BASIC THEORY
AND APPLICATION
OF ELECTRON TUBES

United States Government Printing Office

Washington: 1952

AGO 2244A

DEPARTMENTS OF THE ARMY AND
THE AIR FORCE
WASHINGTON 25, D. C., *20 February 1952*

TM 11–662/TO 16–1–255 is published for the information and guidance of all concerned.

[AG 413.44 (5 Dec 51)]

BY ORDER OF THE SECRETARIES OF THE ARMY AND THE AIR FORCE:

OFFICIAL:
    WM. E. BERGIN
    Major General, USA
    The Adjutant General

J. LAWTON COLLINS
Chief of Staff, United States Army

OFFICIAL:
    K. E. THIEBAUD
    Colonel, USAF
    Air Adjutant General

HOYT S. VANDENBERG
Chief of Staff, United States Air Force

DISTRIBUTION:

Active Army:

Tech Svc (1); Arm & Svc Bd (1); AFF Bd (ea Svc Test Sec) (1); AFF (5); OS Maj Comd (25); Base Comd (3); MDW (5); Log Comd (2); A (20); CHQ (2); FC (2); PMS&T (1); Sch (2) except 5, 6, 7, 17 (50); Dep (except Sec of Gen Dep) (2); Dep 11 (including Sig Sec of Gen Dep) (20); Tng Cen (2) except 11 (100); Lab 11 (2); 4th & 5th Ech Maint Shops 11 (2); PE, OSD (3); Mil Dist (3); Two (2) copies to each of the following T/O&E's: 11–7N; 11–15N; 11–57N; 11–95; 11–537T; 11–547; 11–617.

NG: Same as Active Army.

ORC: Same as Active Army.

For explanation of distribution formula, see SR 310–90–1.

AGO 2244A

CONTENTS

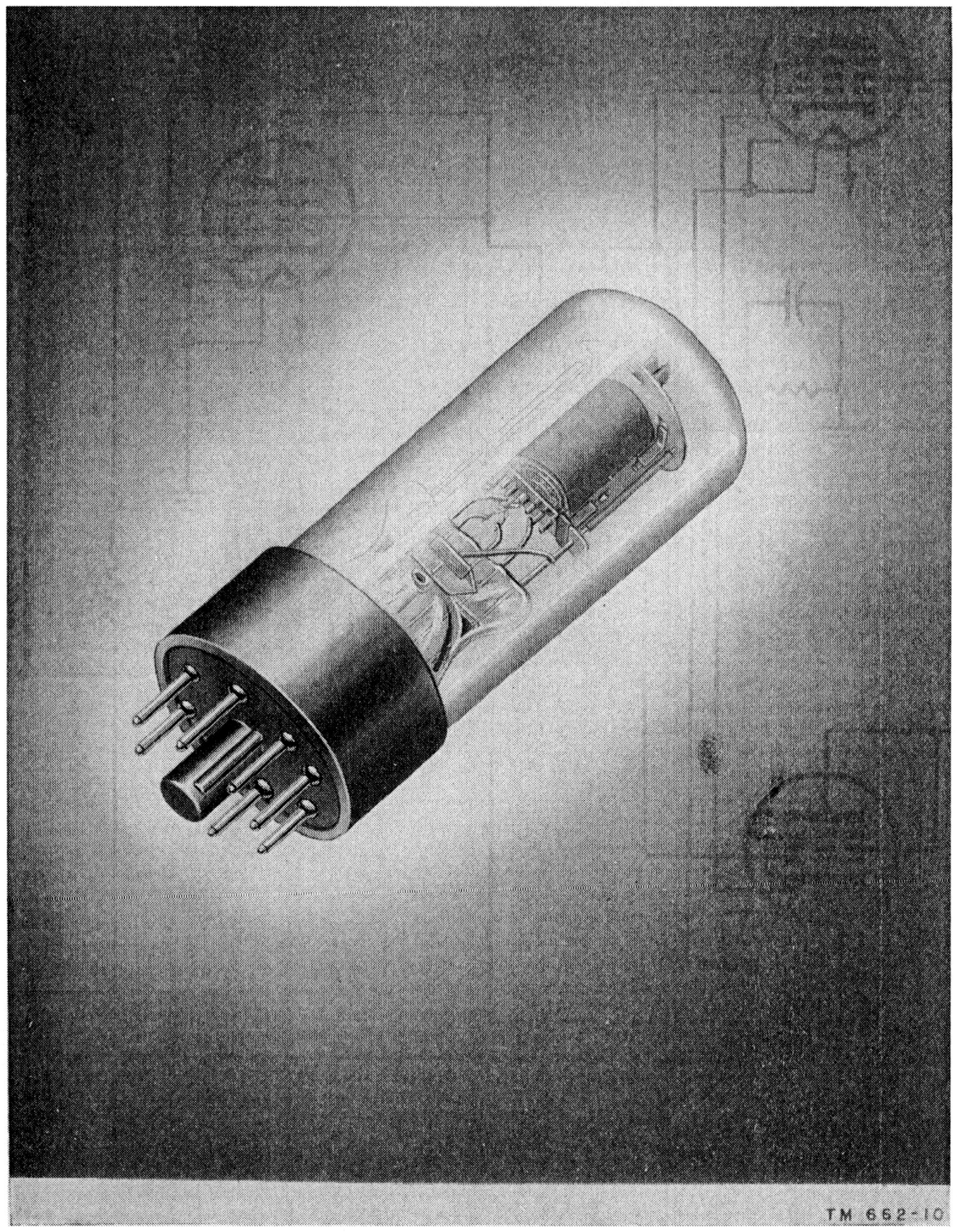

Figure 1. The electron tube is a vital element in modern signal communication facilities.

CHAPTER 1

ELECTRON TUBES—THEIR DEVELOPMENT AND USE

1. Introduction

a. The success of modern military operations, as well as of the peacetime activities of the armed forces, depends in a great measure on signal communication facilities. Whether on the offense or the defense, in the front lines or far behind the lines, exchange of information is vital to the successful movement of personnel and equipment. On the ground, in the air, or below the surface of the seas, electronic devices are in daily use, and in every one of these uses the electron tube is a vital element (fig. 1).

b. Fundamentally, the electron tube is a valve which provides the means of conducting electrons through the enclosed space within a glass or metal container. As the electrons move from the end of one metallic conductor, or electrode, across the intervening space to the other electrode, or conductor, they are controlled more easily than if the same electrons were passing through a wire or any other type of conductor. This power to *control* electrons or electron currents and their associated voltages makes the electron tube the most useful single element in modern signal communication equipment. The present high level of signal communication and its tremendous versatility are due to the existence of the electron tube. Radio communication was utilized prior to the development of electron tubes, but it attained its present versatility only after the electron tube was invented.

2. Importance to Communication

a. In the armed forces, signal communication means many things and a wide variety of equipment. Radio communication is prominent among these, but it is just one method of exchanging *intelligence*, or information, between two points: namely, that method which depends on electromagnetic waves as the link. The intelligence that is transmitted in this fashion may be in code or it may be voice or music intended for specific points or for general reception over a wide area, or it may be both, plus other forms.

b. It is equally important to understand that the telephone, the teletype, and facsimile also are signal communication facilities, even if they use wire as the linking medium. These wired systems frequently are integrated into radio systems wherein the intelligence first spans a distance over wires and then is fed into a radio link for transmission to some remote point or, possibly, to a point not too distant which cannot be reached by conducting wire.

c. Radar, too, in its many forms, is a signal communication facility. It is conceivable that it might be placed in a special category, but even so it is embraced by the broad meaning of the communication art. Although public address systems and intercommunication systems are removed from radio systems, they are part of military signal communication facilities, as are many other equipments which transmit intelligence.

d. In all of these communication devices, the electron tube is the vital component and its unique feature is an extraordinary versatility. Here is a device which, at first view, appears to be rather fragile and capable of only limited types of service. But appearances are deceptive: The electron tube is unrivaled in sensitivity by any other electrical or mechanical device, since it can discriminate between quantities in terms of millionths of a second of time; yet it can handle electrical energy in terms of thousands of watts of power. Following are some of the types of service in which these tubes play an important part:

(1) *In social life.* The long-distance telephone, national and world-wide radio

broadcasting, sound motion pictures, public address systems, and television are all well known, and their impact on social customs and on national and international relations has been tremendous.

(2) *In commercial communication.* Air, land, and sea communication are carried on by means of radiotelegraphy and radiotelephony. Navigation aids, both aircraft and marine, have lessened the hazards of travel.

(3) *In industry. Electric-eye* control devices for automatically controlling quality, color, and size of manufactured products, and for performing all kinds of counting, sorting, timing, recording, and similar operations, have been invaluable to industry. Induction heating applications for production soldering, welding, and heat treatment of metals in metal fabrications have been instrumental in reducing factory time and costs.

(4) *In medical therapy.* X-ray therapy, diathermy for inducing curative artificial fevers, and high-frequency applications in *bloodless surgery* have provided vital tools for the medical profession.

(5) *In scientific research.* Atomic experimentation, electron microscopy, instrumentation, computer and recording devices have become practical realities.

(6) *In military communications.* Radar, fire-control apparatus, communication facilities between fixed and mobile units have increased the mobility and the power of land, sea, and air forces.

e. Finally, so numerous and varied are the applications of electron tubes that a new branch of engineering has been created, that of *electronics*, which encompasses the widespread applications of electron tubes.

3. Early Experimenters

a. The development of the electron tube and its associated communication circuits as they exist today was not the work of any one scientist. Rather it was the cumulative result of the researches, discoveries, and inventions of numerous investigators. Actually, to trace the very beginnings of certain important discoveries, which later led to fruitful results, would necessitate a discussion beyond the scope of this manual. However, for all practical purposes, modern radio art may be said to have had its beginnings toward the end of the last century when, in 1883, Thomas A. Edison was experimenting with his newly invented incandescent lamp.

b. The Edison incandescent lamp may be regarded, in a sense, as the forerunner or prototype of the modern electron tube. Edison noticed that the carbon wire filament of these first incandescent lamps burned out at the point at which the filament entered the glass bulb. Looking for an explanation, he inserted a second conductor or *plate* into the lamp (this is basically the structure of the *diode*, or two-electrode tube of today), and recorded in his notebook that this *dead end* wire or *plate*, when connected through a current meter to the positive side of the battery, showed a flow of current (fig. 2) across the space between the filament and the *plate*. Normally, such an arrangement constituted an open circuit; therefore, current flow, according to the knowledge of electrical circuits at that time, was regarded as an impossibility, for here was an *open circuit*. Edison could find no satisfactory explanation for this phenomenon, which became known as the *Edison effect.*

c. An accurate and epoch-making explanation of the *Edison effect* was advanced in 1899 by a British scientist, Sir J. J. Thomson. He presented the theory that small, negative particles of electricity, called *electrons*, were emitted by the filament in Edison's lamp as a result of operating it at incandescence or white heat. He said, further, that these electrons, because of their negative charge, were attracted to the positively charged plate. Thus, as long as the filament was heated to the proper temperature, electrons would flow from it to the plate. This movement of electrons constituted a flow of electron current, and the electron stream was the means by which the gap was bridged across the intervening space between the filament and the plate, thus closing the circuit.

d. Thomson's findings came to be known as the *electron theory.* Briefly, this theory views

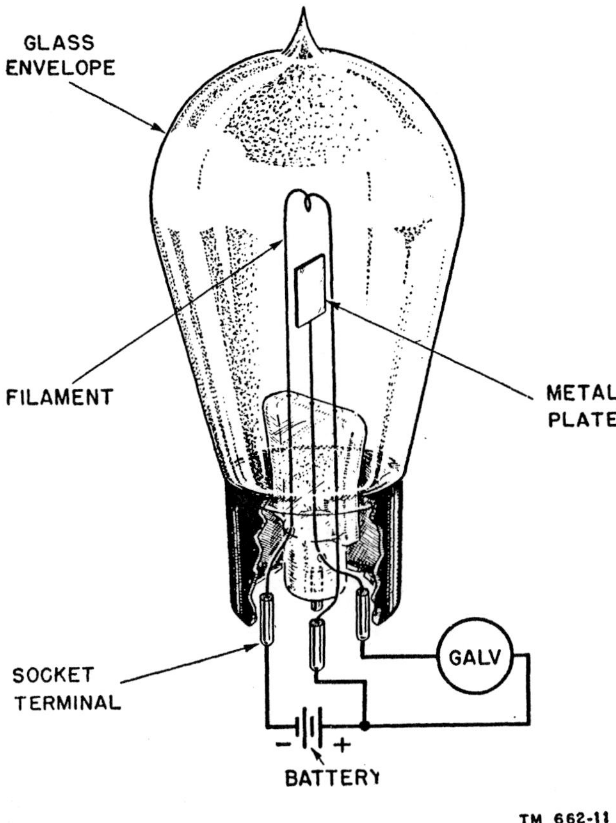

Figure 2. Circuit of Edison's two-electrode tube.

the atoms of all matter as being composed of infinitesimally small, individual negative particles, or electrons, held within the atom by the attraction of a central nucleus of positively charged particles called *protons*. Under suitable conditions, as by the application of heat to a substance, some of the electrons within the substance could be liberated. This extremely important theory gave great impetus to subsequent research, and led to great developments in electron tubes.

e. Equipped with this knowledge, other scientists explored further. The next significant development of far-reaching importance was the work of J. A. Fleming, an English scientist, who designed the first practical electron tube. Fleming observed from Edison's work that, when the plate connection was made to the negative rather than to the positive side of the battery, the current was zero (fig. 3). This property provided the basis for the operation of the electron tube as a *rectifier*, that is, as a device for the conversion of alternating current

into direct current. Fleming, calling his modified version of Edison's two-electrode lamp a *valve* (the term still used for the electron tube in England), thereby provided a superior *detector* to supplant the comparatively insensitive crystal detectors then being used in radio receivers in Guglielmo Marconi's system of wireless telegraphy. The crystal detector was a parallel development, which, following the experiments of numerous predecessors, Marconi made a reality in 1901 when his historical signals of the letter *S* (three dots) spanned the Atlantic Ocean. With the advent of Fleming's valve, the two major lines of discovery and invention, from which the radio art evolved, were joined.

f. Fleming's valve was a two-electrode tube. For several years it was the only electron device in use. At this point it seemed that the progress of wireless communication had reached its practical limit, a limit determined by the existing methods and devices used for transmitting and receiving radio signals. The most powerful

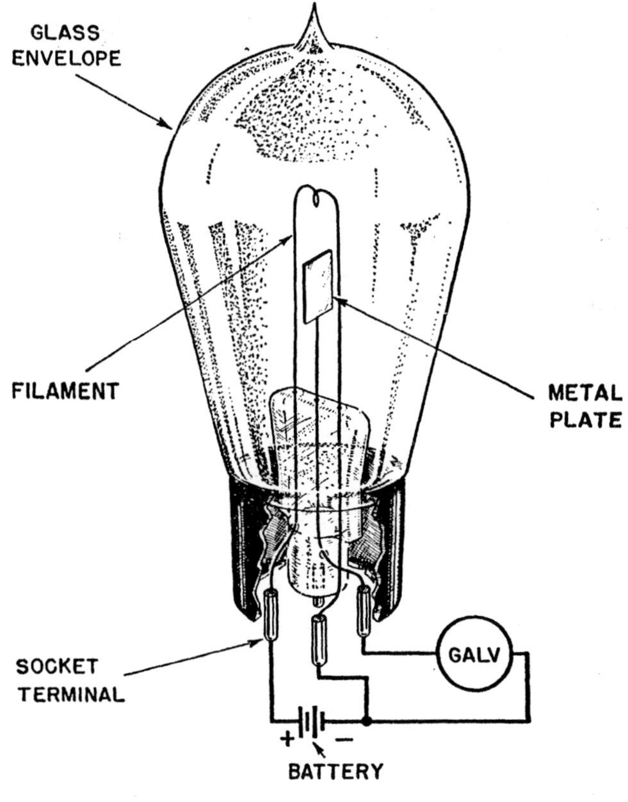

Figure 3. With plate connected to negative side of battery, current through tube is zero.

transmitters could transmit signals to receiving sets more than several hundred miles away, but the reception of such signals was undependable. The range and the dependability of radio communication could be increased only by the development of some method by which the weak signal could be amplified. The vacuum tube developed by Lee DeForest in 1907 supplied this needed means of amplification. Later improvements of this tube have made possible the reception of radio signals millions of times too weak to be audible without amplification.

g. DeForest, by inserting an extra electrode in the form of a few turns of fine wire between the filament and the plate of Fleming's valve, made the tube an *amplifier.* DeForest called the third electrode the *control grid.* It provided the desired amplification by virtue of the fact that relatively large plate current and voltage changes could be controlled by small variations of control-grid voltage without expenditure of appreciable power in the control circuit. DeForest called his three-electrode tube an *audion,* a designation superseded in present-day usage by the term *triode* (fig. 4).

4. Tube Types

a. Each kind of electron tube is capable, generally, of performing many different functions, and therefore initial classification of these tubes is not based on function but upon their physical construction (fig. 5). The envelopes or housings are made of glass or metal or, in a few isolated cases, of both materials. The absence or presence of air or other gases in the envelope distinguishes the two fundamental classes of electron tubes. In the *vacuum* tube, all gases have been removed; in the *gaseous* tube, after all air has been removed, a small amount of mercury vapor or inert gas is placed within the envelope.

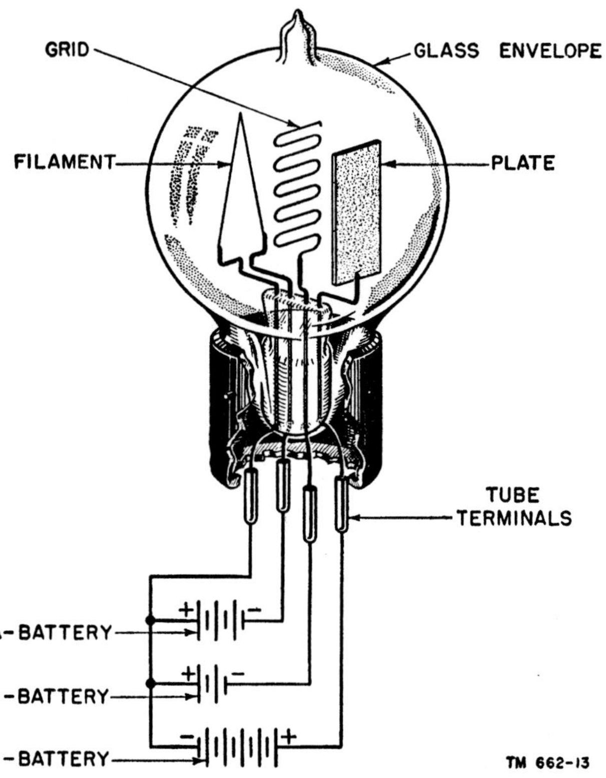

Figure 4. Construction of DeForest's three-element tube, or triode.

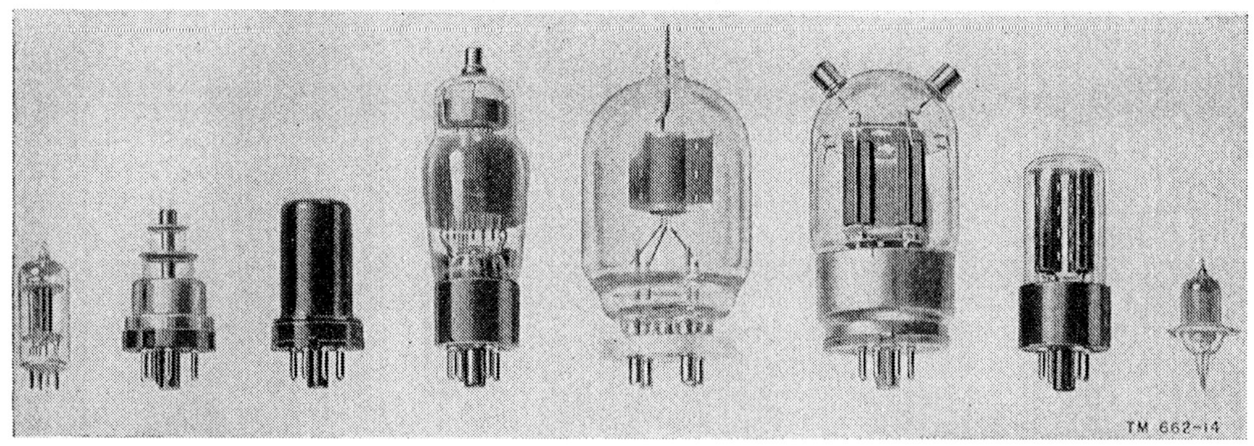

Figure 5. Representative electron tubes.

b. In all electron tubes, one electrode, called the *cathode,* is the emitter of electrons, and it must be heated to cherry red or to incandescence before the electrons are freed from its surface to move across to the second electrode, or *anode.* In vacuum tubes, sometimes called high-vacuum or hard-vacuum tubes, the cathode is in almost every case (the photoelectric tube is excepted) heated by some external source of power, and these tubes, therefore, are not distinguished as to type of cathode. In gaseous tubes, sometimes called soft tubes, a further subclassification is made into *hot-cathode* and *cold-cathode* types. In the first type, the cathode is heated to the proper temperature for emission by some external source of power; in the second type, the gas within the tube is ionized, then the cathode is bombarded by positive ions which raise the cathode to the correct emission temperature.

c. Both vacuum tubes and gaseous tubes of either the hot- or cold-cathode type are further classified as to the number of active elements or electrodes contained inside the envelope. The simplest of these, described above, contains two elements and is known as a *diode.* A tube which contains three elements is known as a *triode,* and a tube with four elements is called a *tetrode.* If it contains five elements it is a *pentode.* In each instance, the type classification indicates the number of elements which make up the tube. In the following chapters each of these tubes is illustrated and the differences between them fully explained.

5. Tube Functions

The functions performed by electron tubes are many and varied, but for convenience these functions may be consolidated into a few general groups. Each function is determined not only by the tube type but also by the circuit and its associated apparatus.

a. A general capability of the electron tube is to alter an ac (alternating current) so that it becomes a pulsating dc (direct current) (fig. 6). Associated apparatus can smooth out the variations in the current, and the system as a whole can be said to change an alternating current into a constant-amplitude direct current. This function provides convenient sources of d-c voltage when the only available primary source of electrical energy is an a-c power line or an a-c generator. These d-c voltages may be as low as a fraction of a volt and as high as tens of thousands of volts. The action of changing an alternating current into a pulsating direct current is referred to in general terms as *rectification,* and the electron tube which does this is called a *rectifier.*

b. Another significant and useful capability of the electron tube, and perhaps its most important function, is described as *amplification* (fig. 7). A stronger signal voltage may be obtained from the tube than is fed into it. In effect, the tube is a signal voltage magnifier. A signal equal to 1 volt fed into the input system of the amplifying tube may appear as 20 volts at its output. Different arrangements provide for different amounts of signal amplification.

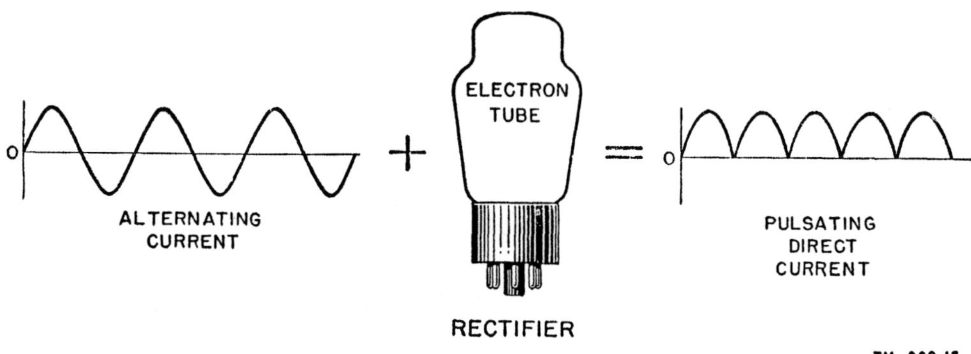

Figure 6. Function of tube as rectifier of alternating current.

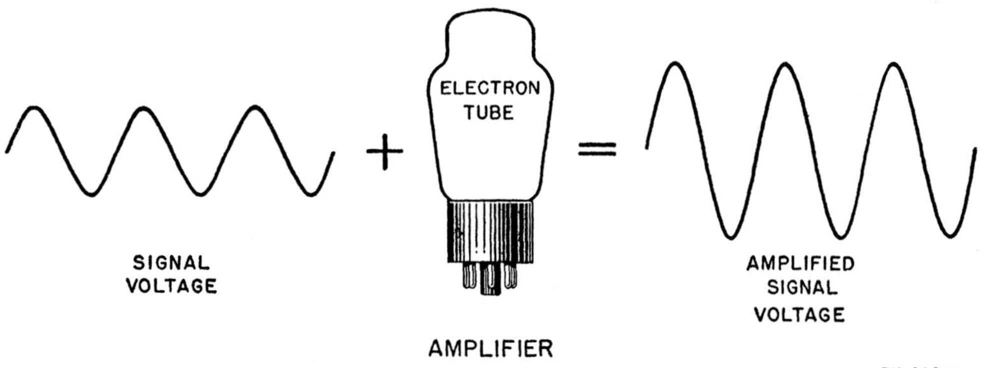

Figure 7. Function of tube as amplifier.

(1) This amplification ability of the electron tube accounts for the advanced development of modern-day communication. It is the basis of all long-distance telephony, because repeaters (electron-tube amplifiers) compensate for the energy losses encountered in the wires. Also, by means of public address systems, the electron tube is used to amplify the voice of an individual so that thousands of people gathered together may hear it clearly.

(2) Amplification by electron tubes makes radar possible because it strengthens the echo signal received from the target so that it can be made visible on a special screen. It is responsible for teletype operation. Television would be impossible without it. Amplification is essential in radio transmitters and receivers of all kinds to build up electrical energy to proportions necessary for proper operation of the various circuits of the equipment.

c. Still another extremely important facility offered by the vacuum tube is the conversion of electrical energy existing as direct current and voltage into alternating current and voltage (fig. 8). Used in this manner, the tube draws energy from a d-c source and, in conjunction with suitable apparatus, *generates* high-frequency oscillations. This function has been responsible for innumerable developments in the communication field.

(1) The principle of oscillation underlies the operation of virtually every type of radio transmitter, large or small, fixed or portable. As a generator of high-frequency oscillation, the electron tube replaces ponderous rotating machinery. Even more important is that specialized oscillators opened up the very-high- and ultra-high-frequency and microwave regions for operation. These extend from approximately 30 mc (megacycles) to tens of thousands of megacycles. The use of these frequencies has helped to overcome the

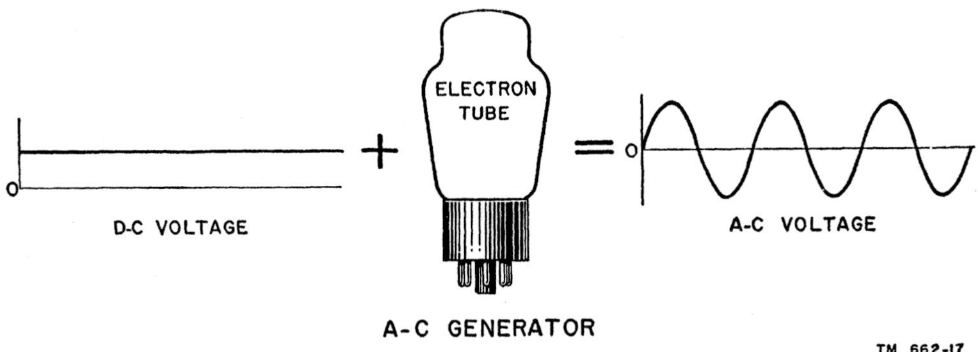

Figure 8. Function of electron tube as generator of alternating current.

communication limitations caused by changing seasons and the effects of weather. It helped create new techniques, among which are radar, television, and radio facsimile.

(2) Availability of these high-frequencies has made possible the convenient use of low power and the design of small receivers, transmitters, and antennas. All of these are important contributions and make feasible radio contact between planes and tanks, or landing forces and ships.

d. The electron tube can modify the shape of electric current and voltage waveforms; that is, it can change the amplitude of these quantities relative to time. Voltage and current shaping (figs. 9 and 10) are vital to the operation of numerous electronic devices. It is used in code transmission, the timing of circuit actions in radar, in the production of television pictures, and in the operation of teletype equipment. Electronic computers could not operate without waveshaping of the currents and voltages present in the equipment.

e. A singular type of electron tube is capable of converting light energy into electrical energy. This is the *photoelectric tube*, also called the *phototube*. In contrast to the conventional electron tube, the phototube does not employ an incandescent electron source. Instead, electrons are liberated from a specially prepared surface inside the tube when radiant energy strikes the material. The stream of charges thus developed is an electric current and gives rise to a voltage in the system. The net result is a voltage output for visible or invisible light input. Although not as versatile as the conventional electron tube, the phototube has many uses. It affords a means of using light energy to control devices or systems or to operate many kinds of equipment. The reproduction of sound from a photographic image on a film is one of many applications of the phototube.

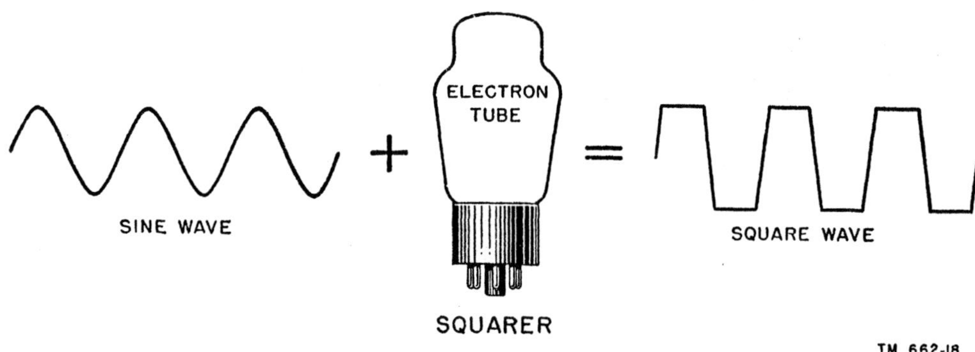

Figure 9. One type of waveshaping accomplished by electron tubes. The input signal is a sine wave; the output is a square wave.

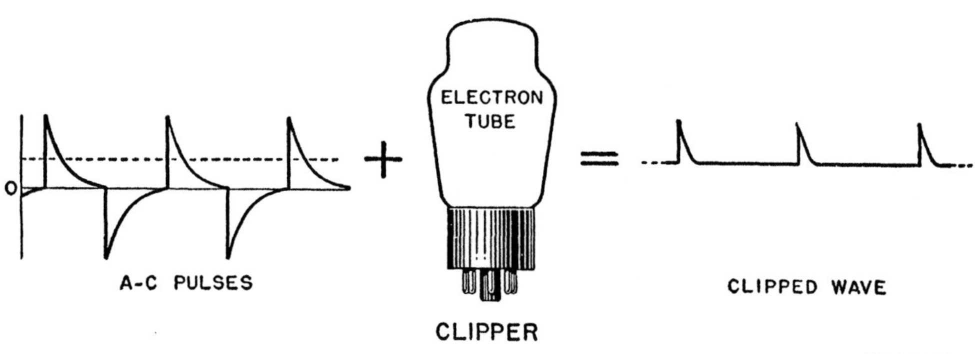

Figure 10. Another type of waveshaping. Only parts of the positive peaks of the input signal are present in the output.

6. Summary

a. The term *electron tube* is a general one under which are grouped vacuum tubes, photoelectric cells, phototubes, cathode-ray tubes, and many other types. The operation of all electron tubes is based on the controlled flow of electrons within the tube structure.

b. In their various forms, electron tubes are found in devices that perform some necessary function in almost every phase of social, industrial, scientific, and military life.

c. Probably the very first important step in radio development was the discovery of the *Edison effect.* Many other scientists contributed important discoveries and inventions. Thomas, Fleming, DeForest, and Marconi are among these early pioneers.

d. DeForest's original three-electrode tube was called an *audion,* a designation now superseded by the term *triode,* and Fleming's original diode was termed a *valve,* the designation that still is used in England instead of the term *tube.*

e. According to the number of elements they contain, electron tubes are classified as diodes, triodes, tetrodes, or pentodes.

f. Among the many special functions which they are capable of performing, that of amplification of electrical energy is perhaps the outstanding feature of electron tubes. The diode, however, does not provide amplification, but it performs another basically important function—rectification.

7. Review Questions

a. What are the two classifications of electron tubes as distinguished by the presence or absence of gas within them?

b. List and describe briefly 12 or more general uses for electron tubes.

c. What are some of the various services included under the general term *communication?*

d. Name some types of signal communication facilities used by the armed forces.

e. Explain the Edison effect.

f. Explain briefly the contributions to the radio art by Fleming, Thomson, Marconi, and DeForest.

g. What are the names for two-, three-, four- and five-element tubes?

h. What is the linking medium in radio broadcasting, facsimile, teletype, telephone, radar?

i. Name and describe briefly some of the special functions which electron tubes can perform.

j. In what sense is an electron tube an amplifier?

k. How does a phototube differ from an ordinary electron tube?

l. What special capabilities of electron tubes differentiate them from electrical and mechanical devices designed for similar types of service?

CHAPTER 2
ELECTRON EMISSION

8. Basic Theory of Electron Emission

a. The operation of all electron tubes depends on an available supply of electrons. Electron emission can be accomplished by four methods, the most important of which is *thermionic emission.* Thermionic emission is the liberation of electrons from a metallic emitter by the application of sufficient heat. Other methods are secondary emission, photoelectric emission, and cold-cathode emission. All of these are discussed later in this chapter.

b. In the *vacuum* type of electron tube, as much air as possible is withdrawn from the envelope and all the electrons needed for operation of the device are obtained from the emitter. In the other general category of electron tubes, known as the *gaseous* type, all air is removed from the tube and a small amount of mercury or some inert gas (neon, argon, xenon, etc.) is placed within the envelope. The emitter furnishes the primary supply of electrons. The emitter output provides electron bullets for ionization of the gas atoms. This process is explained in a later chapter of this manual.

c. The *atomic theory,* which maintained that the atom was the fundamental *building block of matter,* gave way to the *electron theory,* which revealed that the electrons and protons which comprised the atoms were actually the *primary* particles of matter. Accordingly, all electron-tube and electrical phenomena now are explained in terms of the electron theory.

d. The electron theory states that all matter consists of two basic electrical charges: *positively charged protons* and *negatively charged electrons.* These charged particles are the principal and fundamental building blocks which form the atoms comprising the 90-odd elements constituting all matter. Recent investigations have disclosed neutrons, positrons, mesotrons,

and photons, but their study concerns the chemist and the nuclear physicist. The early electron concept, as suggested by Nels Bohr, is satisfactory to students concerned with electronics and provides a useful physical picture necessary for the study of the properties of electricity.

e. In any material, confined in a given volume, the atoms or molecules are in a state of random motion around a mean position. The extent of this motion is determined by the nature of the substance and its temperature. In solids, this random motion is restrained the greatest amount. In liquids, the restraint is less, and in gases, it is the least, almost none. This explains why a solid substance holds its shape unless forces are applied to change it, why a liquid changes its shape to suit its container, and why a gas attempts to fill the space in which it is liberated.

f. Inasmuch as the item of concern is the emission of electrons from metals which may or may not bear coatings of certain chemical elements, it will be well to discuss the action of atoms and molecules. It is significant to note that the amount of random motion in atoms and molecules is a function of temperature. If the temperature of a metal is reduced, its amount of random motion is reduced, and its resistance to the flow of electric current is reduced. This action can be explained by saying that the electrons which comprise the current meet less opposition and experience fewer collisions with the atoms, because the atoms are restricted in their movements.

g. On the other hand, if the material is heated, energy is added to the energy already possessed by the atoms and the molecules in motion, and they perform greater movements at greater velocity. When a fuse blows, the action is simple. The high current overload heats the fuse material so that the kinetic energy of the atoms and molecules is sufficient to disrupt the molecular organization of the metal completely, and it passes instantly through the stage of liquefaction and turns into vapor. Consequently, it becomes a part of the surrounding air.

h. In like fashion, the free electrons in a substance held at normal temperature perform random motion in their travel between atoms, as illustrated in exaggerated form in figure 11. This takes place at a relatively high velocity. Since these electrons move around so rapidly, one is tempted to ask the natural questions: Why don't they move out of the material—for instance, a piece of wire? Why don't they break through the surface and get into free space? Some do just that, but they are so few in number as to be of no importance. The generally accepted description of what happens is that there is no liberation of electrons until special conditions are created deliberately. The reason for such behavior was first explained by O. W. Richardson in 1901. Richardson's theory concerning conditions at the surface of a hot metal has subsequently been accepted by scientists.

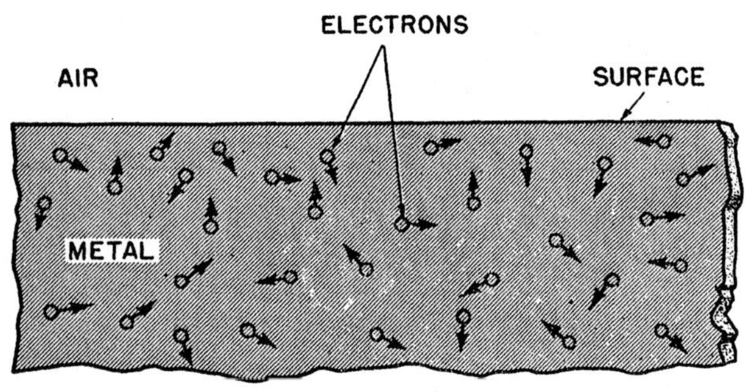

Figure 11. Free electrons having random motion at normal temperature within a metal.

i. Richardson advanced the idea that the boundary of a substance exerts a force in a direction toward the inside of the substance, and so prevents the electrons from leaving the material by penetrating through its surface. For an electron to break through the surface, it is necessary that its velocity, as it approaches the surface, be greater than a critical amount, in order that kinetic energy be sufficient to overcome the barrier at the surface which tends to keep it within (fig. 12). This behavior of different materials is described as the *work function* of the substance. As a matter of convenience, this constant usually is expressed in *electron volts of energy* or simply *electron volts*. Consequently, if an electron advances from a point of zero potential to a point that is 10 volts positive, it has acquired 10 electron volts of energy. Under such conditions, the velocity of the electron is described as 10 electron volts or, for convenience, 10 volts. *Potential barrier* is another name for the action that occurs at the surface of a material and is also expressed in electron volts.

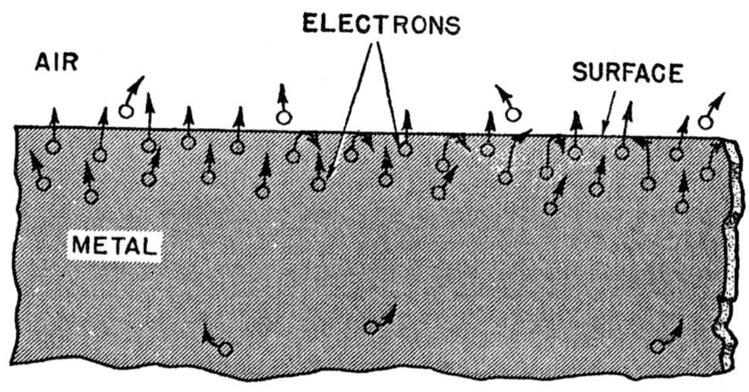

Figure 12. Free electrons escaping into air and to surface of metal.

j. Various substances have certain work functions. A substance like thoriated tungsten, which is commonly used in electron tubes as the source of electrons, has a work function rating of 2.63 volts. On the other hand, nickel coated with barium oxide has a work function rating of about 1 volt. The lower this rating is the more easily an electron can penetrate the surface and leave the material. Consequently, oxide-coated substances often are used as emitters of electrons in electron tubes when a large electron flow is desired. In terms of electron velocity, the lower the work function rating, the less the velocity of the electron needs to be in order to leave the surface. This gives rise to the use of different kinds of electron-emitting substances in electron tubes.

k. The statements made so far show that the emission of electrons by a substance is possible by giving the electrons such velocity that they will overcome the restraining forces present at the surface. This means adding energy to the amount already possessed by the electrons. Several methods of doing this will be described presently.

9. Thermionic Emission

a. Thermionic emission is one means of securing an adequate supply of electrons for the operation of both vacuum and gaseous kinds of electron tubes. By definition, *thermionic emission* is the emission of charged particles from a heated cathode or emitter. The acceleration of the electron to the velocity required for it to leave an emitter is accomplished by means of heat applied by any one of a number of processes. Consequently, the name *thermionic tubes* sometimes is used to describe electron tubes.

b. In thermionic emission, heat is the form of energy that is used to liberate electrons from the substance. When heat is applied to a metal and the temperature is raised sufficiently, some of the heat energy is transferred to the electrons. The electrons then move with greater velocities than previously. If the temperature is high, so that the motion of the electrons reaches sufficient velocity, they will escape through the potential barrier of the emitter into space. When this action is controlled, as in the case of electron tubes, it can be made to provide a continuous stream of electrons.

c. For a given type of emitter, there is a definite rate of thermionic emission at each temperature. The rate depends on the type of emitter material used and the temperature of the emitter. This process is fully described in chapter 3.

10. Other Types of Electron Emission

a. Thermionic emission sometimes is referred to as *primary emission* to distinguish it from another type of electron emission called *secondary emission*. In this comparison, primary emission signifies that emission takes place directly from an emitter substance by the application of heat, or other means; in secondary emission, electrons are detached from a body as the result of its being *bombarded* by electrons emitted from a primary source (fig.

13). When, for example, a stream of high-velocity electrons strikes a metallic substance, these bombarding electrons impart sufficient energy to the electrons within the metal to enable them to break through the potential barrier. Although in the figure only one secondary electron is shown to be released for each primary electron, in actual practice the number of secondary electrons may be more, depending on the material from which the body is constructed. In some vacuum tubes, secondary emission takes place in the normal tube operation, but usually it is undesired and provision is made to prevent it. However, this condition is created purposely in other tubes to obtain special operating characteristics.

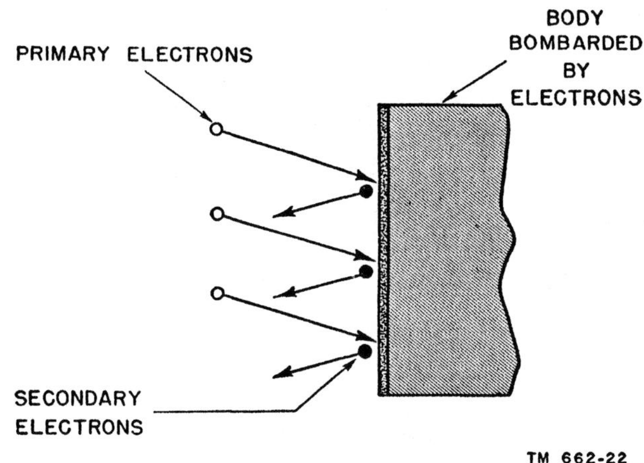

Figure 13. Secondary emission caused by electron bombardment.

b. Another method of producing electron emission is by the application of light. Light, an electromagnetic radiation, is a form of energy. The theory by which electrons are liberated from substances when electromagnetic waves of the proper frequency impinge on them is very complex and is beyond the scope of this book, but it is an accepted fact that electron emission does take place. Consequently, when energy in the form of light strikes a *photosensitive* metal, electrons are liberated from this metal under impact of the energy of the light rays (fig. 14). This is known as *photoelectric emission*. The photoelectric current is directly proportional to the intensity of illumination. All photoelectric tubes, or phototubes used in countless applications for control and detection depend on photoelectric emission for their operation.

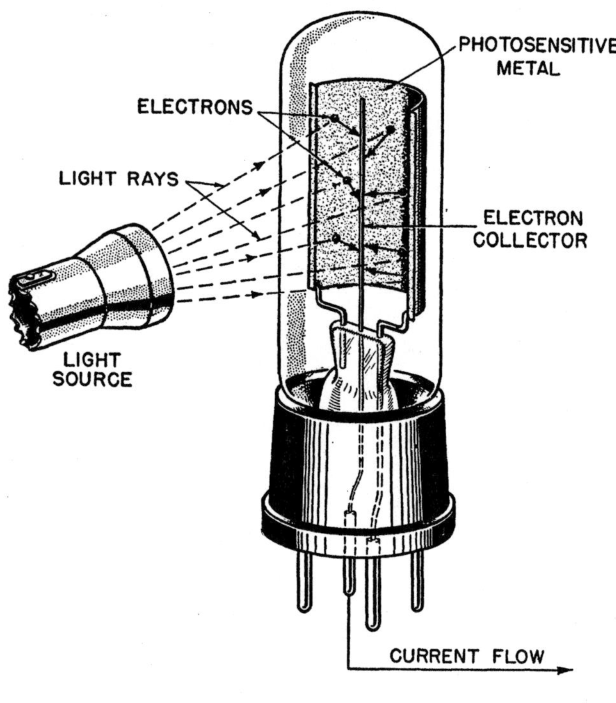

Figure 14. Photoelectric emission of electrons.

c. Still another method is called *cold-cathode emission.* In this type of emission, electrons are virtually pulled out of a substance by the power of an extremely strong electric attracting force. Since high voltages which raise many problems are required in this method, cold-cathode emission is not commonly used.

11. Types of Emitters

Electron emitters are classified according to the method used to heat them. In the *direct method*, the emitter is called a *filament* or *directly* heated cathode, and the electric current is applied directly to the cathode. In the *indirect method*, the emitter is called an *indirectly* heated cathode and the electric current is sent through a separate *heater element* which is located inside the emitting cathode and transfers its heat energy to it by conduction. Both methods can use either alternating current or direct current.

a. DIRECTLY HEATED ELECTRON EMITTERS (fig. 15). A directly heated electron emitter or *filament* is usually of the general construction shown in A. The radio symbol for the filament in a vacuum tube is shown in B. Physically, the filament usually is shaped either in the form of an inverted V or an inverted W. The filament voltage required to produce electron emission is applied across the filament prong terminals located in the base of the tube. When current flows through the filament circuit, the filament emits electrons when it reaches emission temperature. Filament materials are tungsten, thoriated tungsten, or metals that have been coated with alkaline-earth oxides. In the latter case, the electron-emitting material is the coating; the metal core is used to carry the heating current. Some directly heated oxide-coated filaments require comparatively little heating power, and for this reason they are extensively used in tubes designed for operation from batteries and in portable equipments. An added advantage of the directly heated electron source is the rapidity with which it reaches electron-emitting temperature. Since this is almost instantaneous, many equipments that must be turned on at infrequent intervals but must be instantly usable use directly heated tubes. Usually these are of the oxide-coated variety.

b. INDIRECTLY HEATED ELECTRON EMITTERS (fig. 16). For the indirectly heated type of electron emitter, shown in A, the cathode electrode is the emitting element. B represents the radio symbol as it appears in vacuum tubes.

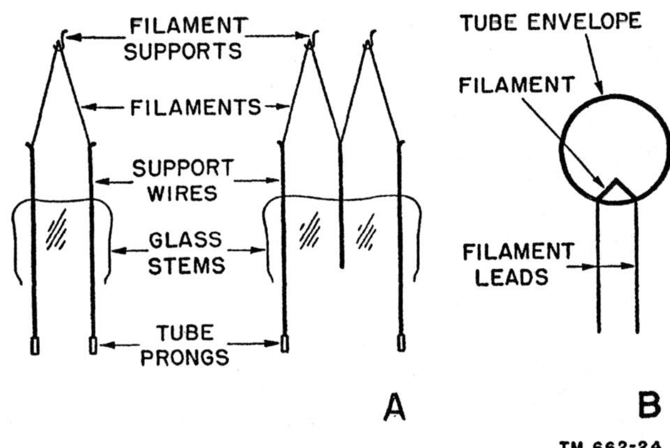

Figure 15. Directly heated filaments and schematic symbol.

The electron source is an oxide coating on the cathode electrode. It is heated to emission temperature by the heat from the heater. The electric current flows through the heater, located within the cathode that surrounds it. Heat energy produced in the heater by the electric current is conveyed by conduction to the cathode. The majority of electron tubes are of the indirectly heated type, because of the practicability of operating the tube from alternating-current supply lines. Variations in heater current do not cause a fluctuating output in electron emission (for all practical purposes) because the temperature of the cathode remains fairly constant when the a-c input reverses its direction. The constant temperature is caused by the inability of the cathode to cool off quickly when the alternations are approximately 60 cps. This is not true for directly heated emitters, and, consequently, special circuits are necessary to adapt these tubes to an a-c filament supply.

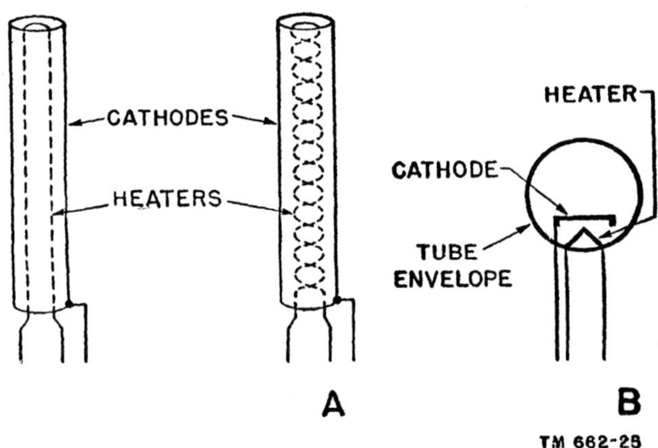

Figure 16. Indirectly heated emitter and schematic symbol.

c. FACTORS DETERMINING EMISSION. The amount of electron emission that can be obtained from an incandescent filament or a cathode depends on various factors. Essentially, it is determined by the temperature of the emitter, which in ordinary vacuum-tube operation is determined directly by the amount of current applied to the filament. In general, the higher the temperature of the emitter, the higher is the resultant emission. However, a practical limit must be considered. The substances used in the manufacture of emitters are designed for operation within definite temperature limits to assure maximum operating life and efficiency. *Excessive* temperature causes extremely rapid deterioration of the emitter, which reduces its useful life. This applies equally to all types of tubes using heated electron emitters.

12. Materials Used

a. Very high temperatures are required to produce satisfactory thermionic emission. As previously mentioned, the materials best suited for the purpose are tungsten, thoriated tungsten, and oxide-coated core materials. Platinum and nickel are examples of oxide-coated core materials. Of all these substances, tungsten possesses the greatest durability, and therefore, is used in tubes which may be subjected to heavy overloads. Similar service is also performed by thoriated-tungsten filaments. Both of these generally are found in equipments such as transmitters, since they can withstand high voltages and rigorous conditions.

b. Thoriated-tungsten filaments are manufactured by mixing thorium with tungsten. The thorium coating behaves as a profuse emitter of electrons and gradually evaporates during use. As it boils off, it is replenished from inside the tungsten filament wire. At the same time, a gradual evaporation of the wire occurs; consequently, it becomes thinner with time. Eventually one part of the filament becomes too weak to carry the current and it burns out.

c. Thoriated-tungsten filaments usually are operated at temperatures of about 1,900° C. At this point, the filament becomes bright yellow. Tungsten, on the other hand, is operated at approximately 2,200° to 2,500° C. and glows with a white light. The evaporation of tungsten is like that of thoriated tungsten; consequently, longest tube life is attained by keeping the voltage constant across the filament and allowing the current to adjust itself in accordance with the changes in filament resistance as the wire becomes thinner and thinner.

d. The most efficient electron emitters are the oxide-coated filaments and cathodes. These coatings usually are barium or strontium oxides. The electron emission takes place from the oxide coating; the core carries the heating current. The operating temperature is approximately 800° to 1,150° C.

13. Summary

a. According to the electron theory, all matter is composed of two fundamental electrical charges—namely, protons and electrons. Different combinations of both form atoms. A molecule consists of one or more atoms, depending on the particular element.

b. Free electrons in a substance at normal temperature perform random motion in their travels between atoms.

c. Some types of electron emission are *thermionic, secondary, photoelectric,* and *cold-cathode.*

d. Electron emitters are either directly or indirectly heated by an electric current. Because alternating current is the principal source of electrical power available, the indirect means of heating electron emitters is most widely used.

e. Oxide-coated materials for cathodes are used most commonly because of their high emission rate. Where higher voltages are involved, mechanical strength is a requirement, and therefore, tungsten or thoriated tungsten is used.

14. Review Questions

a. What are the building blocks of matter?

b. Name the two fundamental particles found in the atom.

c. State three methods by which electrons can be liberated from a material.

d. What enables electrons to leave a material?

e. What is meant by the terms *work function* and *potential barrier?*

f. Name three materials which are used as electron emitters in electron tubes.

g. Which type of substance emits electrons most profusely?

What is meant by indirect heating of an emitter? By direct heating of an emitter?

h. What is the name generally applied to an indirectly heated emitter? To a directly heated emitter?

i. What is the effect of operating emitters at excessive temperatures?

j. Why are tungsten and thoriated-tungsten emitters operated at constant voltage instead of constant current?

k. What is the difference between primary emission and secondary emission?

l. Relative to the source voltage, what advantage does indirect heating offer over direct heating of emitters?

m. Which operates more rapidly, the indirectly heated emitter or the directly heated emitter?

CHAPTER 3

DIODES

15. Construction

a. The simplest type of electron tube is the *diode.* It consists of two elements or electrodes, one of which is an *emitter of electrons* and the other a *collector of electrons.* Both elements are enclosed in an envelope of glass or metal. Although this discussion revolves around the vacuum diode from which most of the air has been removed, it should be understood that gaseous diodes also exist. The term *diode* refers to the number of elements within the tube envelope rather than to any specific application. In this connection, the complete *electron-emitting system* is treated as one element. Different names are applied to the diode to indicate the specific function of the tube in any particular electrical circuit. As one example, the diode can change alternating current into direct current; it is then a *rectifier,* and the tube is named accordingly. Therefore, when discussing the basic diode, reference is made to the tube as a type, rather than to any of its applications.

b. The electron collector is called the *plate* and the electron emitter is called the *cathode.* Although the latter term more specifically applies to the indirectly heated type of emitter, whereas the directly heated type of emitter is referred to as the *filament,* the term *cathode* usually is used regardless of the method of heating. This usage is not so odd as it may seem, since the majority of tubes in use today are of the indirectly heated type.

c. In directly heated tubes, the filament is of the general construction illustrated in figure 15, showing two typical filamentary cathodes. The type shown at the left is known as an *inverted V,* and that on the right as an *inverted W.* The filament is held in place within the tube envelope (glass or metal) by means of suitable metal supports firmly resting in the glass stem of the tube. It is suspended from the top of the tube by a metal support which allows for the expansion of the filament wire when heated. The filament voltage is applied across the prong terminals of the filament in the tube base. Figure 17 is a cross-sectional view of a simple diode tube, showing the internal construction, tube base, and wiring.

d. In indirectly heated tubes, the cathode-heater design can be either of the two common types shown in figure 16. The heater wire is usually either U-shaped, as shown on the left,

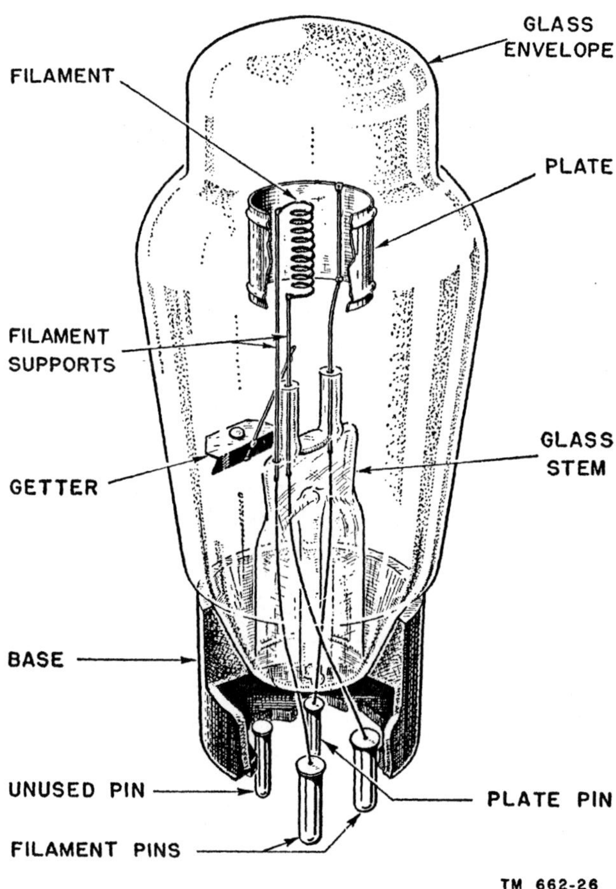

Figure 17. Cross-sectional view of simple half-wave diode of filament type.

in A, or it can be twisted throughout its length, as on the right, in A. In indirectly heated tubes, the cathode is an oxide-coated cylindrical sleeve, usually of nickel, which encloses the heater wire. The heater is insulated from the nickel sleeve by an Alundum coating on the heater wire or by passing the heater wire through fine parallel holes in an Alundum tube. In directly heated high-power tubes, the cathode heater is constructed of tungsten and thoriated tungsten, since high voltages tend to destroy oxide-coated cathodes.

e. The plate is usually of cylindrical construction, although frequently it has an elliptical form. In modern tubes it surrounds its associated emitter, as in figure 18. The metals used for the diode plate (and the plates of most other tubes) usually are nickel, molybdenum, monel, or iron. A tube which contains one emitter and a single related plate is identified generally as a *half-wave rectifier,* for reasons which will be explained later. Another name for tubes of this kind is simply *diode.*

f. Another structural organization of diode tubes utilizes two diode sections inside a single envelope, each section consisting of an emitter with its own heater and its related plate (A of fig. 18). A tube that contains a pair of diode sections is called a *duo-diode,* or a *full-wave rectifier.* The choice of term depends on the application of the tube. Rectifiers usually are associated with circuits which change alternating current to pulsating direct current at reasonably *high* levels of electrical power, whereas the name diode or duo-diode is applied to tubes which perform the same function in connection with radio signals. These represent appreciably lower levels of electrical power. The differences will become much clearer later.

g. The circuit symbol for an electron tube with two cathodes and two plates is shown in the lower part of A. The two heaters, one for each cathode, are shown joined in series. The same heater current flows through both. The two extreme terminals are arbitrarily labeled H1 and H2. In some tube types, a center tap is provided in the heater circuit. This is shown by the dotted line labeled H3. The two cathodes, K1 and K2, are independent of each other. The plates associated with these cathodes are des-

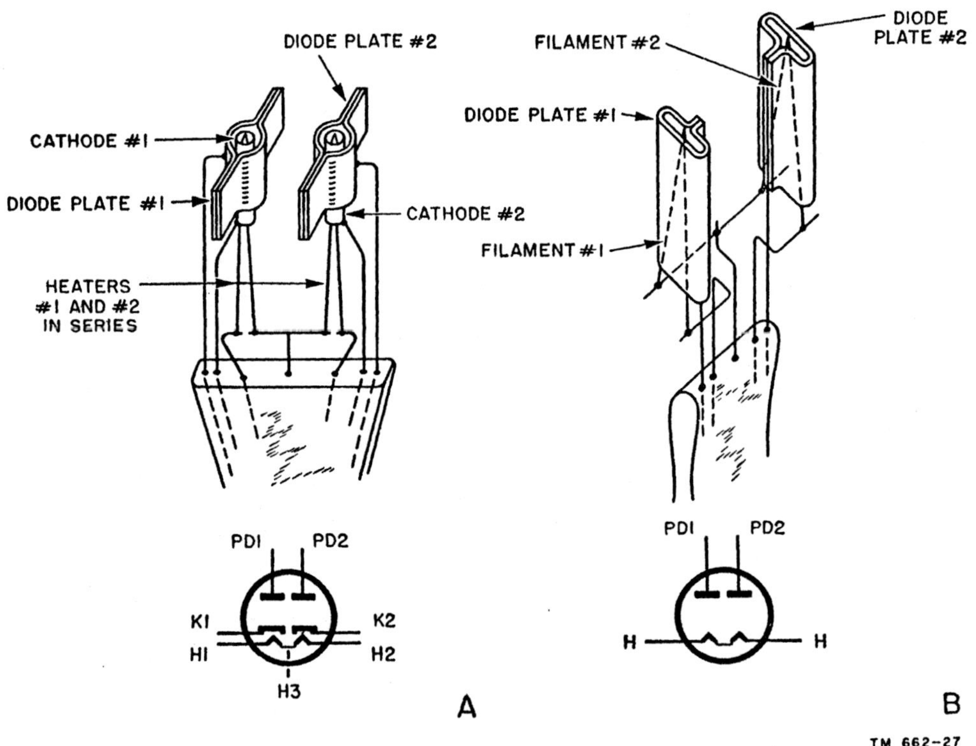

Figure 18. Construction of duo-diode and its schematic representation.

ignated appropriately PD1 and PD2, corresponding to the markings for the cathodes. The circle around all of the electrodes symbolizes the tube envelope.

h. B shows a different physical arrangement of two diodes within a single envelope. This is a directly heated tube in which oval-shaped plates surround the inverted V-shaped filament emitters. These filaments are connected in series, as shown in the related circuit symbol. Again, the two plates are independent of each other and each bears its own identifying designation.

i. Modifications of tube structure occur among the many diode tube types. Two of these are illustrated in figure 19. One heater can serve a single cathode, which has sufficient emission to serve two independent plates, as in A, or two series-connected heaters can serve two cathodes joined to each other internally and terminated at a single tube base pin, as in B. The plates are separate from each other for connection to individual circuits.

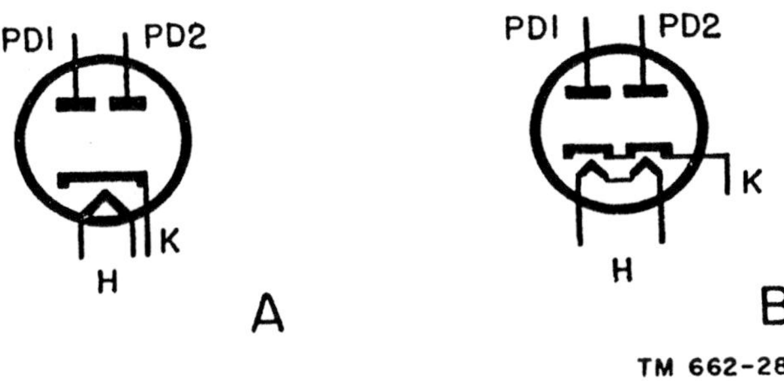

Figure 19. Other schematic representations of duo-diodes.

j. One of the essential requirements of a *vacuum* tube is that it must remain practically free of internal gases during operation. Every precaution is taken during manufacture to pump all the air from the tube and bring it as close as possible to a fully evacuated state. When a tube is not highly evacuated (because of some defect in manufacture, or deterioration during use), it is said to be *gassy.* The flow of current through a gassy tube may be erratic, and this can cause erratic operation. It is important to note that some gases remain after evacuation, being imbedded in the metal parts. To remove these imbedded gases, the tube elements first are brought up to red heat; then a *getter,* generally magnesium or barium, is *flashed* within the tube to absorb the gases released by the heat. The getter condenses on the inside surface of the tube envelope as a silver or reddish coating. The slightest air leak caused by a cracked glass envelope will result in a chemical reaction between air and the getter flash, turning it milky white. The tube interior is evacuated so completely that it approaches a nearly perfect vacuum. Fewer than half a trillion air molecules per cubic inch is considered a very efficient vacuum for receiving tubes.

16. Operation

a. It is necessary to review briefly several fundamentals concerning the behavior of the electron before proceeding with the explanation of the operation of the diode. The electron is a particle of negative electricity. It is attracted by a charge of opposite polarity and repelled by a charge of like polarity. Such behavior is in accordance with the basic laws of electricity which state that like charges repel and unlike charges attract. What is true about charges is true also about bodies or plates which are *charged positively* by the removal of electrons or *charged negatively* by the addition of electrons. Consequently, if a source of voltage, E, is connected between two facing plates, N and M, as in A of figure 20, with the positive terminal to N and the negative terminal to M, a difference of potential is established between N and M. This difference in potential equals the voltage source, E; in addition, plate N is positive relative to plate M, or plate M is negative relative to plate N. Finally, as the consequence of the redistribution of charges on plates N and M, an *electrostatic field* is created in the space between the plates.

b. As explained in TM 11–661, the direction of a line of force is the direction in which a positive test charge tends to move when placed in an electric field. Thus, because like charges attract and unlike charges repel, a positive test charge tends to move toward a negative charge and away from another positive charge, or, from positive to negative. For this reason, the direction of the line of force is shown as leaving a positive charge and entering a negative charge. The choice of a positive charge as the test charge is conventional. Actually, if a negative

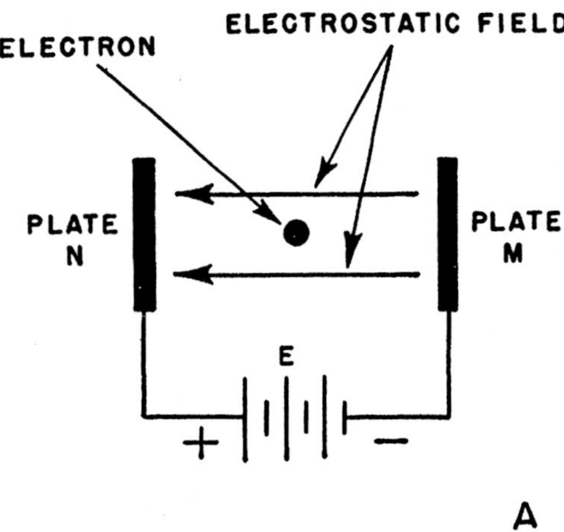

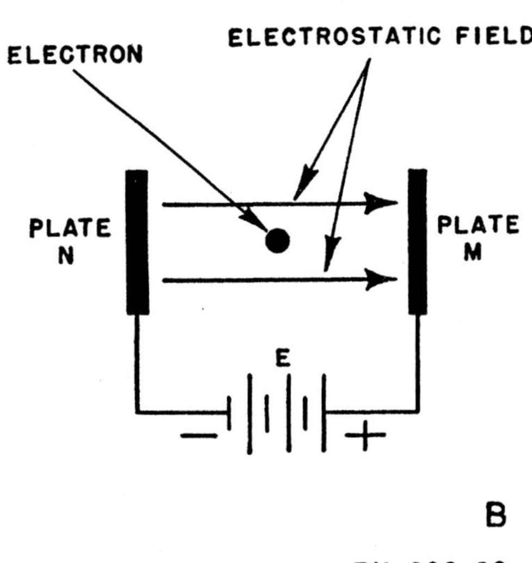

Figure 20. Electrostatic fields resulting from two plates and a battery.

test charge is chosen, the direction of the line of force will be reversed; that is, a negative test charge, when placed in an electric field, tends to move toward a positive charge and away from another negative charge, or, from negative to positive. Actually, it makes no difference whether a positive test charge or a negative test charge is selected because, in either case, the choice is arbitrary. The only difference is in the direction of the arrows on the lines of force in the illustrations. Throughout this manual, the use of a negative test charge is assumed and the direction of arrows on all lines of force in the illustrations are shown from negative to positive (A of fig. 20). This choice is a logical

one because the study of vacuum tubes is concerned largely with the behavior of the electron in motion, and the electron, being a negative charge of electricity, always tends to move from negative to positive. Also, the electron tends to move in the same direction as the arrows on the lines of force, the latter being the graphical representation of the electric field.

c. No matter where the electron is placed in the field between N and M, it travels toward N along the lines of force. Only two lines of force are shown as a matter of convenience in the drawing; actually, the entire space between both plates is full of these imaginary lines of force. Although they issue from point charges, the radial arrangement of electrostatic fields shown for isolated charges is not used here. Great numbers of these charges are located on the two parallel plates and the effect of the forces caused by these charges within the space bounded by the two plates can best be shown by *parallel* lines.

d. If the connections of the voltage source are reversed so that plate M is joined to the positive terminal and plate N to the negative terminal, as in B of figure 20, the reverse condition is established. The electrostatic field still exists in the space between the two plates, but now the direction of the lines of force of the field is reversed. Plate M is positive relative to plate N and the electron advances along the lines of force toward M, the positively charged plate. The action of an electron in an electrostatic field of this kind can be described simply by saying that it advances along the lines of force to the most positive point in the field. In A of figure 20, this is plate N; in B of figure 20, it is plate M.

e. One factor that determines the intensity of the force of attraction and repulsion acting upon the unit test charge is the distance between plates M and N. The smaller the separation between M and N, the greater is the force of the field for a given value of applied voltage. If the applied voltage is doubled, the force of the field is doubled, and this has the same effect as reducing the distance between the plates by half. (Plate separation has the inverse effect with reference to the applied voltage.) Actually, the intensity of the electrostatic field can be increased by three methods: an increase in voltage applied to the plates, a reduction in the separation between the plates, and a combination of both of these methods. It follows that any increase in the intensity of the electrostatic field results in a strengthening of the force acting upon the electron. In effect, the greater this force, the more powerfully is the electron either attracted or repelled in its travel between the plates.

f. Another point of interest that might be mentioned is that, regardless of the number of electrons advancing under the influence of the field, their velocity per unit time is a function of the difference of potential established between the limits of the field. The greater the voltage on plate M relative to plate N, the faster the electrons travel through the field. If a great many electrons start from some point near N, the greater will be the number that will be able to advance to M in a unit time. The lower the difference of potential between the two plates, the lower the velocity of the electrons, and consequently, fewer electrons will span the field per unit time.

g. All of the preceding leads to the most important part of the review. It is evident that the direction of the electron advance is determined by the polarity of the field. If a source of electrons is assumed somewhere in the field, near plate N, and their behavior is examined relative to plate M, it is seen that the electrons will feel a force *attracting* them to plate M when that plate is *positive,* and will feel a *repelling* force when that plate is *negative.* Under one set of conditions, electrons will *advance* to plate M and under another set of conditions they will *not go to plate M.* The relationship between electron movement and the polarity of the field is basic to the operation of the diode and to all electron tubes.

h. As a preliminary explanation at this time and without attempting to deal with all of the pertinent points, reference is made to A of figure 21. This is an elementary version of a diode circuit. The emitter is indirectly heated. The voltage required to drive current through the heater is secured from a source normally identified as *heater voltage supply.* The polarity of the filament battery can be reversed, since its only purpose is to heat the filament. In the circuit shown it is a battery, but it can just as

readily be a transformer of appropriate voltage and current rating operated from the a-c power line. The latter is the usual arrangement.

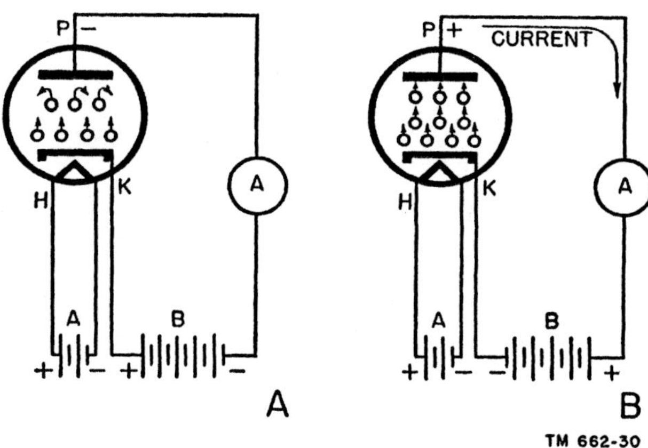

Figure 21. Different conditions of current flow through diode.

i. Apart from the heater circuit, the operating system consists of the emitter, in this instance the cathode, K, plate P, and the external circuit consisting of a *plate-voltage* supply source, B, and a current meter, A. The voltage applied to the plate is connected between the plate and the emitter. The purpose of the meter, which is connected in the circuit, is to indicate the behavior of the current. The plate battery is so connected as to make the plate *negative* with relation to the cathode.

j. At this point, it is important to understand a certain action inside the tube. Regardless of what may be connected between cathode and plate, or if no apparatus of any kind is joined to these electrodes, an electron cloud appears near the cathode as soon as the cathode is heated to electron-emitting temperature. This electron cloud is located in the *space* between the cathode and the plate. What happens to these negative charges? The answer is found in the discussion of charged plates earlier in this chapter. With plate P negative relative to cathode K, the emitted electrons are prevented from advancing to the plate; the direction of action of the electrostatic field between K and P repels the electrons. In the meantime, the continued heating of the cathode results in continued emission. The net result is a cloud of electrons in the space near the cathode, as illustrated by the tiny circles in A of figure 21. The absence of electron flow between cathode and plate makes the interelectrode space the equivalent of an open circuit; consequently, there is no current flow through the tube circuit formed by the cathode, the plate, the current meter, and the plate-voltage supply source.

k. In B of figure 21 the system has been modified by changing the polarity of the voltage applied to the plate, making the plate *positive* relative to the cathode. What happens to the emitted electrons under such polarity conditions? The previous discussion concerning the movement of negative charges in an electrostatic field furnishes the answer. The positively charged plate now exerts an attracting force on the emitted electrons. They advance across the space between cathode and plate. Since moving charges comprise electric current, the stream of electrons is the equivalent of an electric current without benefit of a wire conductor. The movement of these charges can be viewed conveniently as being from the cathode to the plate, thence through the meter, A, through the plate-voltage source, B, and back into the cathode, completing the circuit.

l. If it is assumed that the cathode is kept at an emitting temperature and the plate is maintained positive relative to the cathode, the movement of electrons continues through the circuit for as long as the aforementioned conditions prevail. Because the current flows to the plate in the tube and through the circuit elements joined to the plate, it is identified as *plate current.* The electrons shown in B of figure 21, now are seen to fill the space between the cathode and the plate.

m. This behavior of a diode leads to a number of important and far-reaching conclusions:

    (1) Current flow in a diode occurs only when the plate is positive relative to the electron emitter.

    (2) Current will not flow in a diode when the plate is negative relative to the electron emitter.

    (3) Current flow in a diode can be in *one* direction only, from cathode to plate, never from plate to cathode. This action is called *unidirectional or unilateral* conduction.

    (4) A diode can behave like a control valve, automatically starting and stopping current flow.

n. These capabilities lead to many different applications of the diode, several of which are detailed elsewhere in this manual. One of these can be appreciated here. If the plate battery in B of figure 21 is replaced by an a-c voltage source, which alternately changes the polarity of the plate and the cathode, the tube will alternately conduct and cease conducting. Each time the plate is made positive, current will flow through the system and continue flowing until the polarity of the plate is changed to negative, at which time, current flow through the plate circuit will cease. This is the principle of *rectification.*

o. The fact that two diode systems are contained within a single envelope does not modify the behavior. Each diode behaves as outlined. The use of two diode sections is a matter of accomplishing certain results conveniently. These will be explained later.

17. Space Charge

a. The statements made thus far in regard to the electrons emitted from a cathode (or filament) and the statements made concerning the plate voltage as well as the plate current may well be considered more critically. How much electron emission is possible? How much plate current is possible or permissible? How high can the voltage on the plate be? These questions deserve answers because they influence the understanding of the over-all operation of the device. This leads to the subject of *space charge.* In order to make this manual easy to read, the cathode and filament types of electron emitters will be treated in like manner. Their physical differences are taken for granted, but it is necessary to select one or the other term for easiest reference to emitters. Accordingly, the word *cathode* will be used to denote the electron emitter in all cases, unless otherwise specifically stated. It will have meaning only as a source of electrons, rather than as a term which distinguishes indirectly heated emitters from directly heated emitters.

b. The number of electrons which traverse the space between the cathode and the plate in a diode per unit time (in other words, the plate current), is determined by four important factors:

    (1) The temperature of the cathode;
    (2) The emission per unit time;
    (3) The voltage of the plate relative to the cathode;
    (4) The space charge.

c. Items (1) and (2) are related to each other because the temperature of any cathode determines the number of electrons emitted per unit time. Logically, a supply of electrons is necessary in order that plate current flow may exist.

d. Item (3), the plate voltage, is another important factor. It is the force attracting the emitted electrons. At first thought, it would seem as if a very low value of positive plate voltage would immediately attract all the electrons which have been liberated by the cathode. This does not happen because of item (4), the space charge. The space charge is responsible for the control of plate current. As a matter of fact, plate current in diodes generally is described as being *space-charge limited,* and an analysis of the behavior of the space charge explains why changes in the value of voltage applied to the plate change the value of plate current. The term *space charge* refers to the *effect* of the electrons which have been liberated and which accumulate in the neighborhood of the cathode, with or without plate-current flow in the tube.

e. Imagine for a moment a tube arranged as illustrated in A of figure 22. The cathode is heated to an electron-emitting temperature and the plate is joined to the cathode. Since no voltage is applied to the plate, it exerts no attracting force on the emitted electrons. If the cathode is emitting electrons, what happens to them? As a beginning, it is necessary to make several assumptions. The heat applied to the cathode imparts sufficient velocity to the electrons to enable them to overcome the work function of the emitter material. They break through the potential barrier. But most of the energy possessed by the electrons is given up in passing through the potential barrier and very little is left after they leave the emitter. This is the first assumption: The electrons are considered as leaving the cathode at a relatively low velocity. A few emitted electrons can possess sufficient velocity to complete the excursion to the plate, without any attracting pull from the plate. These are so few in number that they can be disregarded in terms of normal functioning of the tube.

f. The second assumption is that the shape of the emitting surface of the cathode is a straight side or a plane surface. The third and final assumption is that the electrons leave the cathode in an orderly manner and have orderly motion, something like the movement of troops in columns. This is not really true because emission is haphazard and in random directions away from the surface, but the assumption does no harm. An idea of what is meant is illustrated in B; the disposition of the electrons is shown more accurately in A. The orderly emission aids the understanding of what follows. Let us start with the emission of the first row of electrons, in B. They have nothing in front of them to impede their progress. However, their low velocity of emission is a limiting factor, and they do have to contend with the second row in back of them, the third group in back of the second, the fourth group in back of the third, and so on. Since the basic law of charged bodies states that like charges repel one another, the electrons back of the first group are experiencing a repelling force exerted by the electrons in front of them, and the direction of this force is such as to retard the motion of the electrons in their advance away from the cathode.

g. Layers after layers of electrons feel these forces acting on both sides of them and, while they do move under the velocity of emission, the net result is a *cloud of electrons near the cathode.* The density of this cloud is greatest near the cathode, and less and less dense away from the cathode, as in A. To complete the picture, it is necessary to consider two other points. Since each charge in space exerts an influence upon each neighboring charge, the accumulation of the charges in space gives rise to an accumulative effect of great importance. This is the condition created in the proximity of the cathode. It is spoken of as the *space charge.* Actually it is an electrostatic field that exists between the boundary of the electron cloud facing the cathode and the emitting surface of the cathode, as shown in C.

h. One limit of this electrostatic field is the number of charges and the other is the emitting

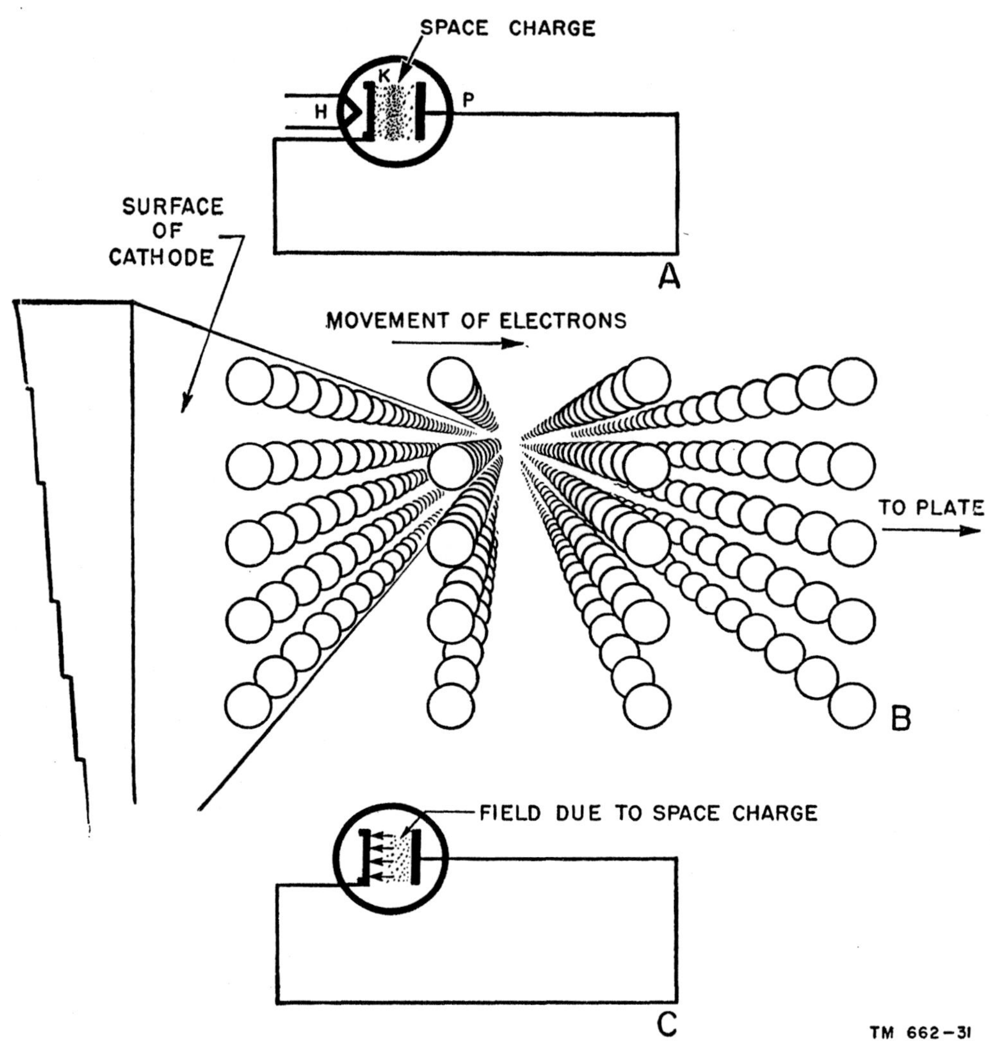

Figure 22. Electron accumulation in diode.

surface of the cathode. The direction of the lines of force in terms of emitted electrons is away from the electron cloud. The cathode surface, on the other hand, is assumed to be the positive boundary of the field, because as each electron leaves it during emission, the cathode contains one additional positive ion (an atom lacking one or more electrons) and tends to attract electrons back into itself. It might seem that the density of the space charge will increase as long as the cathode is emitting electrons. Such is not the case. For any one given temperature of the cathode, the emission of electrons is definitely limited, as is the *density* of the space charge. The reason for this is the repelling action of the space charge. Having acquired a critical density, the space charge develops a field of such intensity as to *repel back into the cathode one electron for every electron which enters the electron cloud.* This condition of equilibrium in the space charge is called *emission saturation.*

i. Every value of cathode temperature has a condition of emission saturation. This is understandable because an increase in cathode temperature increases the velocity of the emitted electrons and these new electrons will enter the space charge and make it more dense. After the density has been increased to the point where the field strength can offset the increased velocity of the emitted electron, a new level of emission saturation is reached for the new cathode temperature. It is evident, therefore, that the space charge has a *controlling influence upon the emission of electrons from the cathode.* The control exerted by the space charge on the plate current can be seen by reference to figure 23. The positive plate voltage is applied to that electrode, giving it a positive charge. The plate is positive relative to the cathode, but in between these two electrodes is the space charge with its retarding influence on the emitted electrons.

the field created by the voltage applied to the plate. The field between the space charge and the cathode is treated separately. The movement of an electron between cathode and plate can, therefore, be viewed as being first into the space charge and then out of the space charge to the plate. In this way equilibrium or constant electron density is maintained in the space

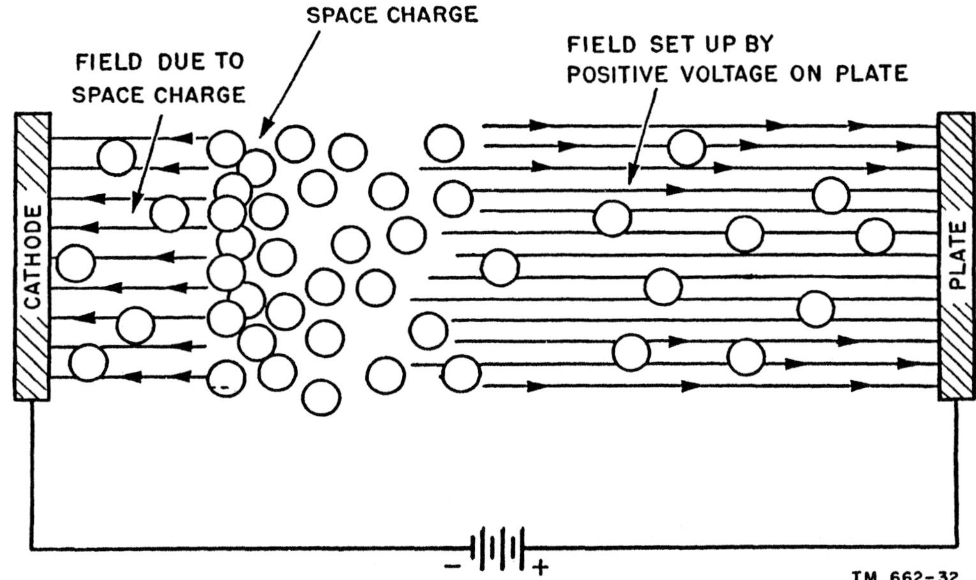

Figure 23. Electrostatic fields present between cathode and plate as result of space charge.

j. Two electrostatic fields are, therefore, present between the cathode and the plate: one between the cathode and the space charge and the other between the space charge and the plate. Because of the presence of the space charge, it is entirely proper to view the attracting force of the positively charged plate as acting on the electrons in the space charge, rather than directly on the electrons being emitted.

k. The reason for this is as follows: The lines of force going to the positively charged plate represent a certain number of excess positive charges on that plate, and, naturally, a deficiency of a certain number of negative charges. The location of sufficient negative charges to serve as terminations of the lines of force going to the positive plate is unimportant. Although it is true that the plate voltage is applied between the plate and the cathode, the electron cloud (space charge) located in space between these two electrodes furnishes the number of negative charges required as terminations for

charge. If the equilibrium in the space charge is disturbed by removing electrons from the portion that is nearer the plate because of the attracting power of the positive voltage on the plate, it results in a reduction of the magnitude of the field of the space charge. Its ability to repel electrons emitted from the cathode is, therefore, reduced, and consequently only that number of electrons required to restore equilibrium will be added to the space charge in the zone near the cathode. More than this number will not be added to the space charge, because the state of equilibrium must be maintained. If electrons equal to 10 ma (milliamperes) of current are removed from the space charge per second, that many electrons will enter the space charge from the cathode each second. The movement of electrons is from the cathode into the space charge, through the space charge, and out of the space charge to the plate *without altering the density of the space charge.*

l. A directed movement of electrons in this manner can be described as being the flow of plate current between the cathode and the plate, thence through the remainder of the system. It is permissible to view the space charge as a reservoir of electrons, from which are drawn the negative charges which are required to equal the instantaneous plate current. The cathode is the supply which replenishes the space charge. This is similar to a water-supply system in which the consumers draw water from a local reservoir, and a stream or lake automatically supplies the equivalent amount of water required to maintain a predetermined level.

m. For a given value of plate voltage and a constant cathode temperature, a fixed number of electrons are drawn from the space charge per unit time. This results in a plate current of a certain value which is measured in amperes. A *change* in plate voltage *changes* the rate of electron flow and the *plate current* likewise changes to an extent determined by the direction of change in plate voltage. At a sufficiently high positive plate voltage, all the emitted electrons enter the space charge because that many leave the space charge. Then the *plate current* is equal to the *emission current*, by which is meant the total number of electrons emitted by the cathode per unit time.

n. The space charge performs a useful and necessary function. In most tubes, the cathodes (or filaments) are designed to emit a surplus of electrons so that more of them are available than are actually required during each unit time. If there were no space charge, even a very low positive plate voltage would result in very high values of current and short emitter life. Some kind of control is, therefore, desirable, and this is exerted by the space charge. Hence the statement that plate-current flow in diodes (and other electron tubes) is *space-charge limited.*

18. Vacuum-tube Characteristics

To study more adequately the various factors affecting diode and other tube types, this discussion of vacuum-tube characteristics is of fundamental importance.

a. In a vacuum tube, a definite relationship exists between plate voltage and plate current —that is, between the magnitude of the attract-ing force exerted by the positively charged plate and the number of electrons attracted to the plate. There is also a relationship between emitter temperature and electron emission—more specifically, filament emission and filament current—and between cathode emission and heater current.

b. The manner in which these variable factors influence a vacuum-tube operation establishes the characteristics of the tube, and this is most clearly illustrated by means of graphs known as *characteristic curves.* As will be shown, these curves provide useful reference data for studying tube performance, and also for predicting performance under particular conditions of operation.

c. For simplicity, characteristic curves can be regarded as being charts of *cause and effect.* Characteristic curves are plotted within the boundaries of two reference lines: one, a vertical reference line called the *vertical axis,* the *Y-axis,* or the ordinate; the other, a horizontal line called the *horizontal axis,* the *X-axis,* or the abscissa (fig. 24). These two reference lines serve as scales upon which can be indicated the units of measures (volts, ohms, amperes, and so on), representing variations in the dependent variable (effect), and the units of measure (volts, ohms, amperes), representing variations in the independent variable (cause).

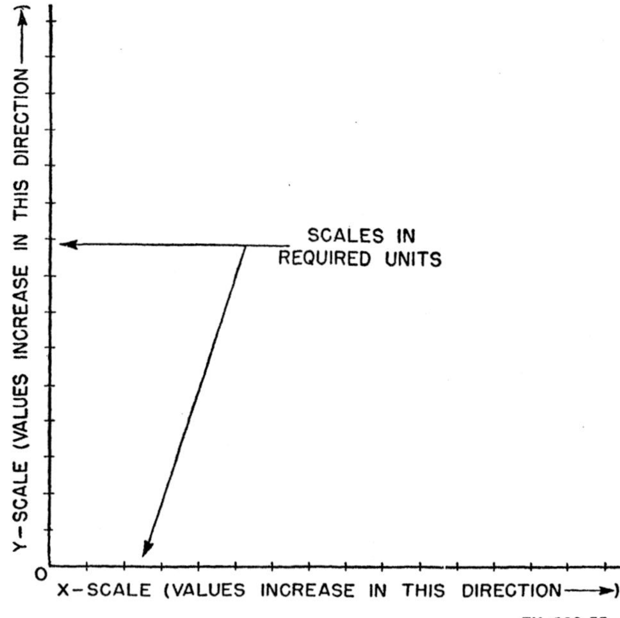

Figure 24. Organization of chart on which tube characteristics can be drawn.

d. Generally, changes in the magnitude of the *cause* (X-axis) are shown on the abscissa, and changes in the magnitude of the *effect* (Y-axis) are shown on the ordinate. These scales of magnitude usually, though not always, start at the junction of the two axes, the 0 point. The units indicated depend on what information is to be conveyed.

e. The characteristic curve (or curves, for there can be more than one), appears inside the space bounded by the two axes (heavy line in fig. 25). The characteristic of a device also is shown in the figure. Its description is unimportant; all that matters is that the characteristic curve displays cause and effect: The current (in milliamperes) through the device varies for different applied voltages (in volts). Current is effect, and voltage is the cause. The effect is shown on the Y-axis in steps of 1 ma and the cause is shown on the X-axis in steps of 2 volts. What is the current corresponding to a voltage of 10 volts? A projection upward from the 10-volt point on the voltage scale meets the Y-axis or current scale at 5.6 ma. This is the amount of current corresponding to 10 volts.

f. To establish the effect of a change from a known reference value, assume that the current corresponding to 24 volts is required. Repeating the previously described process, but starting at the 24-volt point on the voltage scale, leads to a new current value of 10.6 ma. Consequently, a change of 14 volts (from 10 to 24 volts) means a change of 5 ma.

g. Finally, simultaneous projections can be made from both axes, for various magnitudes of whatever is indicated upon the abscissa and the ordinate. These intersect, establishing points from which a characteristic curve *can be drawn* that will identify the relationship between the quantities indicated upon the reference axes. Several examples appear later.

h. In general, characteristic curves involve two variable factors. These have already been indicated, in terms of *cause* and *effect*. The characteristic curve for this relationship is termed the *plate-current plate-voltage characteristic* curve, or the I-E curve. Similarly, if the curve illustrates the cathode temperature for varying heater current, it would be called the *cathode-temperature heater-current* characteristic curve, since it is the change in heater current that causes the variation in cathode temperature.

i. Characteristic curves operate in both directions. If the curve is viewed as the expression of the cause and its resultant effect, the effect resulting from a change in the cause can be

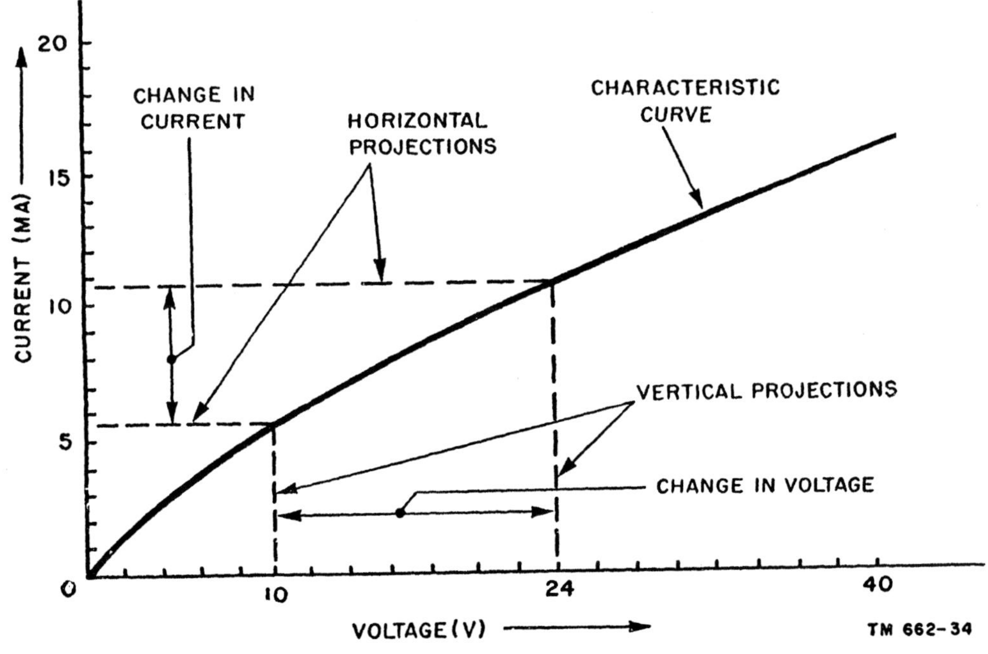

Figure 25. Plot of simple characteristic curve.

noted. It is also possible to regard this curve in another fashion. It may be desired to know what change must be made in the primary variable in order to produce a definite condition in the second variable. Consequently, a plate-current plate-voltage curve can be examined to ascertain the plate voltage that is required to produce a certain plate current, rather than the effect upon the plate current by a certain change in plate voltage. In other words, the curves operate with equal facility in two ways, one being the resultant change in effect as the cause is varied, and the other being the required change in cause to produce a certain change in effect.

j. Although characteristic curves illustrate the relationship between two variables, a third quantity usually is associated with them. For example, in the I-E curve, it is necessary to state the temperature of the cathode, the current through the filament or heater wires, or the voltage applied to the filament or heater wires. If no information is given regarding the cathode temperature, current flow, or voltage, the characteristic could not be identified as applying to any *particular* condition. For, while the curve can be representative of behavior, the exact values of plate current indicated mean nothing unless the exact conditions producing the state of behavior are identified. At least three values are indicated on or required for a vacuum-tube characteristic curve: the two variables illustrated in the curve, and a third, each indicating the exact condition contributed by the other factors to the information illustrated in the characteristic curve.

19. Linear and Nonlinear Characteristics
(fig. 26).

a. Two terms often encountered in descriptions of characteristic curves are *linear* and *nonlinear.* The meaning of the first term is graphically illustrated in A of figure 26. For example, if a d-c voltage is applied across a fixed resistor, the relationship between the voltage (one quantity) and the current (the second quantity) is a straight line. The resultant change in the second quantity (current) is *directly proportional* to variations in the first quantity (voltage). Such a relationship is said to be *linear*, and it appears as a straight line over the full range of variation. In this case, the term *curve* is not truly descriptive of the straight line. However, not many characteristic curves of vacuum tubes are linear throughout their length.

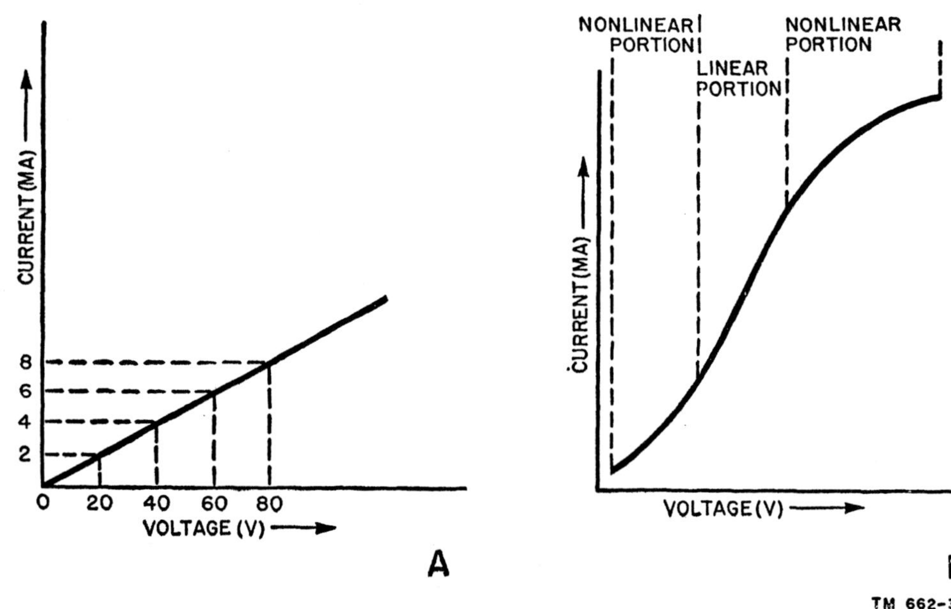

Figure 26. Linear and nonlinear characteristic curves.

b. B of figure 26 shows a tube characteristic curve which resembles the type found frequently. The characteristic is *nonlinear* over a region, representing a range of applied voltage, then becomes linear over another range of voltages, and again becomes nonlinear. Just why this is so is unimportant; the importance lies in recognition of the meaning of linearity and nonlinearity. The reason for this is frequent reference to operation over the linear portion of a tube characteristic. These are specific conditions of use.

c. At all times, linear operation means *direct proportionality* between a cause and an effect, as illustrated in A of figure 26, where a two-time increase in voltage, from 20 to 40 volts, causes a two-time increase in current, from 2 to 4 ma. Only a straight-line characteristic is a linear characteristic. This can be the entire characteristic curve or a part thereof, as indicated in B of figure 26. The completely nonlinear characteristic is shown in figure 27. At no point is one variable directly proportional to the other.

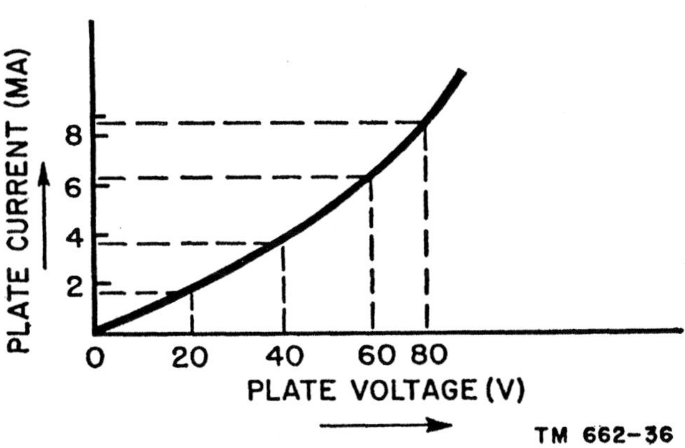

Figure 27. Curve illustrating completely nonlinear characteristic.

20. Static and Dynamic Characteristics

Linear and nonlinear characteristic curves can be of either the *static* or the *dynamic* type. Both static and dynamic characteristic curves exist for each vacuum tube. They differ in shape as well as in the actual values they represent. A simple explanation of the difference between these two types of curves is that, in *static* characteristics, the values are obtained with different d-c potentials applied to all the tube electrodes, and the organization of the system is not typical of actual operation. The *dynamic* characteristics are the values obtained when the organization of the system conforms with typical operation. This is explained more fully later.

21. Family of Curves

a. Several characteristic curves usually are shown together, to illustrate the relationship between the same two quantities under different conditions. A group of such curves is known as a *family.* Figure 28 is a family of different plate-current plate-voltage curves for different values of cathode temperature. For a given value of plate voltage, the plate current will be maximum at curve A. Therefore, curve A indicates the highest cathode temperature, curve B shows a decrease in cathode temperature, and the temperature is the least for curve C. Numerous families of curves appear in later sections of this book.

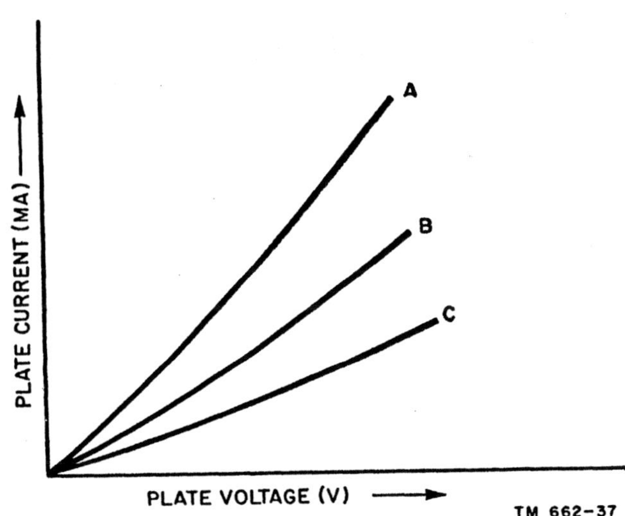

Figure 28. Family of plate-current plate-voltage curves for three values of cathode temperature.

b. The characteristic curve, or family of curves, usually encountered in vacuum-tube literature relates to the flow of plate current in the tube. The most common, and perhaps the most important of vacuum-tube characteristics, is the one illustrating the magnitude of the electron stream arriving at the plate for various values of positive voltage applied to the plate—namely, the plate-current plate-voltage characteristic. Most vacuum-tube applications are judged in terms of the plate current, since it is by the flow of plate current through the tube and the different electrical devices externally connected to it that the various capabilities of the tube can be utilized.

22. Plate-current Emitter-temperature Characteristic
(fig. 29)

To investigate the manner in which the emitter temperature and the space charge influence plate current in a diode, a circuit is set up in which different values of emitter temperature can be created by varying the heater or filament current while the plate voltage is held constant.

a. In A of figure 29, a fixed filament voltage source, A and a variable resistor, R, allow changes in the heater current. The filament meter indicates the current in this circuit. I_f indicates the flow of filament current and E_f the voltage across the filaments. The plate voltage is supplied by battery B and the plate current is read on the meter in the plate circuit. I_b indicates the flow of plate current. It might be well to mention that the capital letter I for current indicates constant, average, or rms (root mean square) values by appropriate subscripts. Instantaneous values are indicated by a small letter i for current and e for voltage. Constant, average, or rms values of voltage are indicated by the capital letter E with appropriate subscripts. The subscript letter f in figure 29 indicates the filament, whereas the subscript b near the plate-current meter associates it with the plate. Common usage has established

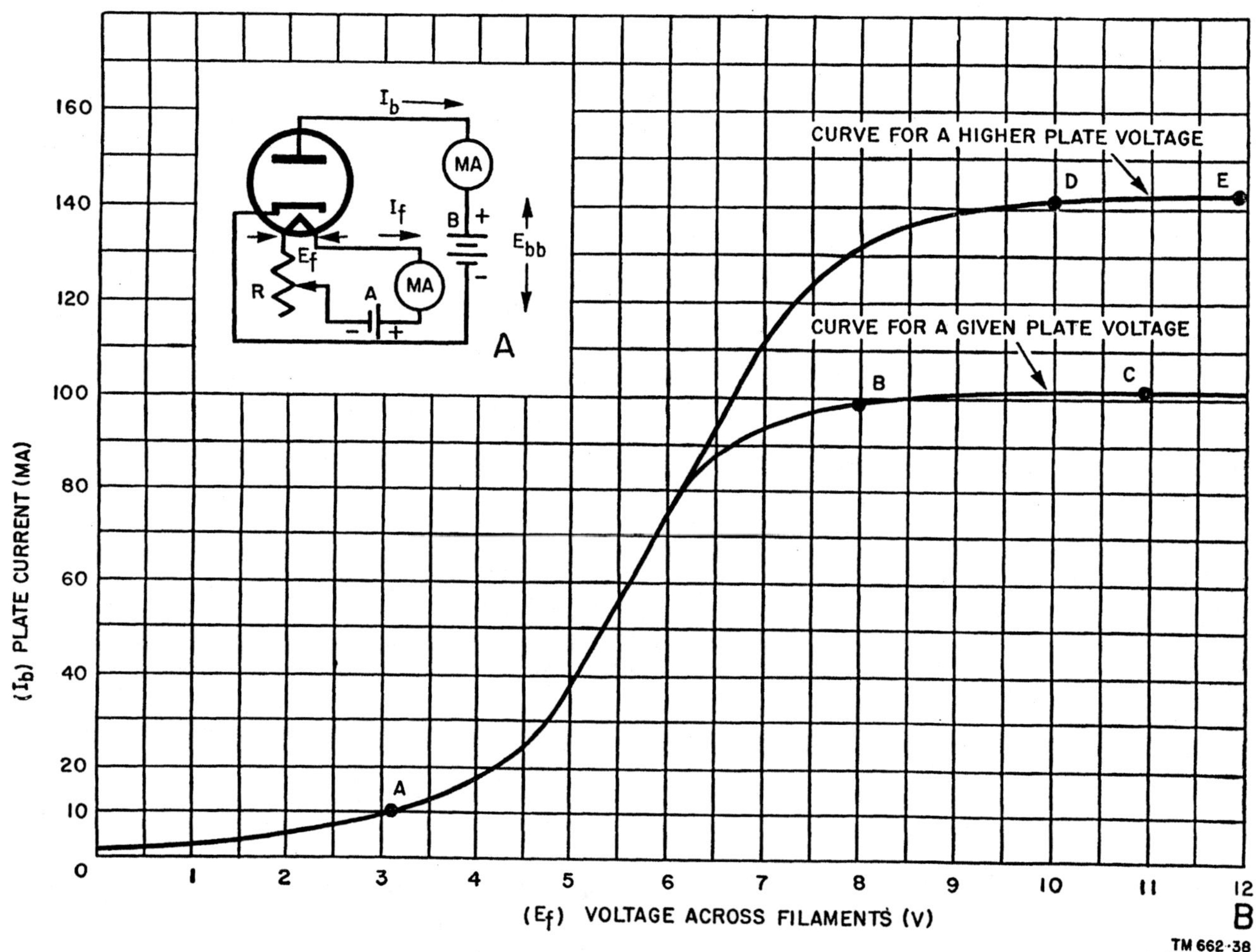

Figure 29. Plate-current filament-voltage characteristic curves.

E_{bb} as the voltage of the plate voltage supply. These subscripts are further explained in chapter 4.

b. The general contour of the $I_b = E_f$ curve can be predicted from what has already been said about the conditions of emission. The curve of filament voltage versus plate current, with fixed plate voltage, would start at 0 corresponding to the 0 value of filament voltage. Then it would rise as the voltage is increased. Eventually it would flatten when the emission saturation points (B and D of fig. 29) would be reached for the applied value of plate voltage. Beyond this value of filament voltage, the increased emission contributes little to the plate current (lines BC and DE), since the number of electrons pulled out of the space charge by the plate voltage is not increased. Only the density of the space charge is increased. As is evident in the graph, a slight rise in plate current occurs with increased filament voltage beyond emission saturation, but this increase is so slight, relative to the change in plate current up to the saturation point, that it can be disregarded for all practical purposes.

c. Referring to curve ABC (B of fig. 29), the span A-B represents current limited by cathode emission for the existing plate voltage. Between points B and C on curve ABC the limitation of plate current is due to the action of the space charge. A means of raising the plate-current limitation is by reducing the effectiveness of the space charge. This can be done most simply by raising the plate voltage.

d. Raising the plate voltage results in a rise of the plate-current curve, and a new value for the point of emission saturation or the start of the space-charge-limited plate-current area. This is shown as curve ADE. Consequently, it can be seen that, for every value of plate voltage, there is a limit to the number of electrons which can be drawn over to the plate in unit time to form the plate current. For the two curves, these are points B and D. The point at which emission saturation is reached is also known as the *saturation point,* and the maximum current flowing (as BC and DE) is known as the *saturation current.*

23. Plate-current Plate-voltage Characteristic

a. Another important relationship is that between the voltage applied to the plate and the resultant plate current when the emitter temperature is held constant. This approaches the conditions under which the tube is used. The circuit for developing these data is shown in A of figure 30. A voltage source, B, applies a voltage between cathode and plate through a current meter, I_b. The voltage applied to the plate of the diode is shown on a voltmeter and is labeled E_{bb}. The subscript letters bb associate voltage E with the source—namely, the battery, B. The meter labeled I_b indicates the plate current flowing in the system.

b. The diode in A of figure 30 is shown without a heater and its associated current source. It is common practice to do this when illustrating the circuity of simple, indirectly heated tubes. The presence of the heater is assumed. If this were a filament-type tube, the circuit would appear as in B of figure 30. Since the emitter temperature is not a variable, the heater or filament circuit is not indicated.

c. If curve ABC shown in C of figure 30 is examined, it will be noted that a fairly rapid rise in plate current occurs as the plate voltage is increased from the 0 value. Then the increase in plate current becomes less rapid, with uniform increases in plate voltage until a point, B, is reached where very little increase in plate current occurs as the plate voltage is increased still more. Point B represents the condition when *all of the emitted electrons* are admitted into the space charge to form the plate current. Between points A and B, the plate current is limited by the space charge; between B and C, it is limited by the emission. Consequently, in the example shown, raising the plate voltage from 0 to 350 volts increases the plate current from slightly above 0 to about 68 ma. Raising the plate voltage from 350 to 500 volts causes an increase of only 2 ma from 68 to about 70 ma. The explanation for the presence of a very slight amount of plate current with 0 plate voltage is the fact that some of the electrons are emitted from the cathode with such high velocity that they advance across the space between the emitter and plate without benefit of any attracting force issuing from the plate.

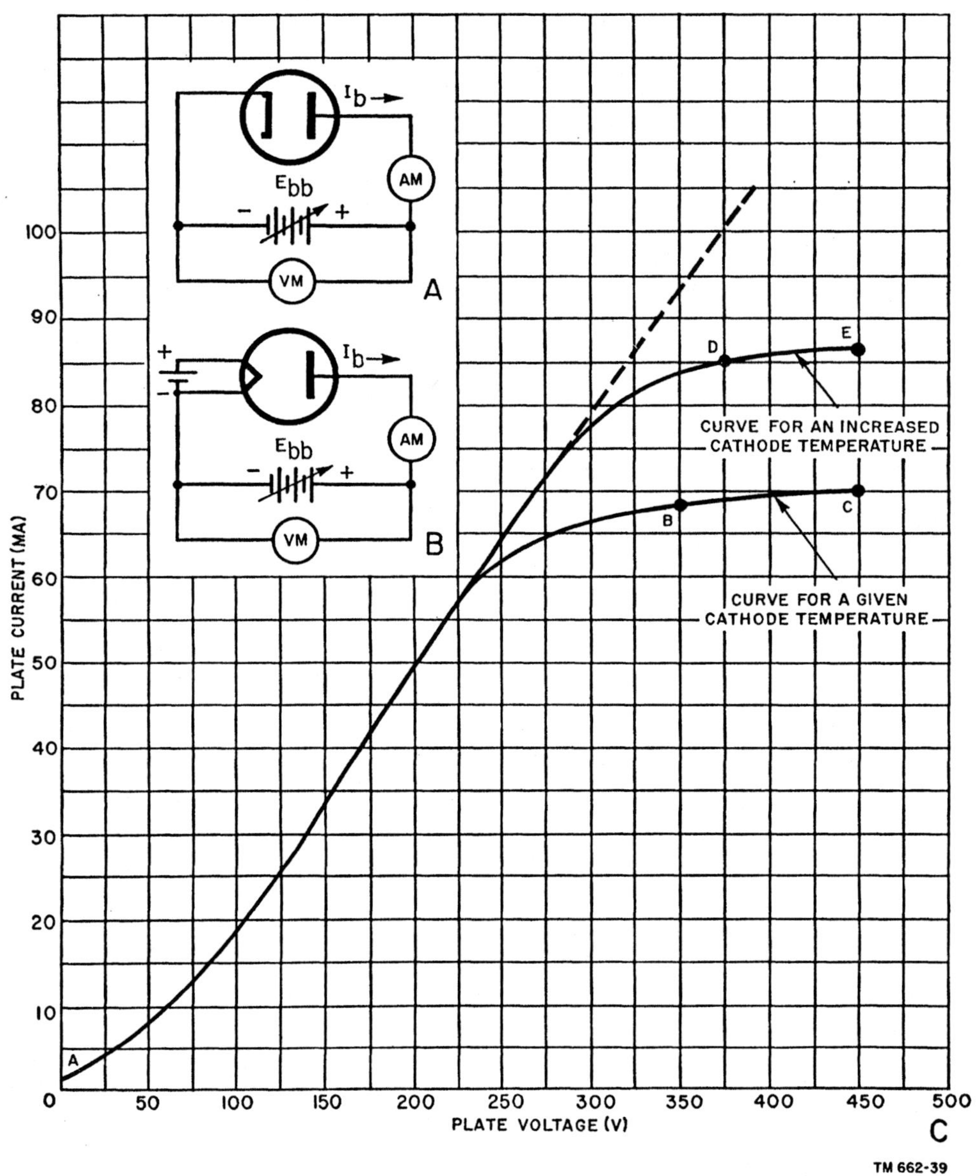

Figure 30. Plate-current plate-voltage characteristic curves.

d. To obtain an increase in plate current for plate voltages in excess of 350 volts, it is necessary to provide a greater flow of electrons. This can be accomplished by raising the emitter temperature. Assuming an increase in cathode temperature, a curve similar to ADE might be obtained. The plate-current saturation point D corresponds to a plate-current flow of about 85 ma (in contrast to the previous 68 ma at point B) and a plate voltage of about 375 volts. Any further increase will not produce a substantial increase in plate current.

e. The discussion thus far reveals that, for every tube operated at a fixed emitter temperature, there is a certain value of plate voltage beyond which it is useless to increase the voltage further, since all of the emitted electrons are moving over to the plate, and consequently, no appreciable plate current increase can be obtained. However, one factor remains: There is a difference in saturation points depending on whether a tungsten, a thoriated-tungsten, or an oxide-coated cathode is used. The curve in C of figure 30 was obtained with a diode having a tungsten filament. Figure 31 shows that the emission from an oxide-coated emitter is considerably larger, since the plate current is larger, than that of the tungsten types. Oxide-coated emitters release so many electrons that it is practically impossible to find plate voltages which will draw off all of these electrons without ruining the tube. Consequently, most modern tubes which use oxide-coated cathodes do not display a plate-voltage saturation characteristic, but rather show continually rising plate-current curves as the plate voltage is increased as shown in figure 31. Note from the curves that as the operating temperature of the cathode increases, the plate current increases. In addition, this increase will have a more noticeable effect on the oxide-coated cathodes since the solid and dashed lines are much farther apart in the higher operating temperatures than in the lower ones.

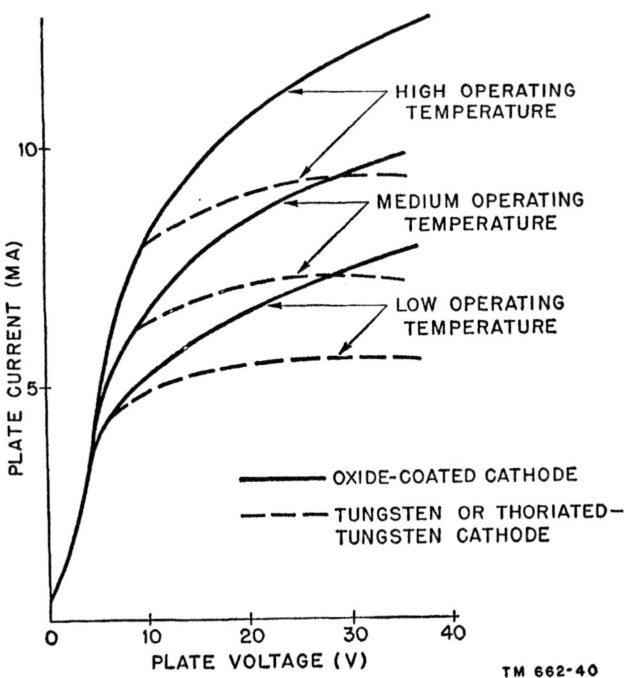

Figure 31. Plate-current emitter-temperature characteristic curves of diode with different types of emitters.

24. D-c Plate Resistance of Diode

a. Since the application of a positive voltage to the plate of a diode results in the flow of plate current, and since some form of control of the magnitude of plate current for different values of voltage exists in the tube, it is entirely proper to view the current control mechanism, what-ever may be its nature, as being the equivalent of a resistance. Electrical fundamentals teach that the control of current in a system issues from some form of opposition to the current flow present in the system.

b. Such opposition exists in all electron tubes, among which is the diode. It stems from many factors, such as the spacing between the electrodes, their physical size, the conditions of emission, the conditions of the space charge, and, in general, the energy wasted while the electrons are traversing the space within the tube. Tube resistance is of two kinds: *d-c plate resistance* and *a-c plate resistance*, each of which will be explained. The former usually is symbolized as R_p and the latter as r_p.

c. The d-c plate resistance of the diode is that opposition to plate current flow offered by the tube when a d-c voltage is applied to the plate. It can be calculated from Ohm's law, $R=E/I$, with the plate voltage and the plate current inserted as the known factors in the equation.

d. Referring to figure 32, B shows a typical plate-current plate-voltage characteristic for a type 6H6 duo-diode tube, using only one pair of the two pairs of elements contained inside the envelope—that is one cathode and its associated plate. With 8 volts dc applied to the plate, in A, the plate current, I_b, is seen to be 10 ma, or .01 ampere. (The notation E_b is used as the varying constant voltage between the plate and cathode. This may or may not be the same value as E_{bb}.) Applying Ohm's law for d-c circuits, these values of plate voltage and plate current represent a d-c plate resistance of

$$R_p = \frac{E_b}{I_b} = \frac{8}{.01} = 800 \text{ ohms.}$$

e. Similarly, with 20 volts dc applied to the plate, the curve shows a plate current of 40 ma, or .04 ampere. This corresponds to a d-c resistance of 20/.04 = 500 ohms. With 28 volts dc applied to the plate, the plate current on the curve is 66.4 ma or .0664 ampere, and the d-c resistance is 28/.0664 = 422 ohms.

f. Examination of these figures, and also of the shape of the characteristic curve, shows that the resistance offered by the diode to the flow of plate current is *not* constant, as is ordinarily the case when resistance is present in a conventional d-c circuit. The characteristic shows that the resistance of the diode *decreases*

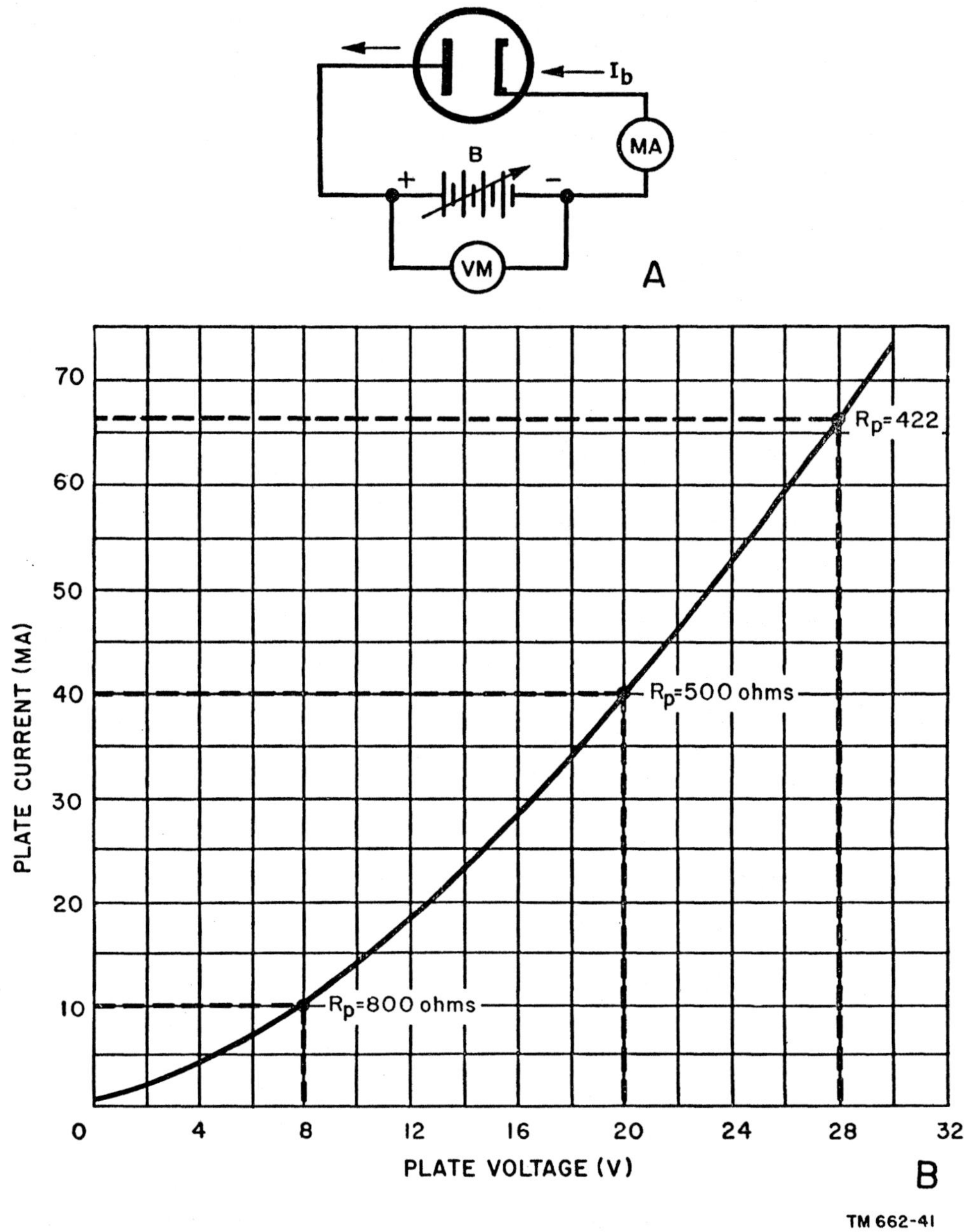

Figure 32. Plate-current plate-voltage characteristic for type 6H6 tube, using only one diode section.

as the plate voltage increases, and *increases* as the plate voltage is decreased. It has a nonlinear behavior. If the relationship of plate voltage and plate current were linear over the entire range of plate voltage, the plate-current curve would be a straight line instead of having a curved characteristic. A straight line would indicate that the d-c plate resistance remained constant over the full range of plate voltage.

g. Another significant fact in connection with the $E_b = I_b$ characteristic curve is the condition resulting when the two sections of this duo-diode are connected in parallel, and the two cathodes also are joined in parallel. The plate current in this case becomes twice the amount obtained when the same value of plate voltage is applied to a single pair of elements. Accordingly, the values at any point along the plate-current curve are doubled when both sections of the duo-diode are paralleled. The plate resistance, therefore, becomes half of that existing when one pair of the two diode portions is used. This will be found to be true in all multiple-section diodes.

h. The absence of plate-current saturation points similar to those appearing in figures 29 and 30 is due to the use of oxide-coated emitters in this tube. An application as low as 32 volts to the plate results in a fairly high value of plate current, almost 70 ma. Judging by the shape of the characteristic, the saturation point is a long way off. The tube probably would be damaged long before the plate-current curve would flatten because of the application of sufficient plate voltage to attract all the emitted electrons. This does not mean that diodes are not used at plate voltages exceeding 32 volts; they are used with very high voltages, many tens of thousands of volts, but when so used, diodes are specially designed.

25. A-c Plate Resistance of Diode

a. The a-c plate resistance of the diode is defined as the resistance of the path between cathode and plate to the flow of *an alternating current* inside the tube. Such is the definition meant when the term *resistance* is used without qualification for the diode and all other electron tubes. As will be seen later, it is an important term and is closely related to operating conditions.

b. The a-c plate resistance is the ratio of a small change in plate voltage by the corresponding change in plate current. Expressed in an equation

$$r_p = \frac{\triangle e_p}{\triangle i_p}$$

in which r_p is the a-c plate resistance in ohms, $\triangle e_p$ represents a small change in plate voltage, and $\triangle i_p$ represents a corresponding small change in plate current caused by the plate voltage change. The Greek letter $\triangle$ signifies a *small change of*.

c. Figure 33 shows a plate-current plate-voltage characteristic curve identical to the one in B of figure 32 and obtained from the circuit in A of figure 32, where one section of a 6H6 diode is used. To establish the *a-c resistance* at any point along the curve, the plate voltage is varied on both sides of it at some definite value. B of figure 32 shows that the d-c plate resistance, R_p for this diode is 500 ohms when the plate voltage is held constant at 20 volts dc.

Suppose it is desired to find the a-c resistance when the mean value of this assumed a-c voltage is 20 volts, and the variation or *swing* is arbitrarily chosen at 1.6 volts on either side of the 20-volt point; that is, when $\triangle e_p$ equals 3.2 volts (fig. 33).

d. To compute r_p, the change in plate current for the given change in d-c plate voltage first must be found from the curve. At 21.6 (20 plus 1.6) volts, the plate current is 45 ma or .045 ampere. At 18.4 (20 minus 1.6) volts the plate current is 35 ma or .035 ampere. These provide the two limits of voltage and current. Consequently,

$$r_p = \frac{\triangle e_p}{\triangle i_p} = \frac{21.6 - 18.4}{.045 - .035} = \frac{3.2}{.010} = 320 \text{ ohms.}$$

e. The value of 320 ohms given above is the a-c plate resistance for a plate voltage which varies between 18.4 and 21.6 volts, whereas the same point is equal to a d-c plate resistance of 500 ohms when the plate voltage is held constant at 20 volts. Using the same method, 278 ohms a-c plate resistance is obtained at the 28-volt plate-voltage point for a change in plate voltage of 1.6 volts each side of the mean value of 28 volts, whereas the d-c resistance is 422 ohms when the plate voltage is maintained constant at 28 volts. Similarly, r_p is shown equal to 527 ohms when the 8-volt plate-voltage point varies *1* volt on either side of the 8-volt mean value. If 1.6 volts is used in the latter case instead of 1 volt, r_p still remains relatively close to 527 ohms. However, it must be kept in mind that the smaller the change in plate voltage, the more accurate are the results. This is because of the nonlinearity of the characteristic curve about this small change. In the former two cases, the curve is fairly linear and 1.6 volts are used arbitrarily. Normally, for linear operation, the operating point for this curve is chosen with an r_p equal to about 320 ohms.

f. It can be seen that an appreciable difference exists between d-c plate resistance and a-c plate resistance, the latter being approximately one-half of the former. This is, in general, true for all types of vacuum tubes. Furthermore, as figure 33 shows, the a-c resistance also is related to the plate voltage, *decreasing* as plate voltage is *increased*, and *increasing* as plate voltage is *decreased*. The *exact value of a-c resistance depends on the point of operation se-*

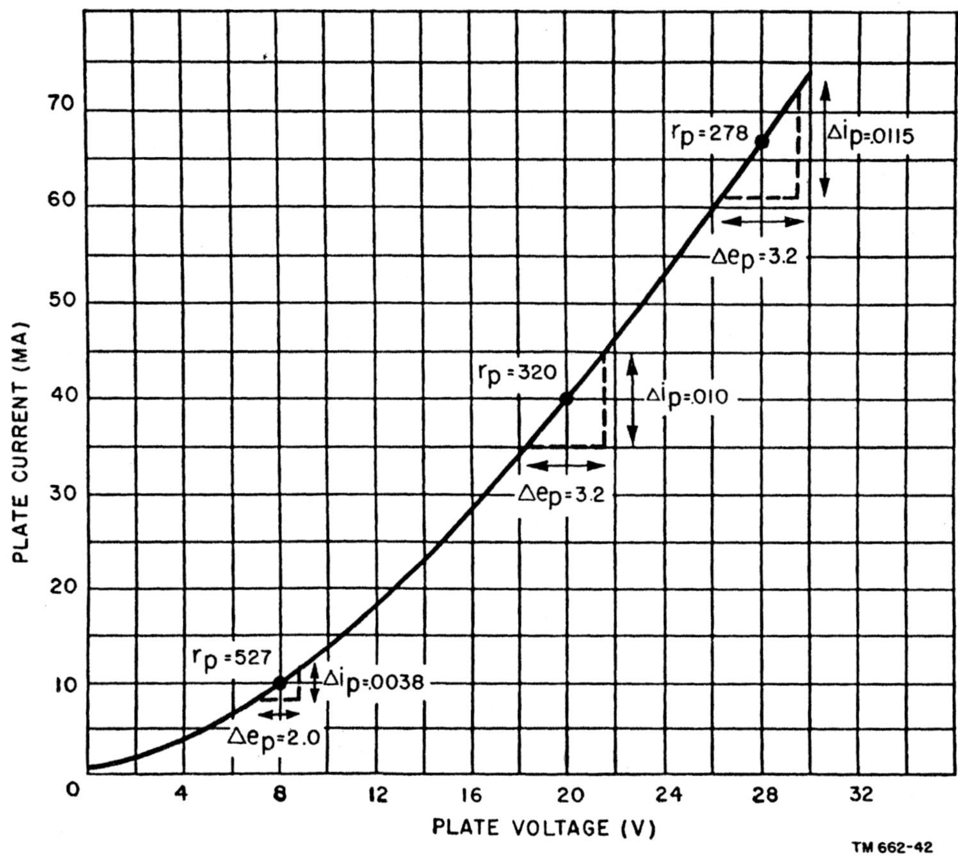

Figure 33. Plate-current plate-voltage characteristic which can be used to determine the a-c plate resistance of a diode.

lected on the characteristic curve. Consequently, if the point of operation is at a plate voltage of 20 volts, the a-c plate resistance is greater than when the operating point chosen corresponds to a mean plate voltage of 28 volts.

26. Static and Dynamic Diode Characteristics

a. The discussion of diodes thus far has been for static conditions and not for actual operating conditions. For a diode, or any tube, to be able to perform its normal function, its external circuit must contain a *load*. It is through this load that the diode current flows outside the tube, and the *voltage drop developed across this load then represents the output of the tube.* With such a load, represented by a resistance R_L (fig. 34), the operating characteristic of the tube is changed materially. The plate-current plate-voltage curve is altered noticeably, and it represents the *dynamic* characteristic rather

than the *static* characteristic which applies when there is no load.

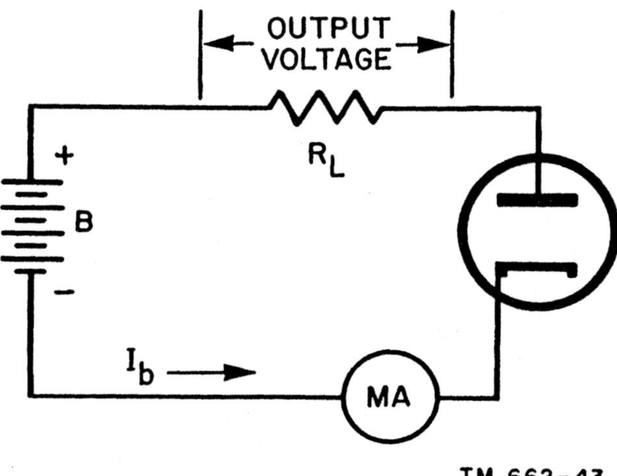

Figure 34. Adding a load resistor, R_L, to a diode circuit to obtain an output voltage.

b. With no load in the circuit, or when R_L equals 0, the resistance in the circuit is the resistance of the tube; also, the small resistances of the battery and meter are present, but these can be neglected. But when an external resistance, the *load resistance,* is added (fig. 34), the total opposition to plate-current flow includes that of the tube itself and also that of the load. When the load resistance is many times the value of the internal resistance of the tube, the tube resistance offers only a negligible percentage of the opposition to plate-current flow. Consequently, if the external load resistance maintains its value regardless of the amount of current flowing through it, the plate-current plate-voltage characteristic of the system is changed from a curved line into a substantially straight line.

c. These various conditions can be noted on a separate graph (fig. 35). The dashed line represents the static characteristic—that is, when there is no load resistance. Solid line 1 illustrates the voltage-current relationship when the load resistance is 1,000 ohms. This line possesses some curvature, but it is considerably less than that of the static line. Solid line 2 is obtained with a load resistance of 10,000 ohms. This line is almost straight; some curvature appears in the region of low plate voltage where the internal resistance of the tube is highest. A line representing a value of 100,000 ohms is so straight throughout its length that it lies too close to the horizontal axis, and it cannot be drawn clearly on the graph. Therefore, the higher the load resistor value, the straighter is the dynamic curve, and the lower is the amount of plate current flowing in the circuit. This is not a disadvantage, provided the load resistance is maintained within reasonable limits.

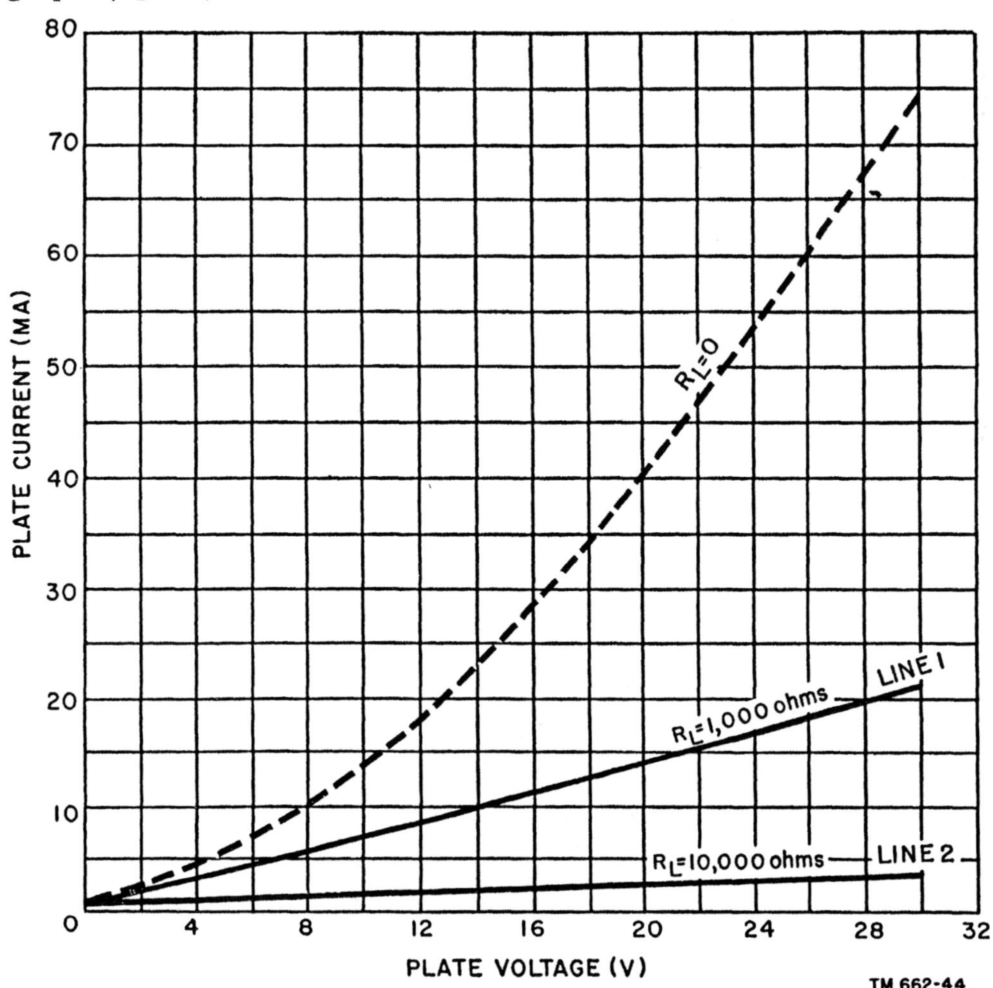

Figure 35. Plate-current plate-voltage characteristic curves of diode having different load values.

d. A linear dynamic characteristic is desired for the diode, principally because it assures proportionality between applied plate voltage and output current. This affords freedom from distortion in many circuits that use the diode, although, in other applications, distortion is of secondary importance. The most important consideration is the use of a load; a *load* is necessary if a diode is to be of practical use.

27. Uses

a. Since current flow in the diode is unidirectional, one of its most prominent uses is as a rectifier or converter of alternating current into direct current. In figure 36, an alternating voltage is used as the plate voltage. During the positive alternations of the input voltage, the plate is positive relative to the cathode, and plate current i_p flows through the tube and load resistance R_L. The current develops a voltage drop across the load. At every instant the output voltage (in volts) is:

$$e \text{ output} = (i_p \times R_L) - (i_p \times r_p)$$

where R_L is the load resistance in ohms

r_p is the a-c resistance of the diode in ohms

i_p is the instantaneous current in amperes.

voltage. The absence of conduction in the tube during the two negative alternations, B and B', of the input voltage is caused by the diode plate being negative with relation to the cathode during these periods. If the input voltage has a frequency of 60 cps (cycles per second), the output current pulses will occur 60 times per second, each lasting for a period equal to one-half of each input cycle.

d. Both single- and dual-section diodes have many other uses, developed later in this manual. All of these depend on the principle of unilateral conduction through the tube.

28. Types

High-vacuum diode tubes are available in various types (fig. 37), differing principally in physical dimensions and shape. They differ internally as well but not in the fundamental arrangement of the electrodes. The physical dimensions are related to the magnitudes of voltages applied to the plate and the currents flowing through the tube during operation. Tubes used in circuits which handle signal voltages of relatively low values are known as signal diodes. Another general category is that of power diodes or vacuum-tube rectifiers which handle high values of voltage and current. Each

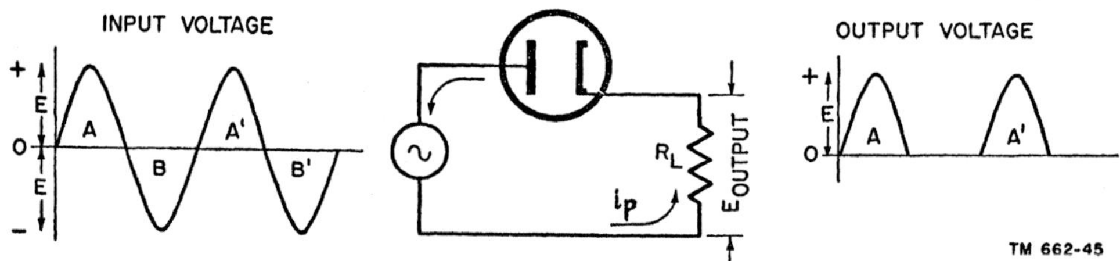

Figure 36. Diagram illustrating unidirectional current flow in diode.

b. If load resistance R_L is very much greater than r_p, the internal tube resistance, the effect of the latter is negligible and i_p times r_p of the preceding formula can be neglected.

c. Concerning the action in figure 36, it can be seen that the changes in current amplitude follow the instantaneous changes in applied plate voltage. This becomes evident when the input plate-voltage alternations, A and A', are compared with the output voltage pulsations, A and A'. These pulsations have the same shape as the two positive alternations of the applied

of these types is intended for special application and they are not interchangeable. Signal diodes are restricted to signal circuits. Power diodes generally are used in power circuits, except where certain signal systems entail relatively high voltages, and circuit operation demands an action similar to rectification.

29. Electron Transit Time

a. The time required for an electron to traverse the distance from the cathode to the plate of a diode, or other kinds of electron tubes, is

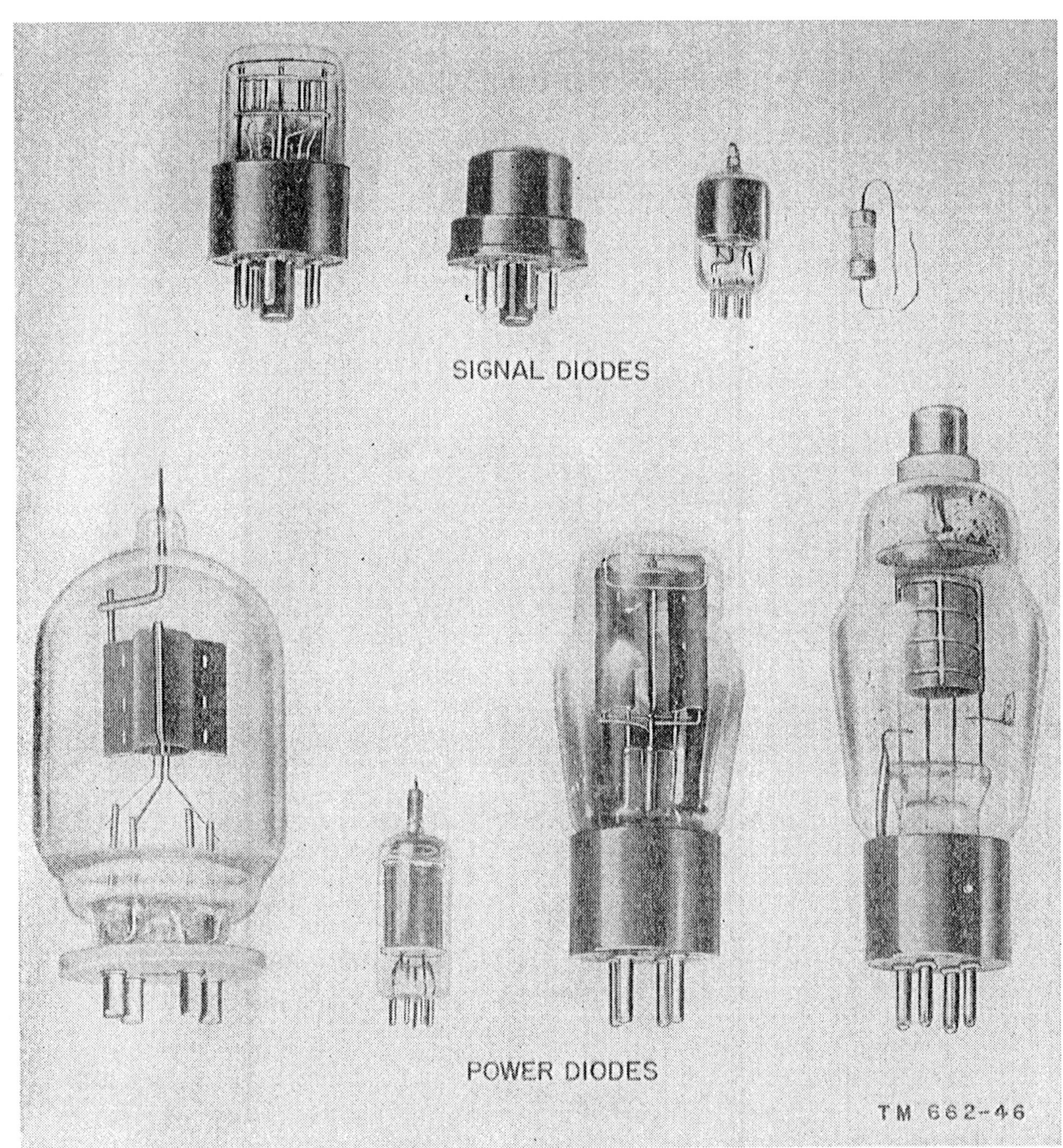

Figure 37. Various types of high-vacuum diodes.

called *electron transit time*. At operating frequencies below 200 mc, transit time introduces no particular problems. Under such conditions, the movement of the electron conforms with the changes in instantaneous amplitude and polarity of the plate voltage. This can be described more broadly by saying that the output plate current conforms with instantaneous changes in tube electrode voltages.

b. As the frequency of the electrode voltage rises, above 200 mc, for example, the transit time becomes an appreciable part of the cycle. The elapsed time for the completion of 1 cycle of a 200-mc voltage is one two-hundredth of a microsecond. The positive alternation of such a voltage is completed in one four-hundredth of a microsecond, or .0025 usec. For example, suppose that the applied plate voltage and the physical arrangement of the electrodes are such that .005 usec is required for the electron to travel from the cathode to the plate. Then if a 200-mc signal is applied to the plate, the electron undergoes a change while traveling to the plate because the polarity of the plate voltage has reversed from that which existed a moment earlier.

c. A *lag* of this kind cannot be tolerated because it interferes with the operation of the system. Therefore, diodes (and other electron tubes) are designed specially for use at very high frequencies. As a rule, these design differences are found in the physical spacing between electrodes and in the value of the applied voltages. Moreover, such tubes bear special designations indicating their suitability for very-high- and ultra-high-frequency applications.

30. Summary

a. The diode is a two-electrode tube containing an emitter of electrons and a collector of electrons. The emitter is known as the filament or cathode. In modern usage, the term *cathode* is used to indicate either type. Cathodes are of two types, directly heated and indirectly heated. The collector is commonly known as the *plate* and sometimes is referred to as the anode.

b. The flow of electrons in a diode constitutes the plate current. Plate-current flow in a diode is unidirectional, and the tube is, therefore, a unilateral conductor. This underlies the use of the diode as a rectifier.

c. The direction of plate-current flow is identical to that of electron flow, that is, from negative to positive.

d. An electron in an electrostatic field moves toward the positive limit of the field, its velocity per unit time being a function of the difference of potential existing between the limits of the field.

e. The important features of a vacuum tube can be obtained from its characteristic curves. A number of curves drawn on the same graph are known as a family of curves.

f. Static characteristics are obtained with different d-c potentials applied to the tube electrodes, whereas dynamic characteristics are obtained with an a-c potential applied to the control grid with different values of d-c potentials applied to the other electrodes.

g. One of the effects interfering with maximum plate-current flow in a tube is that resulting from the presence of the space charge. The space charge is caused by an accumulation of electrons in the immediate vicinity of the cathode. In this position, it tends to repel electrons back to the cathode and prevents them from flowing toward the plate.

h. The state of equilibrium in which the maximum density of the space charge for every fixed temperature of the cathode is reached is known as *emission saturation,* or the *saturation point.* The value of plate current flowing at that time is referred to as the *saturation current.*

i. Two types of resistance are present in a diode, the d-c plate resistance and the a-c plate resistance. This also is true of other types of tubes.

j. The phenomenon of electron transit time is not a factor of importance in medium- and high-frequency operation of vacuum tubes, but it is of very special concern in ultra-high-frequency operation.

k. The principal uses of the diode are as a rectifier and as a detector.

31. Review Questions

a. Describe the construction of a diode and draw its schematic.

b. By what other names are these identified: the emitter of electrons and the collector of electrons in a tube?

c. Describe two principal methods for heating electron emitters. Which is more commonly used?

d. Name the two types of rectifiers using diodes. What feature of diode operation makes possible its use as a rectifier?

e. What is a duo-diode and how does its construction differ from that of a regular diode? Illustrate schematically.

f. When an electron moves in an electrostatic field between two charged surfaces, what determines (a) its direction of travel, (b) its velocity of motion?

g. In a vacuum tube, why do electrons flow from the cathode to the plate but not in the reverse direction?

h. In what respect does a diode act like a valve when comparing it to the flow of plate current in a diode?

i. What is the *space charge* in a tube and how does it affect the operation of the tube? Is it desirable?

j. Name four factors that control plate-current flow in the diode.

k. Identify the two electrostatic fields present in a diode.

l. How does the plate current of a tube vary as the plate voltage is increased? Show this by a graph. Can the plate current be increased indefinitely by increase of plate voltage?

m. What is a positive ion?

n. What is meant by the *equilibrium* of the space charge?

o. Explain the terms emission saturation, saturation point, saturation current.

p. What is the effect of a load resistance upon plate-current flow and upon the dynamic curve?

q. What are some kinds of important information that can be obtained from characteristic curves?

r. What relationship is brought out in a plate - current plate - voltage characteristic curve?

s. In what sense do characteristic curves *operate in both directions?*

t. Explain and illustrate the terms *linear and nonlinear.*

u. What are the two types of resistances present in a diode?

v. What general information is contained in a family of characteristic curves?

w. What is the difference between the static and dynamic characteristics of a vacuum tube?

x. Name some types of diodes.

y. What action of the electron is referred to by the term *electron transit time?*

32. Control Grid

a. GENERAL.

(1) The invention of the *triode,* or three-element tube, was one of the most important steps in modern electronics. Up to 1907, the diode was the only electron tube used in the primitive wireless communication systems of that time. In that year, DeForest disclosed his *third element,* an electrode which was added to the diode and so formed the triode. Not only did it modify the diode, but it opened a new era in communication facilities. De-Forest's third element made present-day radio communication in all its forms a practical reality.

(2) The emitter and the plate as used in the diode appear also in the triode. They retain their functions as a source of electrons and a collector of electrons, respectively. In the space between them, and located nearer to the emitter, is placed the third element, commonly called the *control grid.*

b. PHYSICAL CONSTRUCTION OF TRIODE.

(1) Figure 38, illustrating the organization of the cathode, control-grid, and plate electrodes of the triode, is an example of the oval-shaped form of grid, and shows how the grid surrounds the emitter on all sides. The plate electrode is seen enveloping the control grid. Other examples of triode construction exist, but they do not differ greatly from this.

(2) The dimensions and the shape of the electrodes used in triodes, as well as the physical spacing between the electrodes, differ in accordance with the

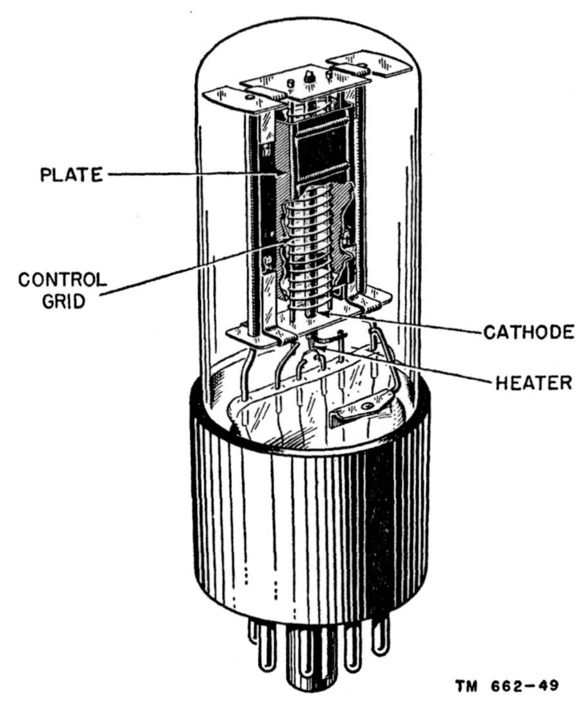

Figure 38. Construction of triode, showing cathode, control-grid, and plate arrangement.

intended uses of the tubes. Since these are related to such details as the values of plate voltage and plate current, the elements of triodes used in transmitters generally are larger than those of tubes used in receivers.

(3) The internal structure of the triode seldom is completely visible through the envelope. When the envelope is made of metal, the reason is obvious. When the envelope is made of glass, the view of the inside usually is obscured because of the opaque coating formed by the *getter* when the tube is *flashed,* that is, heated to a high temperature. The getter is a substance that is placed within the envelope for

the express purpose of absorbing any gases that may be liberated from the electrodes during initial operation. This keeps the vacuum created in the tube as high as possible. Magnesium is widely used as a getter, but other materials, such as barium, zirconium, and phosphorous, also can be used.

(4) The opaque deposit formed by the getter appears as a thin film on the lower half of the inside surface of the glass housing. The top usually remains transparent, and some visual inspection of the electrodes inside is therefore possible. This is the usual means for determining whether the electron emitter is incandescent. In the larger transmitting tubes, the getter coating usually does not obscure the metal parts inside the tube.

c. CONSTRUCTION OF CONTROL GRID.

(1) In appearance, the control grid is a ladderlike structure of metal. In most cases, it has a helical form, consisting of a number of turns of fine wire wound in the grooves of two upright supporting metal structures (fig. 39).

(2) The metals used for grids are usually molybdenum, nichrome, iron, nickel, tungsten, tantalum, and alloys of iron and nickel. The different physical sizes of the control grid, as well as the spaces between the turns that form the electrode, reflect different designs of triodes. The emitters and the plates are similar to those used in diodes.

d. SYMBOL FOR CONTROL GRID. When shown symbolically, the control grid (fig. 40) appears between the symbols used for the emitter and the plate, as either a zig-zag line, in A, or a

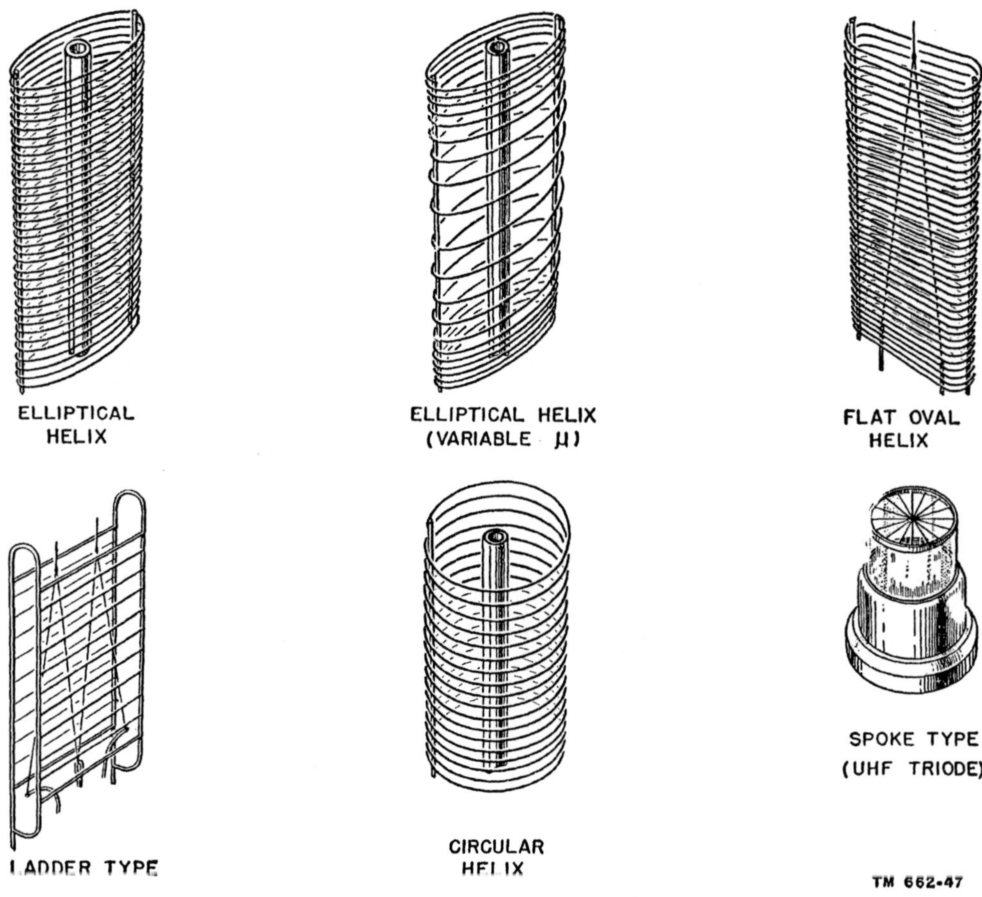

Figure 39. Typical control-grid assemblies.

dashed line, as in B. The symbol using the dashed line is the standard form in Signal Corps schematics. The letter identification of the control grid in the triode electron tube is the capital letter G, although in some instances the letters CG are used.

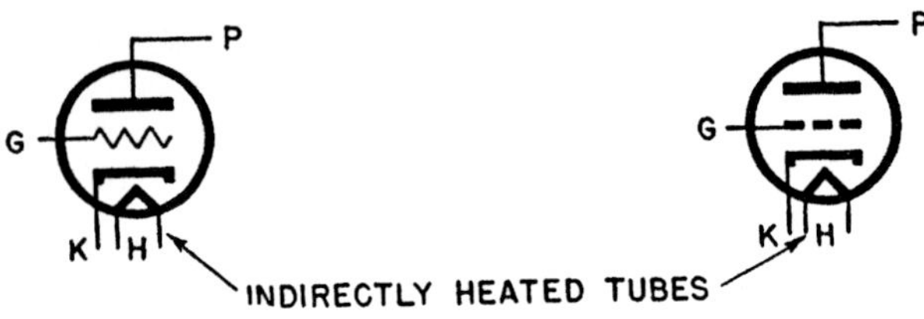

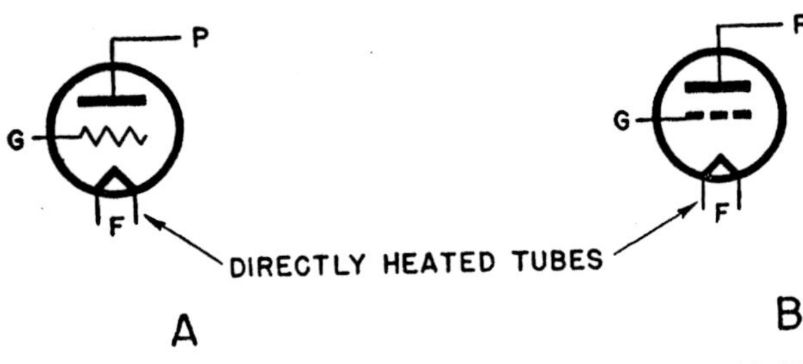

Figure 40. Symbols for directly and indirectly heated triodes.

e. PURPOSE OF CONTROL GRID.

(1) Briefly, the purpose of the control grid is to govern the movement of electrons between the cathode and the plate, thereby controlling the instantaneous plate current flowing through the tube. As previously mentioned, the voltage applied to the plate determines the amount of electrons attracted to it. What, then, is the electrical difference between the triode and the diode?

(2) Essentially, the diode can change alternating current to pulsating direct current. When an a-c voltage is applied to the plate, the instantaneous value of plate current is a direct function of the instantaneous polarity of the applied plate voltage, assuming, of course, that the emitter temperature remains constant. In the triode, a voltage on the control grid can vary the plate current when the voltage applied to the plate of the triode is held constant.

(3) The polarity and amplitude changes in voltage applied to the diode plate result in a unidirectional flow of pul-

sating plate current. In the case of the triode, the voltage applied to the plate of the tube is a d-c voltage obtained from a constant voltage source, but the action of the control grid can, nevertheless, cause variations in the amplitude of the plate current. The statement just made concerning the situation at the plate of a tube is a very important and fundamental point of distinction between the diode and the triode.

(4) The control grid can stop the flow of electrons to the plate. This is the same as saying that the plate current can be cut off despite the presence of a d-c voltage on the plate of the tube.

(5) The positive plate voltage can be as high as 5,000 volts, and, with proper negative grid bias, the control grid can overcome completely the influence of the high voltage. Consequently, it can be said that the control grid controls the current in the plate circuit. This control is external. If current and voltage in a system are representative of electrical energy in that system, then the control grid in the triode is an independent agency affording control of the amount of electrical energy present at any instant in the plate circuit.

(6) The ability of the control grid to govern the flow of plate current is used in many ways. Stopping all plate current is an example of an extreme condition. In addition, the control grid can decrease or increase the plate current instantaneously, even though the positive voltage applied to the plate is constant. It can also create current conditions in the plate circuit which are equivalent to changes in the plate voltage. Finally, the control grid can be made ineffective so that it does not control the plate current at all.

(7) It is natural if the preceding statements lead to the impression that the grid in the triode behaves like a valve by controlling the instantaneous value of plate current. It is common to refer

to the control grid in this manner, although the term valve is not fully descriptive of all the uses of a grid. Nevertheless, the valve action is very important, for it is the basis of the many functions of a triode, especially its ability to deliver a stronger signal than it receives. This process is called *amplification*. It should be pointed out, however, that vacuum tubes actually do not amplify power. In fact, the grid actually controls the flow of power from the plate power supply.

f. Supply Voltages of a Triode.

(1) To operate properly, every triode requires a means of heating the emitter. This is accomplished by a *heater* or *filament* voltage. The voltage source sometimes is referred to as an A-battery (fig. 41). It is connected to the filament in the directly and the indirectly heated triodes, shown in A and B respectively. This d-c voltage commonly is known as the *heater supply* or the *A-supply*. As a matter of simplicity, the discussion of the triode will be carried on using the indirectly heated triode as the basis.

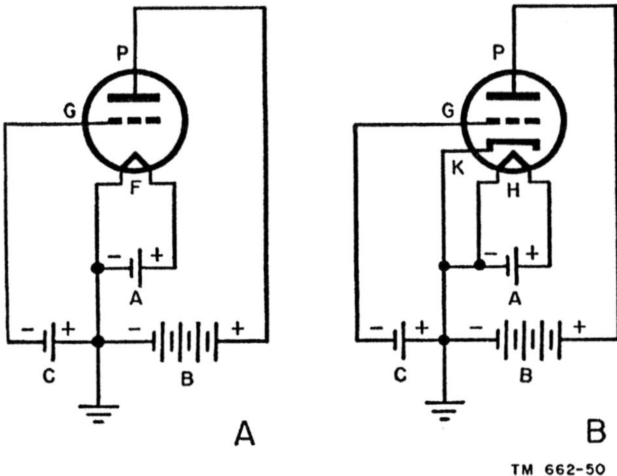

Figure 41. D-c supply voltages of directly and indirectly heated triodes.

(2) Since the plate in the triode functions as an attractor of electrons, it is understandable that a source of *positive* plate voltage (in reference to ground) is required. This source is the B-battery (fig. 41), also called the *B-supply* or *plate-voltage supply*. Electrically, it is connected between the plate and the electron emitter, with the positive terminal of the supply joining the plate of the tube. The junction between the negative terminals of the A- and B-batteries usually is grounded in standard practice.

(3) The third electrode of the triode is the control grid. Discussion of the movement of electrons in the diode (ch. 3) has made clear that control of plate current by means of an added electrode is accomplished by setting up an additional electrostatic field. A field of this kind results from the application of a voltage difference between the grid and another point. Therefore, by all sound reasoning, the control grid should be subject to a voltage.

(4) A third source of voltage, generally assigned the letter C, is connected between the control grid and the emitter (fig. 41). The polarity connections of the C-battery make the control grid negative with relation to the electron emitter. This is standard practice for reasons which will be explained shortly. For the moment, it should not be considered in that light, but viewed rather as a voltage connected between the control grid and the emitter. The reason for this approach is the manner in which the action of the control grid will be analyzed. It will be examined under these conditions: positive relative to the emitter, negative relative to the emitter, and zero potential relative to the emitter.

(5) The circuits in figure 41 are of fundamental triode systems suitable for study. Each diagram shows the voltage sources applied to the electrodes of this type of tube. However, all systems do not use batteries as the source of operating voltages. The voltages can be secured from other sources. The heater or filament voltage, marked A, can be obtained from an a-c trans-

former, or from rotating machinery such as a d-c generator. The plate-voltage source, labeled B, can be a d-c generator, or it can be a device designed for the purpose of furnishing d-c voltages from a variety of primary power sources. Such a device is known as a *power supply*.

(6) The grid-voltage supply, labeled C, also can be secured from a power supply or from special circuits associated with the tube itself, which will be described later.

(7) The circuits of figure 41 warrant additional comments. Previous statements concerning the thermionic type of diode associated that tube with one fixed operating voltage, the heater or filament voltage. The voltage applied to the plate was shown as dc only, for the purpose of explaining the operating characteristics. In rectifier applications, this voltage is alternating and is representative of the a-c power that is being applied to the tube for conversion to d-c power.

(8) In the triode (fig. 41), the situation is different. Here, the three electrodes receive fixed d-c operating voltages. These determine the manner in which the tube functions and performs the duties assigned to it. As long as an electron tube is used as a triode, it bears direct association with a value of heater or filament voltage, a plate voltage, and a control-grid voltage. Sometimes these are referred to as the *operating potentials*.

(9) Another very important detail concerning the operating voltages of the triode (as well as of other vacuum tubes containing three or more electrodes) is the reference point of the applied electrode voltages. In almost all cases it is the cathode, or the common junction point of the voltage sources, shown connected to ground in figure 41 because this is a standard practice in almost all cases. To measure the plate voltage, a voltmeter would be connected between the plate

electrode and the common ground junction. The same is true for the control-grid voltage: the measurement would be made between the control grid and the common ground point.

33. Input, Output, and Cathode Circuits of Triode

a. GENERAL. The application of the triode and other vacuum tubes which contain three or more electrodes is founded upon a division of the electrode circuits into three systems (fig. 42): the *input*, *output*, and *cathode* circuits. A shows an indirectly heated and B a directly heated triode.

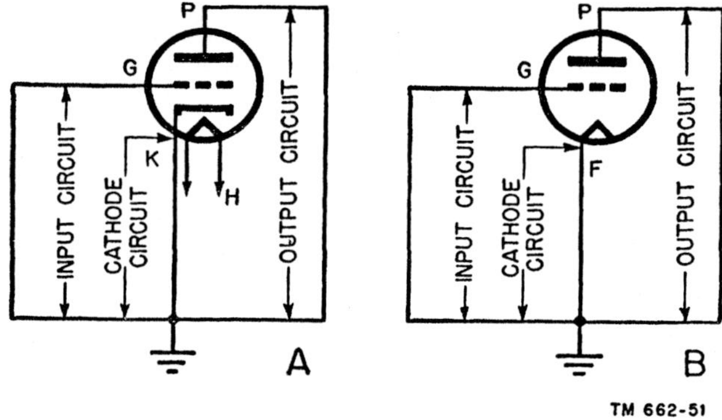

Figure 42. Input, output, and cathode circuits of indirectly and directly heated triodes.

b. DISTINCTION BETWEEN CIRCUITS.

(1) The *input* circuit encompasses all circuit elements between the control-grid electrode as one boundary and the grounded common voltage reference point as the other circuit boundary. The *output* circuit includes all component parts between the plate electrode and the common voltage reference point. Finally, the cathode circuit includes all the elements in the system between the emitter proper and the common voltage reference point.

(2) The circuit division is based on the voltages applied to the electrodes even though the different voltage sources have been omitted. It also is related to the paths of signal currents that flow in the tube circuits. It is too early in this treatment to speak about these,

but they will appear again later in this manual.

34. Electrostatic Field in Triode

a. SPACE-CHARGE CONDITIONS.

   (1) An imaginary view of the electrodes of a triode without any voltages applied to the control grid and the plate, but with the cathode at emitting temperature, is shown in figure 43. The letter K designates the electron-emitting cathode, G is the control grid, and P is the plate. Each grid wire is represented as a small circle. Electrically, these circles are connected. The separation between the grid wires is ample to allow electrons to travel between the wires toward the plate. The space charge is shown by the tiny dots between the cathode and the control grid. The greatest space-charge density is seen to exist nearest the cathode (par. 17). Note the absence of electrons in the space between the control grid and the plate. Some of the emitted electrons have sufficient velocity after emission to advance to the plate, but these are so few in number as to be unimportant in the discussion. Consequently, it is completely satisfactory to show the space between the control grid and the plate devoid of electrons. It is only when current is passing through the tube that electrons are present in this space.

   (2) In figure 43, note the physical locations of the control grid and the plate relative to the space charge. The control grid is very much closer to the space charge than the plate is. In fact, the entire space charge is assumed to be located between the cathode and the control grid. The actual physical separation between the electrodes of an electron tube is a matter of individual tube design. This aspect of electron-tube construction is unimportant in this discussion, although it can be stated that these dimensions usually are in small fractions of an inch, except in the very large transmitting tubes, in which the tube electrodes are farther apart.

b. BEHAVIOR OF CONTROL GRID.

   (1) In figure 44, the cathode is assumed to be emitting an adequate supply of electrons. The B battery applies 100 volts to the plate, making it positive relative to the cathode by that amount. The control grid is at zero potential relative to the cathode, as shown. Mention of zero voltage difference between two points does not imply an open circuit; it means that both points are at the same potential or that there is no difference of potential between them.

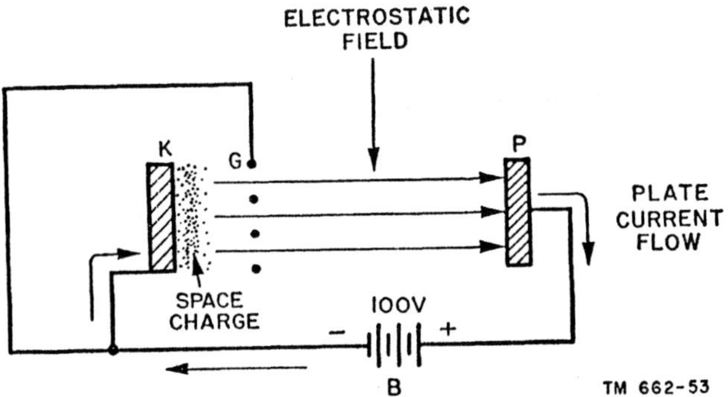

Figure 44. Electron flow in triode with control grid at cathode potential.

   (2) The triode is a thermionically heated electron tube; therefore, a space charge exists in the neighborhood of the cathode, between that electrode and the control grid. This cloud of electrons is inherent in every tube of

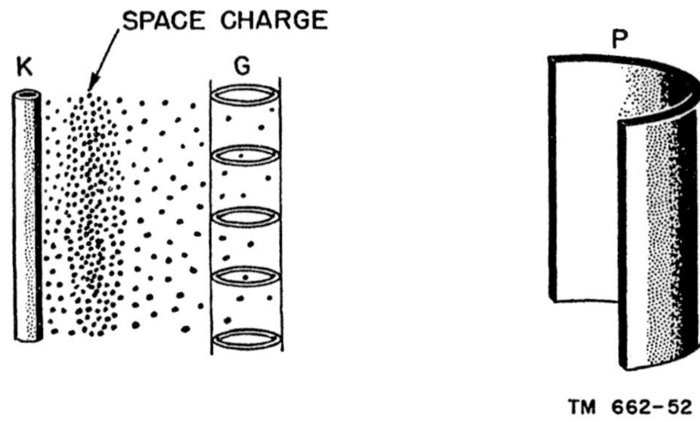

Figure 43. Imaginary view of electrodes in a triode.

this kind, regardless of the number of electrodes.

c. CREATION OF ELECTROSTATIC FIELD. The electrostatic fields created by this organization of electrodes and the voltage relationships are symbolized by the arrows in figure 44. With zero voltage applied to the grid, the latter has no electrostatic field of its own. The positive voltage applied to the plate causes an electrostatic field to exist between the cathode and the plate (par. 17).

d. DIRECTION OF ELECTROSTATIC FIELD. The direction of the field is such as to pull electrons toward the plate. They advance to it through the spaces between the control-grid wires. The amount of plate current flowing through the tube and around the external plate circuit back into the cathode is a function of the voltage applied to the plate, or space-charge limiting that takes place. For the sake of clarity and because the plate voltage has been stated as 100 volts, a plate current of 20 ma is assumed. This value is purely illustrative, and is not intended to imply any plate-current plate-voltage relationships for triodes in general.

e. CONTRIBUTION OF CONTROL GRID. So far, the action of the triode is not much different from that of the diode. The control grid, at cathode potential, contributes very little, if anything, to the behavior of the triode. It is important, however, to consider the *grid current* which flows in the control-grid cathode circuit because of what it means during application of control-grid voltages and its association with the operating voltage conditions at the control-grid electrode.

f. GRID CURRENT.

    (1) The spaces between the control-grid wires allow free movements of electrons to the plate (fig. 44). The control grid is, however, a physical structure; therefore, the grid wires can block the passage of some of the electrons advancing toward the plate. There is no force that tends to make the grid wires attract electrons to it, but since it is in the path of the plate-current electrons, some charges will attach themselves to the control-grid wires. This results in a small amount

of *grid current* flow in the control-grid cathode circuit.

    (2) Slight as it is, this grid current is, nevertheless, undesirable for several reasons. All of them relate to the ultimate performance of the tube. Some of them are associated with the proper performance of the devices which are used with the tube. Since it is too early in the discussion of the triode to explain each of these effects, they will be explained later. However, it can be said that the usual triode operating conditions are such as to prevent the attraction of electrons to the control grid, and therefore, the flow of grid current.

35. Negative Voltage on Control Grid

a. GENERAL. Another set of operating voltage conditions for the triode (fig. 45) is identical with that shown in figure 44, except for the insertion of a 1-volt C-battery which makes the control grid *negative* relative to the cathode by 1 volt.

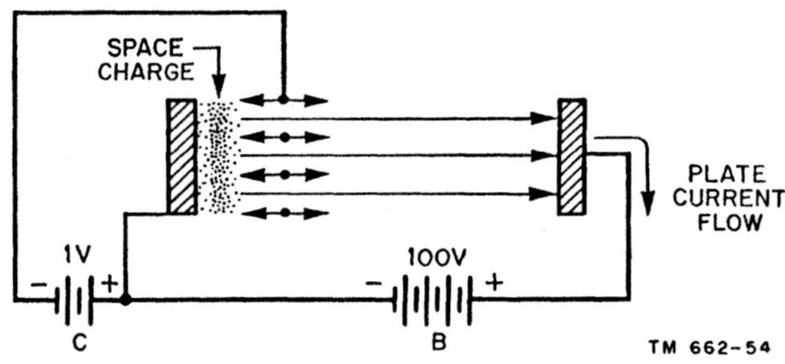

Figure 45. Electron flow in triode with control grid negative in relation to cathode.

b. ELECTROSTATIC FIELD. The voltages applied to the plate and the control grid produce two separate electrostatic fields (fig. 45). The positively charged plate has a field which attracts electrons toward the plate. The negatively charged control grid has a field which repels electrons toward the cathode.

c. PATH OF ELECTRON.

    (1) On the side of the control grid which faces the space charge, the negatively charged grid acts to repel electrons back into the space charge, thereby reducing the number of electrons

which advance to the plate. This tends to increase the density of the space charge, which causes a greater number of the emitted electrons to be repelled back into the cathode in order to maintain equilibrium of the space charge. Immediately, the situation implies a reduction in plate current; however, this is not the complete story.

(2) On the side of the control grid which faces the plate, the direction of the electrostatic field is toward the plate. Consequently, once an electron has traveled into the area between the control grid and the plate, all the forces acting on it attract it toward the plate. Even though this effect takes place, it is not too important in relation to the negative grid.

d. EFFECT OF ELECTROSTATIC FIELD. Figure 45 shows that the lines of force in the plate field penetrate the space between the grid wires and act on the space charge, tending to pull electrons to the plate. At the same time, the lines of force on the cathode side of the control grid tend to prevent electrons from advancing to the plate. The movement of electrons through the openings of the grid to the plate is controlled by the force that predominates.

e. EFFECT OF CIRCUIT VOLTAGES.

(1) It would seem from the simple relation between numbers that the negative 1-volt potential on the control grid and the positive 100-volt potential on the plate would create fields of such relative intensity as to enable the plate field to overwhelm the grid field completely. This does not occur, because the control grid is much closer to the space charge than the plate is. Therefore, a low voltage applied to the control grid can exert as much, or more, influence on the space-charge electrons as a very much higher voltage applied to the plate.

(2) Making the control grid 1 volt negative with relation to the cathode can offset as much as plus 10, 20, or even 50 volts applied to the plate. In some types of triodes the effect may be less, and in still other types can be greater.

For the purpose of comparing figures 44 and 45, the —1-volt grid bias in figure 45 reduces the original plate current by 10 ma, making it 10 ma instead of the 20 ma flowing with 0 grid voltage in figure 44.

(3) In some tubes, —1 volt applied to the control grid can stop the movement of electrons completely, thus cutting off the plate current. Such a condition is known as *cut-off*, and the amount of negative grid voltage necessary to cause this state is known as the *cut-off voltage*. In some instances, cut-off is not reached until perhaps —10 volts, —20 volts, or more are applied to the control grid. The exact value is determined by the design of the tube.

36. Positive Voltage on Grid

What happens if the voltage applied to the control grid is positive, thereby making the control grid positive relative to the emitter? This is not the most frequently used operating condition, but it warrants an explanation because it is found in some applications.

a. ELECTROSTATIC FIELD. The change in polarity of the control-grid voltage modifies the direction of the electrostatic field between grid and cathode (fig. 46). This field tends to pull electrons out of the space charge and accelerate them toward the plate; because of the positive voltage on the control grid, great numbers of electrons pass through the grid openings as they advance toward the plate. The field produced by the positive voltage on the plate also is pulling electrons toward the plate; consequently, the positively charged control grid can be described as aiding the plate voltage, thereby increasing the plate current. It is assumed in the example cited that the plate current is increased 10 ma, so that the current is now 30 ma instead of the 20 ma indicated in figure 44.

b. PATH OF ELECTRONS. Many of the electrons attracted from the space charge by the positive control grid are intercepted in their advance to the plate by the grid wires. This results in grid current which is much larger in value than when the control grid was held at 0 voltage.

c. EFFECT OF PLATE VOLTAGE. The extent to which a positive voltage applied to the control

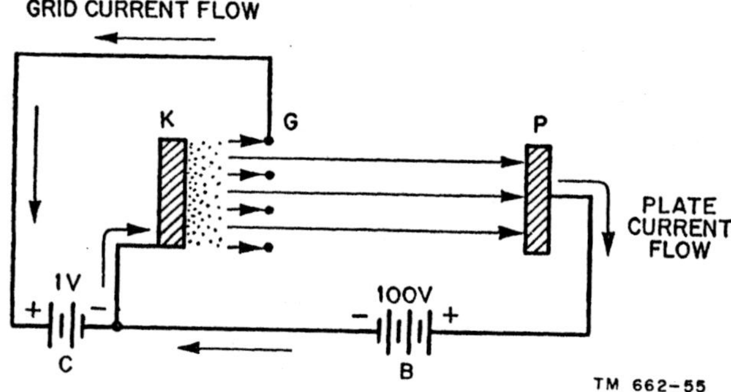

Figure 46. Electron flow in triode with control grid positive in relation to cathode.

grid offsets the space charge and increases plate current is determined by the design of the particular triode. In some tubes, a grid voltage change of a few volts may have the same effect on plate current as changing the plate voltage by 10, 20, or oven 50 volts. In other instances it may be more or less.

d. PLATE-GRID RELATIONSHIPS. The exact quantitative relationship between changes in the control-grid voltage and changes in the plate voltage for equal changes in plate current is dependent upon the individual tube type design. This is a very important *constant* of every triode, and, for that matter, of every type of electron tube which uses more than two electrodes. It determines the suitability of a tube to meet a functional need.

37. Summary of Control-grid Action

a. PICTORIAL SUMMARY. A pictorial summary of the control-grid action associated with figures 44 through 46 is shown in figure 47. Positive and negative polarities and values of control-grid voltage are pictured over a period of time in curves A and C. An imaginary switch mechanism in the grid circuit provides 0, +1, and —1 volts at the control grid. The 1-volt positive and negative limits are arbitrary. The graphs of these voltage variations may assume the shape of a sine or a square wave, as shown in curves A and C. For purposes of easiest correlation with the resultant plate-current changes (curves B and D), each interval of grid-voltage level and polarity is labeled. Corresponding labels appear on the plate-current curve.

b. CONCLUSIONS. The specific values of plate current shown in curves B and D are also arbitrary, but for the sake of simplicity are made to conform with the values stated above during the description of the grid-voltage conditions. Correlating the plate-current changes in curve B with the grid-voltage changes in curve A shows that—

(1) When the grid voltage is held at zero potential (curve A, periods 1 to 2, 5 to 6, and 9 to 10), the plate current is constant at a value determined only by the attractive force of the positive plate voltage. This current is represented in curve B, periods 1′ to 2′, 5′ to 6′, and 9′ to 10′.

(2) A change in grid voltage from 0 to a positive 1 volt (curve A, 2 to 3) results in an increase in the plate current (curve B, 2′ to 3′).

(3) A change in grid voltage from 0 to a negative 1 volt (6 to 7) causes a decrease in plate current (6′ to 7′).

(4) Where the grid voltage assumes a steady positive 1-volt value for a period of time (3 to 4), the plate current also remains constant (3′ to 4′) at the new value created by the positive voltage on the grid.

(5) Where the grid voltage assumes a steady negative 1-volt value for a period of time (7 to 8), the plate current also remains constant (7′ to 8′) at the new value created by the negative voltage on the grid.

(6) The direction of change in grid voltage has a definite relationship to the direction of change in plate current. As the grid voltage changes in the positive direction from 0 (2 to 3), the plate current increases (2′ to 3′). As the grid becomes less positive (4 to 5), the plate current decreases (4′ to 5′). As the grid voltage increases in the negative direction (6 to 7), the plate current decreases (6′ to 7′). As the grid becomes less negative (8 to 9), the plate current increases (8′ to 9′).

(7) The changes in plate current occur in step with the changes in grid voltage.

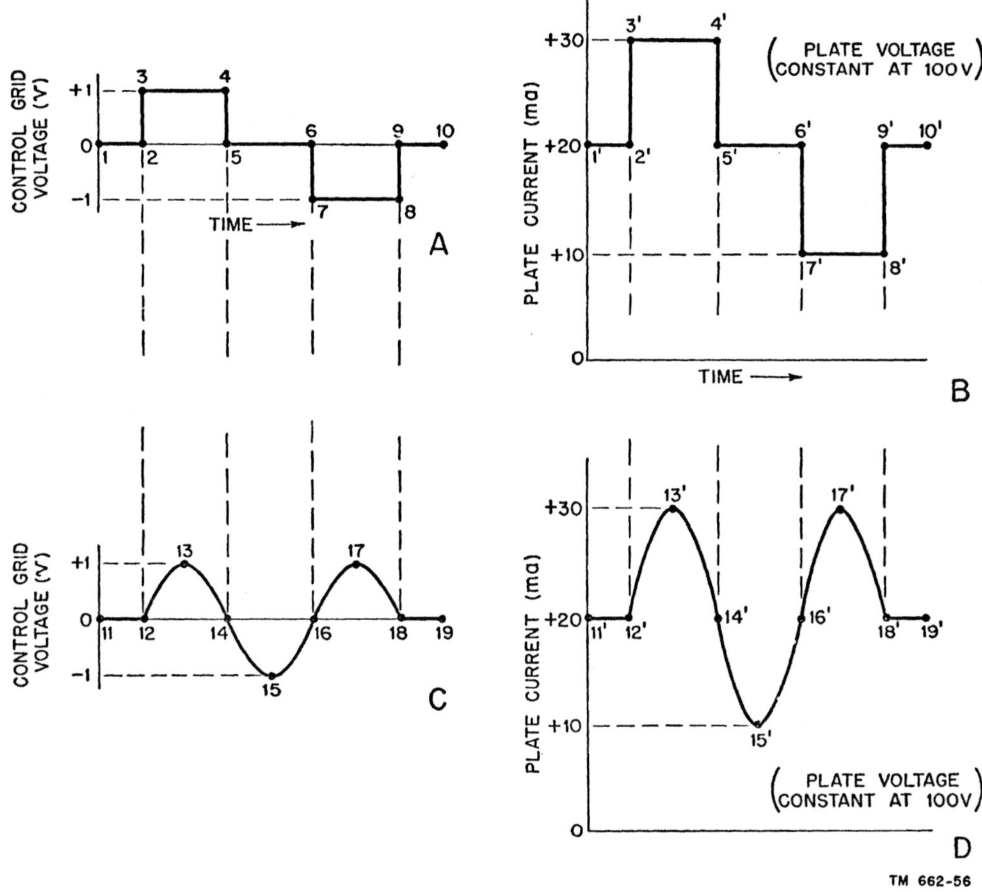

Figure 47. Curves showing relationships between grid voltage and plate current in triode.

The maximum and minimum points of one occur at the same instant as the maximum and minimum of the other. This is commonly referred to as an *in-phase* condition.

38. Additional Characteristics

a. UNIDIRECTIONAL FLOW. Curves A and B warrant some additional discussion. The changes in grid voltage in each direction represent a change in polarity relative to a 0 reference voltage. However, the changes in plate current are not changes in polarity, but rather represent variations above and below a reference value of current. The reference value of plate current in curve B is +20 ma, which corresponds to the grid reference volt of 0 volt in curve A. Curve B shows that even when the plate current is at its lowest value (7' to 8'), it still has a value greater than 0 (+10 ma). The polarity of the plate current always re-mains positive. No other value is possible because of the unidirectional current flow in the tube. The voltage applied to the grid makes the unidirectional plate current greater or smaller in value, in conformance with the instantaneous amplitude and polarity changes of the grid voltage.

b. GRID-PLATE RELATIONSHIPS. Another pertinent detail is that the plate-current grid-voltage relationship is not dependent on sudden changes in grid voltage. The grid voltage can change slowly or very rapidly and the plate current will vary accordingly. This is symbolized by the sine-wave voltages shown in curves C and D. In general, it can be said that the change in plate current usually has the same shape as the change in control-grid voltage. This is not a rigid rule; it is subject to numerous modifications as dictated by the way in which the tube is used. Subsequent explanations will make this point clear.

c. TRIODE OUTPUT. The final point to be made about a varying plate current in the plate circuit of a triode is that it can be looked upon as a steady direct current to which is added an alternating-current component. This results in a current which has a unidirectional flow but which varies in value above and below a steady reference value. In curve C, 11 to 12 and 18 to 19 are the steady (0 grid voltage) reference values. In curve D, 11' to 12' and 18' to 19' show the steady reference value of corresponding plate current, 20 ma. The a-c voltage applied to the control grid (curve C, points 12 through 18) momentarily increases the plate current to a maximum (curve D, points 13' and 17'), 30 ma, and momentarily decreases it to a minimum (point 15'), 10 ma. This plate current can be described as a direct current of 20 ma with an a-c component equal to 10 ma peak.

39. Bias, Signal Voltage, and Plate Current

a. GRID-CURRENT EFFECTS. It has been stated that a value of positive voltage on the control grid results in the presence of grid current between the control grid and the cathode through the system external to the tube. This condition cannot be avoided, because the grid wires intercept electrons which are advancing toward the grid on their way to the plate. The positively charged grid attracts electrons into itself. Disregarding for the moment any applications which permit grid current, the presence of grid current normally is undesirable. It represents the consumption of power and other unwanted effects.

b. BIAS AND SIGNAL VOLTAGES. The signal voltage normally applied to an electron-tube control grid is alternating in character; at least, it is a voltage which varies in amplitude and perhaps in polarity relative to the cathode. During the time that it is negative with relation to the cathode, freedom from grid current is obvious, but when it is positive, grid current is present unless some means are provided to keep the control grid at a negative potential during the positive portion of the input signal. The purpose of the negative control-grid *bias* is to establish this operating condition. Bias may be defined as the d-c voltage between the grid and the cathode. It is represented by battery C in figure 48. The total voltage existing between grid and cathode is the signal voltage plus the bias voltage.

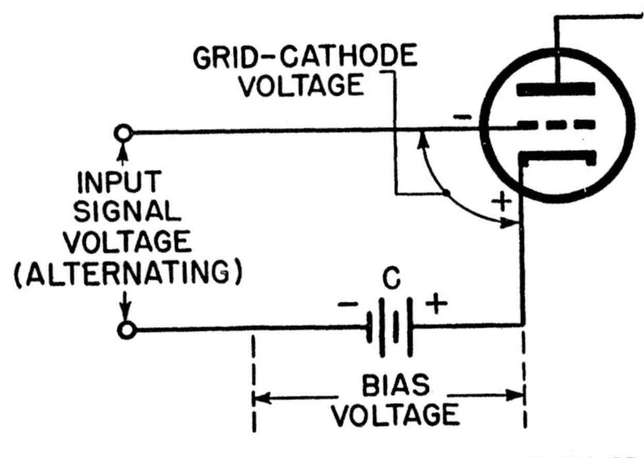

Figure 48. Bias and signal voltages of a triode.

c. BIAS-SIGNAL REPRESENTATIONS.
   (1) The association between the signal and the grid bias is illustrated in figure 49. Curve A represents an input a-c signal of 5 volts peak. It varies between +5 and —5 volts. In order to keep the control grid negative during the entire positive alternation of the input signal, the grid bias must equal, if not exceed, the peak value of the signal. Therefore, the control-grid bias is arbitrarily set at —6 volts, as in curve B. Since the grid is negative with relation to the 0-voltage reference level, it is shown *below* the reference voltage line.
   (2) The resultant of the signal and control-grid bias voltages at the control grid, instant by instant, is shown as curve C in the same illustration. Curve C is the addition of curves A and B. The fixed bias voltage sets up the initial voltage relationship between the control grid and the cathode. This is the no-signal condition as in curve B. It is represented by 1 to 2 and 8 to 9 in curve A and 1' to 2' and 8' to 9' in curve C. The times from 1 to 2 and 8 to 9 in curve A represent the period

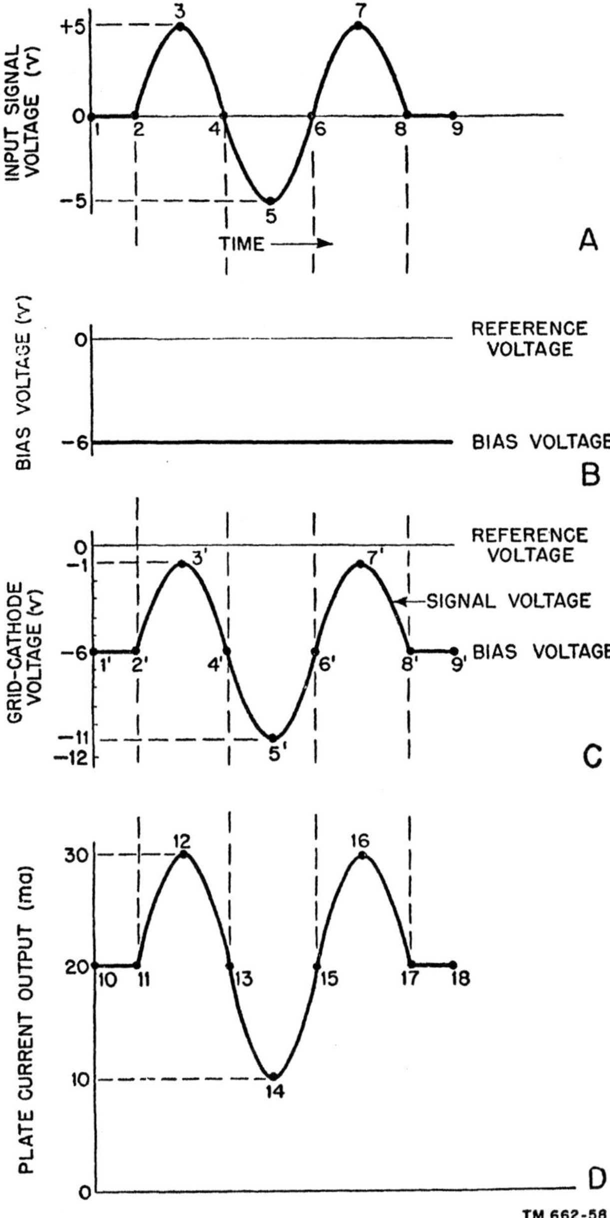

Figure 49. Waveshapes illustrating bias, signal voltage, and plate current.

of 0 signal voltage, during which time the full —6 volts of grid bias is active on the control grid, as shown in times 1′ to 2′ and 8′ to 9′ in curve C.

(3) As the signal voltage starts rising in the positive direction (from 2 to 3), it bucks the fixed negative bias, and the control grid becomes less and less negative until, at the peak of the positive alternation of the signal voltage (point 3), the control grid is 1 volt negative with relation to the cathode, as shown by 3′ in curve C. As the signal voltage decreases in positive amplitude (from 3 to 4 in curve A), more and more of the bias voltage becomes predominant until point 4 is reached, which again corresponds to 0 signal voltage. The control grid again becomes 6 volts negative with relation to the cathode, as shown by point 4′ on curve C. Examining the action during the positive half-cycle of the applied signal voltage, points 2′ to 4′ in curve C, it is evident that a 5-volt change in signal voltage in the positive direction has taken place at the control grid, but the grid electrode remains negative throughout the half-cycle.

(4) During the negative alternation of the signal voltage (points 4 to 6), the signal and the fixed negative bias voltages add. The result is a change in voltage at the grid from —6 volts (4′ in curve C), to a maximum negative voltage of —11 volts (5′) and then a return to —6 volts (6′) again. The control grid remains negative with relation to the cathode by an amount equal to the sum of the instantaneous signal voltage and the fixed grid bias.

d. PLATE-CURRENT REPRESENTATIONS. The plate current varies in accordance with the instantaneous resultant of the signal and fixed bias voltages, as in curve D. Without a signal input, but with the bias applied, the plate current is of constant amplitude (10 to 11 and 17 to 18) at 20 ma. When the signal voltage is applied, the plate current increases and decreases above the no-signal value.

e. GENERAL.

(1) It is interesting to note that curves C and D show changes in voltage and current, respectively; they are in step above and below the no-signal values. Each of these curves represents a combination of d-c and a-c components.

(2) The grid bias may be referred to as being increased. This means that it

is a higher value of bias, making the control grid more negative with reference to cathode. When reference is made to a reduction of bias, it means a lower value of voltage, or one which makes the control grid less negative.

40. Characteristic Curves

a. GENERAL. The relationships between the different voltages applied to the triode and the effects they have on the plate current are very important. As in the case of the diode, they are illustrated by means of *characteristic* curves, except that the curves which display the behavior of triodes are more numerous and present more varied information.

b. TRIODE CONSTANTS. The suitability of a triode or any other kind of electron tube for an application is determined by the *constants* of the tube. An understanding of the meaning of triode constants and the manner in which they are determined is pertinent to proper use of the tube. Accordingly, subsequent paragraphs are devoted to the subject.

41. Triode Circuit Notations

a. GENERAL NOTATIONS. Before explaining triode characteristics, it is necessary to refer briefly to the several triode circuit notations that will appear in the discussion. The electrodes of vacuum tubes are referred to by their common names, such as control grid, plate, filament, and cathode. Voltages and currents related to these electrodes also are treated in the same way: for example, plate voltage, plate-supply voltage, plate current, and control-grid voltage.

b. LETTER NOTATIONS. Such long names can prove unwieldy in text or in illustrations, especially when they need frequent repetition. To overcome this difficulty, it is common practice to assign letter *notations* to indicate certain voltages and currents under different operating conditions. A complete list of these appears in the appendix to this manual. Figure 50 illustrates a simple triode circuit bearing several circuit voltage and current notations which are used often. Only immediately pertinent notations are shown:

(1) E_{bb} = the plate-supply voltage, or the amount of B voltage.

(2) e_b = the instantaneous total plate voltage. (In fig. 50, E_{bb} equals e_b because there is no plate load.)

(3) I_b = the total steady plate current.

(4) E_{cc} = the control-grid supply voltage, the C voltage, or the grid-bias voltage.

(5) e_c = the instantaneous total grid voltage. (In fig. 50, E_{cc} equals e_c because there is no input signal.)

(6) I_c = the total steady grid current.

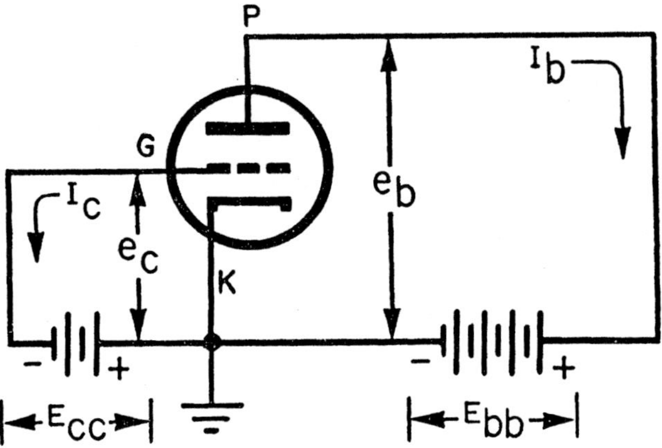

Figure 50. Triode circuit bearing several voltage and current notations.

c. OTHER NOTATIONS. Attention is called to the use of capital letters E and I for *steady* values of voltage and current and to the use of small e and i for instantaneous values of voltage and current. The conditions depicted in figure 50 do not involve instantaneous values of current, and therefore, the small letter i is absent. The subscript letters c and b associate the voltage or the current with the tube electrodes. Only a few circuit notations are introduced at this time. More will follow as the subject unfolds in this manual.

42. Static Plate-current Grid-voltage Characteristics

a. GENERAL. In view of the electrode organization of the triode, three basic factors control the plate current: the emitter temperature, the control-grid voltage, and the plate voltage. The first of these can be disregarded, for, unless otherwise stated, it is general procedure in all vacuum-tube operation to assume that the emitter is being operated at the required temperature. Accordingly, it is necessary merely to stipulate the emitter temperature in terms of heater or filament voltage or current upon whatever tube characteristic is being illustrated.

b. VARIABLE FACTORS. This, then, leaves the control-grid voltage and the plate voltage as the two variable factors (causes) which contribute to variations in plate current (the effect). To treat two variables simultaneously with no constants is impossible; consequently, each of these is considered separately. The effect on the plate current of a change in grid voltage is considered first; this characteristic is known as the plate-current grid-voltage characteristic. Plate voltage must be applied to the triode plate; therefore, the characteristic is developed with a known and stipulated plate voltage. If it is desired, individual plate-current grid-voltage characteristics can be developed for different values of plate voltage. Both kinds of characteristic graphs are treated in this manual.

c. CONSTRUCTION OF CHARACTERISTIC.

(1) The construction of a static plate-current grid-voltage characteristic is a relatively simple procedure. Different values of grid voltage, starting at 0 and advancing first in the positive direction and then in the negative direction (or the reverse) are applied. The plate current flowing at each increment of grid-voltage change is plotted on a suitable graph. If the changes are in steps of 1 volt, the final characteristic shows the plot of plate current over a range of grid voltages from zero to some negative limit, say —6 volts, and from zero to some positive limit, say +6 volts (in 1-volt steps). The graph is laid out in the way described in paragraph 26. The known variable is considered to be the *cause* and is marked off along the abscissa or horizontal axis. The unknown variable, or *effect,* is marked off on the ordinate or vertical axis (A of fig. 51). This is the static plate-current grid-voltage

characteristic for a tungsten filament emitter triode.

(2) The use of the word *static* in connection with the characteristic means that the curve represents tube behavior under no-load conditions. The voltages applied to the electrodes are determined solely by the voltage sources, such as E_{cc} and E_{bb} in figures 50 and B of 51.

(3) The vertical dividing line projecting upward from the 0-voltage point on the grid-voltage axis divides the plate current curve into two parts. The portion left of the 0-voltage dividing line shows the plate-current curve when the grid voltage is negative. The portion to the right of the 0-voltage line shows the plate-current curve when the grid voltage is positive. In all cases, the value of plate current is indicated by the scale on the ordinate, or the vertical axis.

d. EXPLANATION OF CHARACTERISTIC.

(1) The circuit capable of producing the data contained in this graph is illustrated in B of figure 51. Grid-voltage source E_{cc} is so arranged that a variable voltage, either negative or positive, can be tapped to the control grid. An arbitrary choice limits the grid-voltage changes to a maximum of about 7 volts in a positive or negative direction. Appropriate voltmeters VM indicate the applied grid and plate voltages. The ammeter, AM, shows the plate current flowing in the system for each increment of grid voltage.

(2) Referring to the characteristic, it is seen that with the plate voltage fixed at a constant value, plate current starts flowing at about —6 volts on the grid. As the control grid is made less negative by adjusting voltage supply E_{cc}, the plate current increases, but in a nonlinear manner. From approximately —2.75 volts up to about —.75 volt, the plate-current rise is in direct proportion to the reduction in grid voltage. This is the *linear* part of the characteristic, or the straight portion.

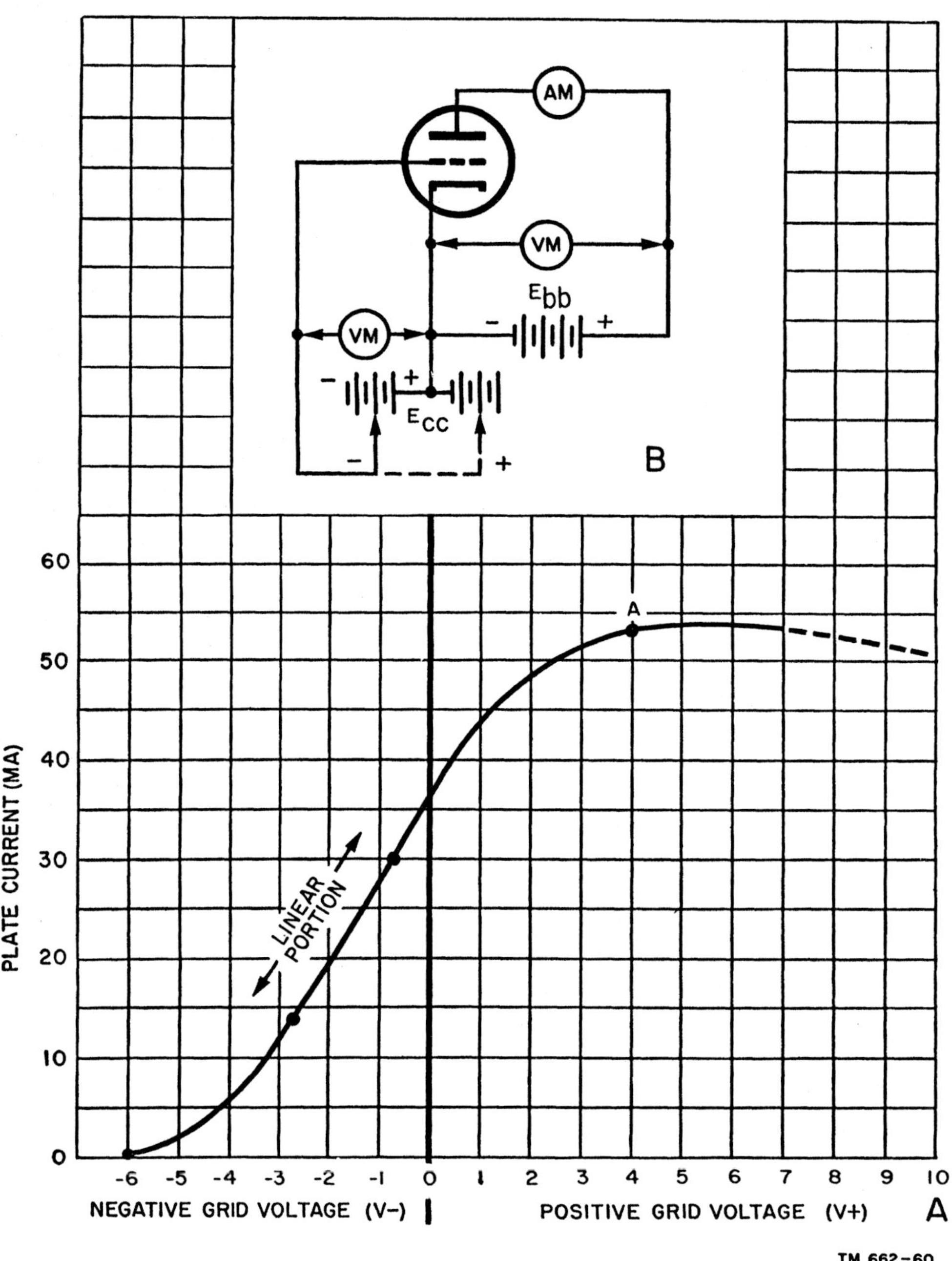

Figure 51. Plate-current grid-voltage characteristic curve for a typical triode circuit.

As the control-grid voltage approaches 0 and passes into the positive grid-voltage region, the plate current continues to increase, but again in nonlinear manner.

e. CONCLUSIONS. Several pertinent conclusions can be drawn from this plate-current grid-voltage characteristic. As the grid is made more and more positive, the plate current continues to increase until the rise in grid voltage produces no further increase in plate current. At this value of positive grid voltage, the plate-current curve flattens off. This occurs at approximately point A, which usually is identified as the plate-current saturation point. It is the value of plate current beyond which no further increase in plate current occurs as the grid is made more positive.

f. OTHER FACTORS.

(1) Plate-current saturation is a peculiarity of tungsten filament tubes. For any one emitter temperature, there occurs a fixed maximum emission. Plate saturation corresponds to that condition when, with the fixed amount of emission from the tungsten filament, all the emitted electrons divide between the grid current and the plate current, and making the control grid more positive does not increase the plate current.

(2) Even this does not remain static. If the control grid is made sufficiently positive, it will *reduce* the plate current, as shown by the dotted-line curve beyond point A in figure 51. The fall in plate current is due to the emission of *secondary* electrons from the plate while under bombardment by the high-velocity electrons which comprise the plate current. These secondary electrons are attracted by the highly positive grid. In a sense, electrons are moving in two directions across the area between the highly positive grid and the positively charged plate. The current flow to the plate consists of those electrons coming from the cathode which pass through the spaces between the grid wires. The higher the positive voltage on the grid, the greater the velocity of these electrons, but fewer reach the plate because of the greater attracting force present at the grid wires. The other flow is from the plate to the grid inside the tube, these being the secondary electrons knocked out of the plate by the electrons which advance from the space charge to the plate. The net result is an over-all reduction of plate current.

g. USING OXIDE-COATED EMITTER..

(1) The plate-current grid-voltage characteristic for an oxide-coated emitter triode, curve ABC (fig. 52), warrants several comments. The absence of plate-current saturation is noted immediately. The oxide-coated emitter is so profuse in electron output that even with a highly positive grid there remains an ample supply of electrons to cause plate current to flow. This means a continually increasing plate current as the grid is made more positive. Raising the plate voltage would have the same effect; that is, the plate current would continue to rise. In the figure, the plate voltage is held constant at 100 volts.

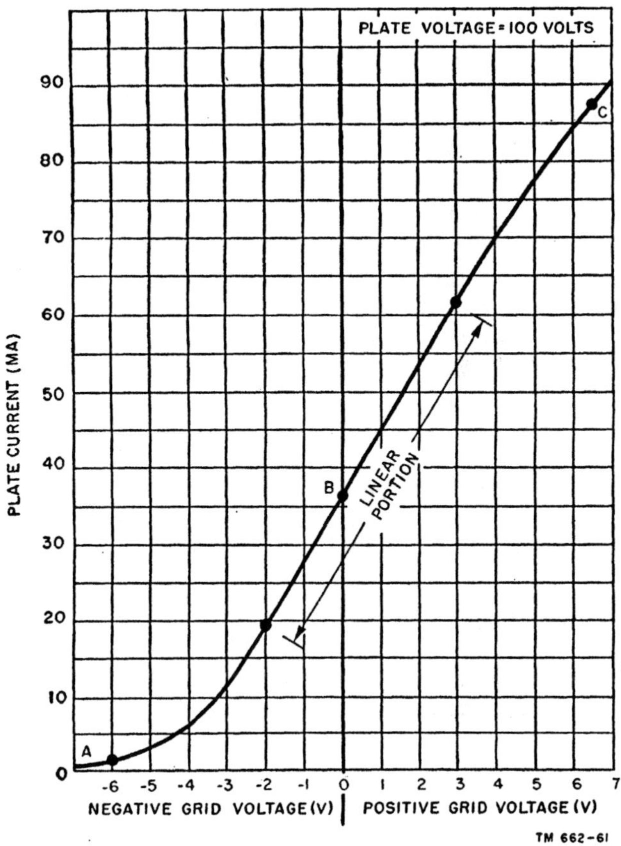

Figure 52. Plate-current grid-voltage characteristic curve for triode using oxide-coated emitter.

(2) The shape of the characteristic in the negative grid-voltage region is very much like that for the tungsten emitter triode. This is to be expected because the fundamental relationships between the tube electrodes are determined only in part by the type of emitter. The grid voltage required for a constant plate current is seen to be much higher than for the tungsten

emitter triode. This is incidental because the specific value of grid voltage required to level off the plate current is determined by the characteristics of the tube. The important point is that every triode is subject to a constant plate current and finally cut off if the grid is made sufficiently negative.

h. GRID-CURRENT EFFECTS. The plate-current grid-voltage characteristic of the oxide-coated emitter in figure 52 is seen to be substantially linear in the positive grid-voltage zone. This does not imply that normal use of the tube is with grid current flowing. In some cases it is used in this manner, but in the majority of instances the control grid is held at some negative d-c value. Whenever the grid is positive, power is consumed in the grid circuit, since the grid then draws current. If the source of grid voltage can compensate for the consumption of power, all well and good, but this type of operation is a special case, and not the usual. As will be seen later, grid-current flow is unusual in receiving system tubes, but it is common in tubes used in transmitters.

i. SUMMARY.

(1) The previous descriptions of tube behavior have the one purpose of making clear the manner in which the various supply voltages acting on the space charge are translated into numbers to form a graph.

(2) Different voltages applied to the plate modify the shape of the characteristic curves shown in figures 51 and 52. All tube characteristic curves have linear and nonlinear portions, although not necessarily to the same degree. The cut-off voltages differ, but every plate-current curve displays the effect because it is inherent in every vacuum tube. Finally, the meaning of plate-current change versus grid-voltage change has been illustrated in a way which leads to further explanations of tube behavior when plate voltages are increased or decreased.

43. Grid Family of Characteristic Curves

a. GENERAL. A single plate-current grid-voltage characteristic curve furnishes important information, but it is limited. A number of such curves shown on the same scale for different values of plate voltage give much more information concerning the effects of the different grid voltages on the plate current. Such curves plotted on a single graph comprise a *family* of characteristic curves, in this instance, the grid family. Another name used is *static transfer characteristics*.

b. GRID FAMILY. As a rule, the grid family or static transfer characteristics do not involve the positive region of grid voltage. This is so because for most triode applications the grid is not driven positive. A direct comparison between figures 51, 52, and 53 is, therefore, not possible. The first two represent arbitrary operating conditions and values, whereas figure 53 treats a specific tube, the 6J5. The circuit used to develop the data shown in figure 53 is similar to that in B of figure 51, except that the plate voltage, E_{bb}, is variable over a substantial range.

c. OBTAINING GRID FAMILY OF CURVES. Each curve in figure 53 is identified by a specific value of applied plate voltage, and is, therefore, the resultant of the stated plate voltage and changes in grid voltage. The curve is formed by noting the plate current as the control-grid voltage is increased in the negative direction, beginning at 0 voltage. These points are joined and form a curve. The —18-volt bias limit is set by that range of plate voltages (100 to 300 volts) considered within the performance capabilities of the tube.

d. ANALYSIS OF CURVES.

(1) A number of facts immediately become evident. There is a close similarity between the general contours of the characteristic curves. Each has a linear and a nonlinear part. Moreover, each plate-current curve has a cut-off point (on the X-axis), and it is seen that as the plate voltage increases, the value of negative bias required to cut off the plate current also increases. For example, —6.5 volts causes cut-off when the plate voltage is 100 volts, whereas —12.6 volts is required when the plate voltage is 200

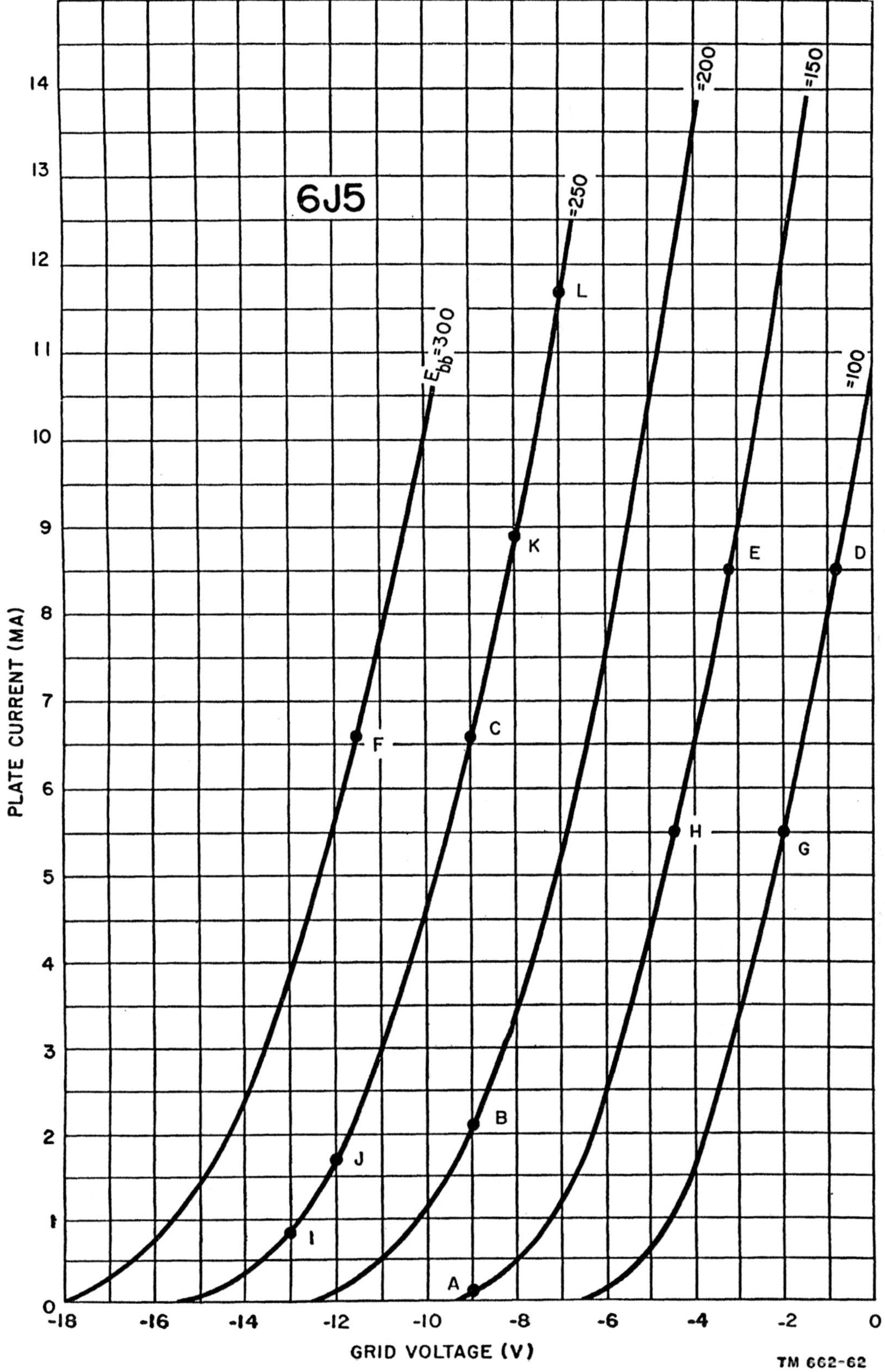

Figure 53. Grid family of characteristic curves for 6J5 triode.

volts and —15.5 volts cuts off the plate current when the plate voltage is 250 volts.

(2) The grid family also discloses an increase in plate current for an increase in plate voltage when the bias voltage is held constant. For example, a —9-volt bias (point A) results in .1 ma plate current with 150 volts applied to the plate, a 2.1-ma (point B) plate current at 200 volts on the plate, and a 6.55 ma (point C) plate current with 250 volts applied to the plate.

(3) The amount of grid voltage required to offset a change in plate current when the plate voltage is changed can be determined from the grid family. This can be restated by saying that these curves indicate the required increase or decrease in bias in order that the plate current be held constant when the plate voltage is changed. For example, point D on the 100-volt plate-voltage curve corresponds to 8.5 ma with a negative bias of .8 volt. If the plate voltage is increased to 150 volts, and the plate current must be held at 8.5 ma (point E), the grid bias must be increased to —3.2 volts. This indicates that an increase of 50 volts in plate voltage is offset by an increase of 3.2 minus .8 or 2.4 volts bias. In similar fashion, points C and F, each representing 50-volt increases in plate voltage, require an increase of 2.5 volts in negative grid voltage in order that the charge in plate voltage be offset and the plate current remain constant at 6.55 ma.

(4) The curves work in the reverse direction as well. They indicate the change in plate voltage required to offset a change in negative grid bias for constant plate currents. For example, assume a starting point G, representing 100 volts on the plate and —2 volts on the grid, and 5.5 ma of plate current. If the bias is increased to —4.4 volts, what is the new value of plate voltage which will result in the same plate current of 5.5 ma? A vertical projection from the —4.4-volt bias point intersects the 5.5-ma current projection along the 150-volt plate-voltage curve (point H). This leads to the conclusion that a 4.4 minus 2 or 2.4-volt increase in negative grid bias demands a 50-volt increase in plate voltage in order to hold the plate current constant.

(5) Still more information is available from the grid family. This is seen between points I-J and K-L along the 250-volt characteristic. It shows that the change in plate current corresponding to a fixed change in bias voltage (grid voltage) is a function of the *operating region* on the plate-current curve. Consider a 1-volt change, between —12 and —13 volts (I-J) on the curve of $E_{bb} = 250$ volts. The plate-current change is .84 ma, going from .86 to 1.7 ma. The same 1-volt grid-voltage change higher upon the curve from —7 to —8 volts (K-L) results in a change of 2.8 ma, going from 11.7 to 8.9 ma.

(6) Examination of the plate-current curve where $E_{bb} = 300$ volts discloses even a greater change in plate current for 1-volt changes in grid voltage. The opposite is true if $E_{bb} = 200$ volts. This can be seen if the points, —7 to —8 volts, are plotted on the plate-current curve of $E_{bb} = 200$ volts. A change of 1.8 ma takes place, from 3.5 to 5.3 ma. This value is 1 volt less than the previous value obtained at 2.8 ma when E_{bb} was equal to 250 volts. This leads to the generalization that the higher the plate voltage applied to any one type of electron tube, the greater is the change in plate current for a given change in grid voltage.

e. OPERATING POINT. The *operating point* along the plate-current grid-voltage curve deserves some emphasis. This term should be understood thoroughly because it is repeated frequently later. It is important to recognize the association between the location of the operating region on the curve and the manner in which it determines the amount of change in plate current per unit change in grid voltage. Proportional changes occur only over the *linear* (straight) parts of the curve. The greatest change in plate current per unit change in grid voltage occurs along the straight, rather than the curved part of the characteristic.

44. Static Plate-current Plate-voltage Characteristics

a. OBTAINING PLATE FAMILY OF CURVES. Figure 54 shows a *static plate family* of characteristic curves obtained by using the same circuit as that used in B of figure 51. In this family of characteristics, the grid voltage is varied in steps of 2 volts, and the plate-current measurements are made over a continuously variable range of plate voltages. For example, with the grid voltage at 0, or $E_{cc} = 0$, the plate voltage is varied from 0 to approximately 135 volts, and the plate current can be noted for this range. Joining the individual plate-current points results in the curve $E_{cc} = 0$. The grid voltage then is increased to —2 volts, or $E_{cc} = -2$, and the plate current again is noted for a new range of plate voltage, starting with that value which is just offset by the grid voltage—in this case 35 volts—up to about 170 volts. These new points of plate current are joined and they form the second curve ($E_{cc} = -2$). In every instance, the plate-current range shown for a particular fixed grid voltage starts at the point along the plate-voltage axis where the grid voltage causes plate-current cut-off. Each curve is drawn over a range of about 135 volts change in plate voltage.

b. USES OF CURVES. In general, the grid family and the plate family furnish the same information, except in somewhat different forms. Whereas the grid family displays the plate cur-

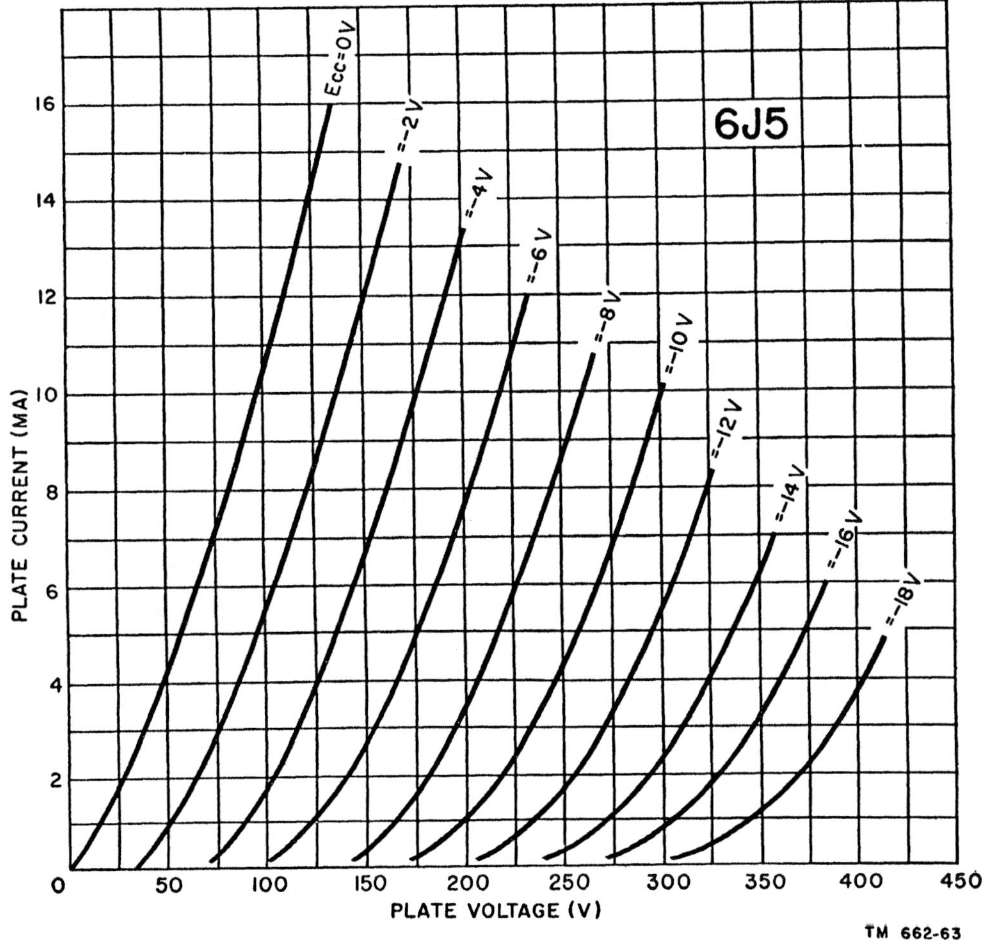

Figure 54. Plate family of characteristic curves for 6J5 triode.

rent for small increments of grid-voltage change and fixed differences in plate voltage, the plate family displays the effects of small increments of plate-voltage and fixed increments of grid-voltage. These family graphs serve to present the relationship between the different triode-tube electrode voltages under static conditions. They also lead to the meaning of tube *constants* or *parameters*, frequently described as tube rating data. Tube constants provide a means of comparing vacuum tubes and determining their suitability for specified functions.

45. Tube Constants

a. GENERAL. The behavior of the plate current in a triode (or in any vacuum tube which contains three or more electrodes), under the influence of different control-grid and plate voltages, does not occur at random. It is a function of the design of the tube—specifically, the geometric organization of the tube electrodes. Examples of these are the separation between the electrodes, the shape and dimensions of the electrodes, and other physical details. It is these factors which determine the maximum voltages that can be applied to the electrodes, the maximum plate current permissible through the tube, the conditions for plate current cut-off, and other similar facts. All of these are expressed by a group of numbers referred to as tube *constants*.

b. NAMES OF CONSTANTS. Tube constants differ from tube characteristics. Whereas the characteristic is a graphical representation of tube behavior under the particular set of conditions shown, the tube constants are individual numerical ratings predicated upon the geometry of the tube. Tubes possessing similar constants demonstrate similar relationships, although the specific values of grid voltage, plate voltage, and plate current necessary to make the tube perform properly may be different for the various tubes. The three primary tube constants are *amplification factor, a-c plate resistance,* and *transconductance.* Each of these will be explained.

46. Amplification Factor

a. DEFINITION. By definition, the *amplifica-*

tion factor or *amplification constant* is the ratio between a small change in plate voltage and a small change in grid voltage which results in the same change in plate current. It is an indication of the effectiveness of the control-grid voltage relative to the plate voltage in controlling the plate current. Expressed in a formula

$$\text{Amplification factor} = \frac{\text{Small change in plate voltage}}{\text{Small change in grid voltage}}$$

or

$$\mu = \frac{\triangle e_b}{\triangle e_c} \quad (i_p \text{ constant}).$$

The amplification factor is represented by the Greek letter μ, pronounced mu. The Greek letter $\triangle$ indicates *a small change in.* It is a pure number without any reference to units. For example, if a tube is said to have a μ of 100, it means that the grid voltage change required to produce a certain change in plate current is 100 times less than the plate voltage change required to bring about the same change in plate current. In other words, the grid voltage is 100 times more effective than the plate voltage in its influence upon the space charge and, consequently, on the plate current. To illustrate, suppose that a plate-current change of 1 ma is produced by a plate-voltage change of 10 volts, and a grid-voltage change of .1 volt produces a 1-ma change in plate current. Then

$$\mu = \frac{10}{.1} = 100$$

and the amplification factor is 100. Emphasis is placed on the fact that it is the *change* in plate voltage and the *change* in grid voltage that are important and not the individual values of plate and grid voltage.

b. FINDING AMPLIFICATION FACTOR.

(1) The amplification factor can be determined by using either the grid or the plate family of curves. Using the grid family (fig. 55), assume that it is desired to determine the μ of the 6J5 with 250 volts on its plate and —8 volts on its grid. This corresponds to point A in the figure. The first step is to locate, on the curve of $E_{bb} = 250$ volts, some convenient point along the straight part of the characteristic, such as point B. Next, project a line parallel to the X-axis from point B to the adjacent curve of $E_{bb} = 200$ volts, which is point C. From point

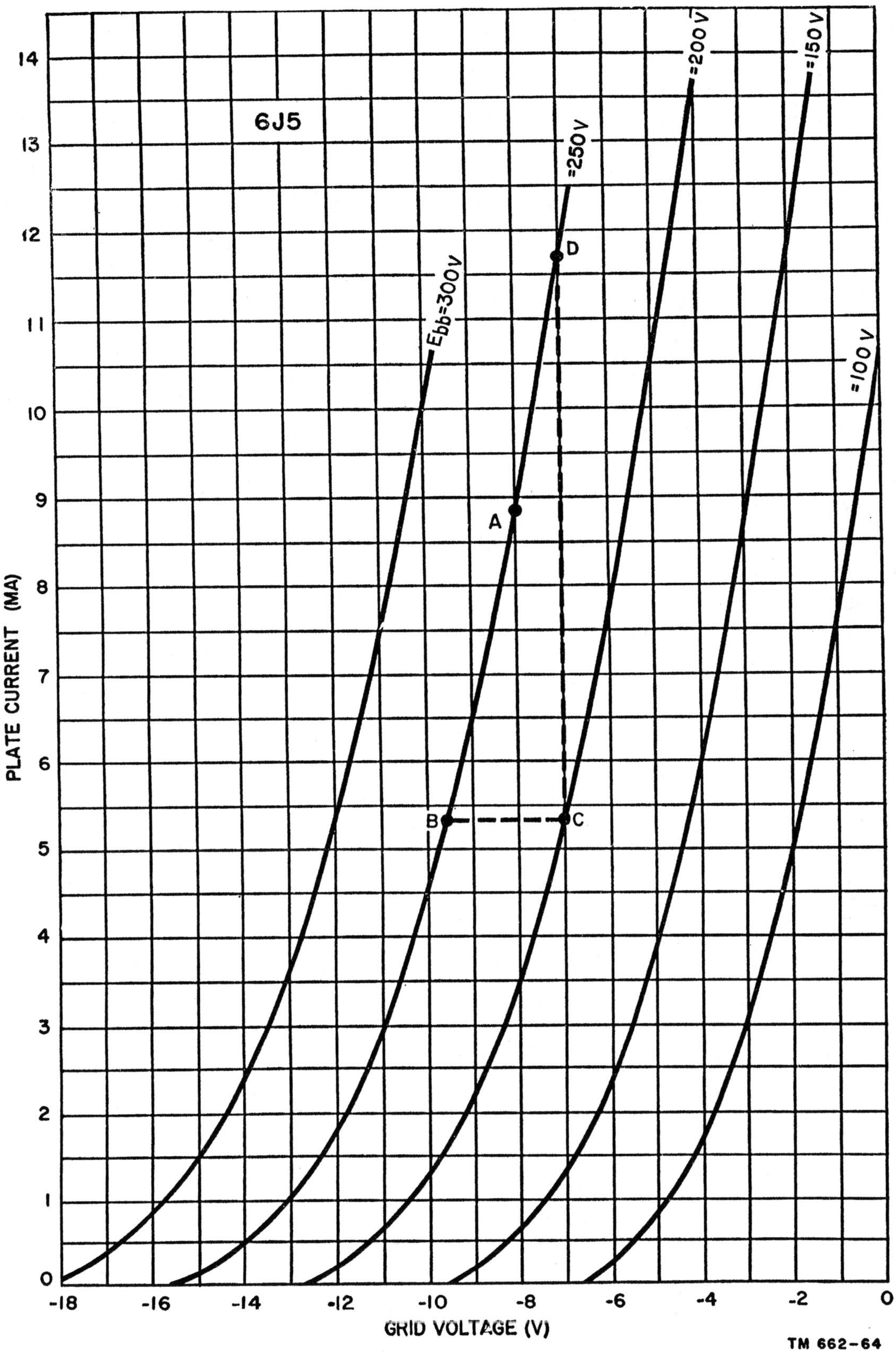

Figure 55. Grid family of curves used to determine amplifiaction factor of 6J5 triode.

C, project a line parallel to the Y-axis upward from the curve of $E_{bb} = 200$ volts until it intersects the curve of $E_{bb} = 250$ volts. This is point D.

(2) The plate current at D is 11.7 ma and at B it is 5.35 ma. Also, point D corresponds to a negative grid voltage of 7, and point B corresponds to a negative grid voltage of 9.6 volts. Consequently, with the plate voltage held constant at 250 volts and the grid voltage varying between points D and B—that is, from 7 to 9.6 volts, or a total change of 2.6 volts—the plate current changes from 11.7 to 5.35 ma or 6.35 ma. As to the plate-voltage change required to produce this same plate-current change, this value is found by holding the negative grid voltage constant at 7 volts and changing the plate voltage from D to C, that is, from 250 to 200 volts. Again, the plate current changes from 11.7 ma to 5.35 ma, or 6.35 ma.

(3) A recapitulation of the action shows that a 50-volt change in plate voltage results in a plate-current change of 6.35 ma, and a grid-voltage change of 2.6 volts produces the same 6.35-ma change in plate current. This can be summarized as

$$\mu = \frac{\triangle e_b}{\triangle e_c} = \frac{50}{2.6} = 19.20.$$

The amplification factor or μ of the tube is 19.2. This means that the grid voltage is 19.2 times more effective than the plate voltage in causing a change in plate current. This μ of 19.2 is only 4 percent less than the figure specified in the manufacturer's literature of the 6J5 triode. For all practical purposes, the μ of this triode can be considered as 20, since the 19.2 figure will vary slightly anyway on the nonlinear portions of the curves.

47. Finding μ from Plate Families

a. GENERAL. The amplification factor of a triode can be determined from the plate family of characteristics as well. Graphical determination of tube constants is accomplished more frequently by means of the plate family than the grid family because the latter is not shown too often in tube literature. A typical static plate family appears in figure 56.

b. DETERMINING AMPLIFICATION FACTOR.

(1) Developing the amplification constant of a triode from the plate family is a simple process. Select a grid-voltage value that is about halfway between the usable limits of the grid-voltage range shown on the graph. Referring to figure 56, —8 volts is a satisfactory grid voltage. Locate a reference point about halfway down the straight portion of the plate-current curve for that grid voltage. This is point A. A horizontal projection parallel to the plate-voltage axis shows point A as being equal to 5 ma. A vertical projection downward to the plate-voltage scale shows A to be equal to 216 volts. Point A, therefore, corresponds to $E_{cc} = -8$ volts, $I_b = 5$ ma, and $E_{bb} = 216$ volts.

(2) Now, project point A parallel to the X-axis to an adjacent grid-bias curve. The direction of this projection is optional. In this instance, it is toward the higher value of negative grid bias. This is point B at $E_{cc} = -10$ volts. A vertical projection dropped to the plate-voltage axis intersects the 257-volt point. Point B can, therefore, be described as $E_{cc} = -10$ volts, $I_b = 5$ ma, and $E_{bb} = 257$ volts.

(3) The next step is to project the higher plate-voltage point on the —8-volt curve; this results in point C. Point C corresponds to $E_{cc} = -9$ volts, $I_b = 9.6$ ma, and $E_{bb} = 257$ volts. The information needed to calculate μ has been obtained. With 216 volts on the plate (point A) to 257 volts on the plate (point C), a change in plate current from 5 to 9.6 ma takes place. With 257 volts on the plate, a change in grid voltage from —10 volts (point B) to —8 volts (point C) causes a change in plate current from 5 to 9.6

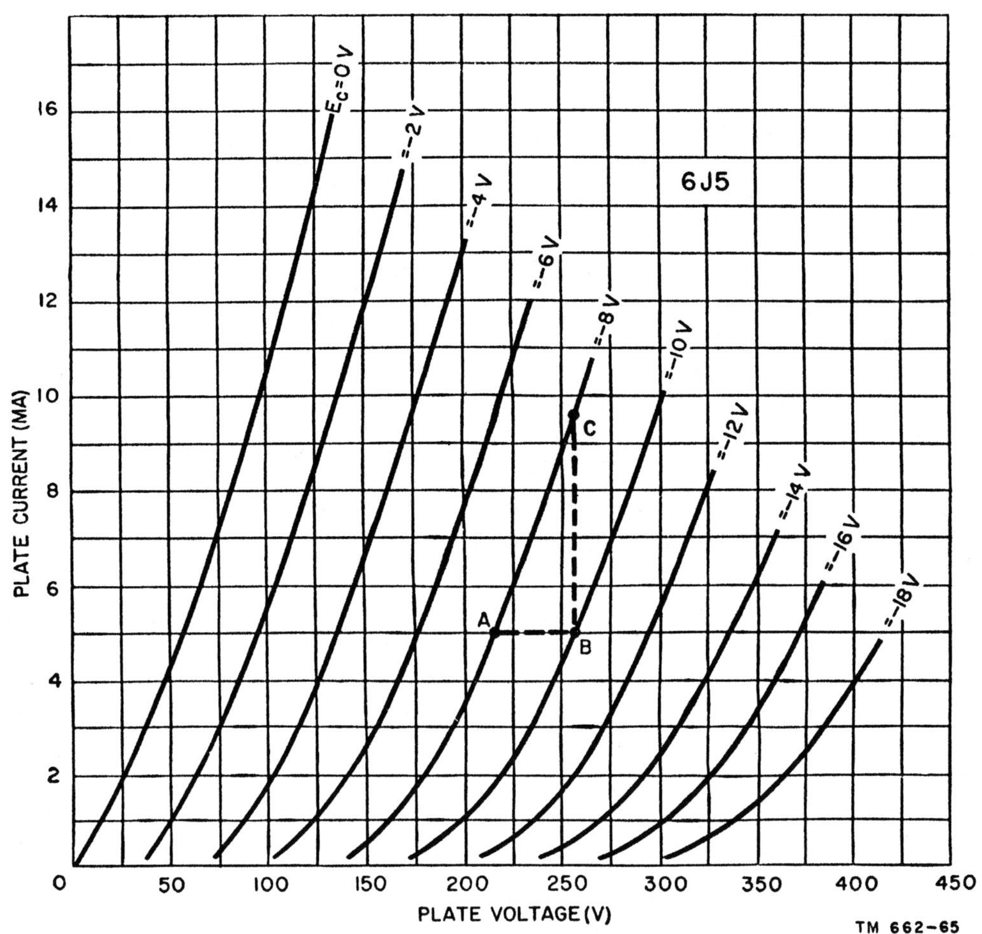

Figure 56. Plate family of curves used to determine the amplification factor of 6J5 triode.

ma. Therefore, the equation

$$\mu = \frac{\triangle e_b}{\triangle e_c} \text{ reads } \mu = \frac{257 - 216}{10 - 8} = \frac{41}{2} = 20.5$$

which is in substantial agreement with the data developed previously from the grid family. The constant determined by means of the plate family is closer to the manufacturer's data because the increments of change are smaller.

c. CLASSES OF TRIODES. There are three general classes of triodes: *low-μ triodes,* in which the amplification constant is less than 10; *medium-μ triodes,* in which it is between 10 and 30; and *high-μ triodes,* in which the amplification factor is 100 or more. A type of tube known as the *variable μ* possesses a changing μ feature.

48. Plate Resistance

a. DETERMINING R_P. The plate resistance is another vacuum-tube constant. It describes the *internal resistance* of the tube or the opposition experienced by the electrons in advancing from the cathode to the plate. This constant is expressed in two ways—the *d-c resistance* and the *a-c resistance.* The former is the internal opposition to current flow when *steady* values of voltage are applied to the tube electrodes, and is determined by the simple application of Ohm's law,

$$R \text{ (ohms)} = \frac{E \text{ (volts)}}{I \text{ (amperes)}}$$

at any point on the plate-current characteristic. The voltage, E, is the d-c plate voltage or E_{bb}, which equals e_b in this case, and the current, I, is the steady value of plate current of I_b. Resistance R, then, is R_p.

b. CALCULATING R_P. Referring to the plate

family in figure 57, point M is an arbitrary point corresponding to a plate voltage of $E_{bb} = 250$ volts, $E_{cc} = -8$ volts, and the plate current $I_b = 8.9$ ma. Applying Ohm's law,

$$R_p = \frac{250}{.0089}$$
$$= 28,100 \text{ ohms}.$$

For any value of plate voltage, the d-c resistance is determined by applying Ohm's law, where the numerator is the steady voltage at the plate, $e_b = E_{bb}$, and the denominator is the corresponding steady plate current. Point N in figure 57, therefore, corresponds to a d-c resistance of 37,500 ohms, since $E_{bb} = 225$ volts and $I_b = 6$ ma. Point 0 corresponds to a d-c resistance of 65,000 ohms. The various triangles in figure 57 can be neglected in establishing the d-c plate resistance; they are needed for finding the a-c plate resistance.

c. CALCULATING R_P.

(1) Finding the a-c plate resistance from the plate family in figure 57 is somewhat more involved. The initial point Q on the curve marked $E_{cc} = -10$ volts corresponds to a plate voltage of 265 volts and a plate current of 8 ma. The d-c resistance using Ohm's law is 33,100 ohms. The varying voltage representing a change of plate voltage for a-c resistance determination is obtained by varying the plate voltage above and below point Q. This operating range is shown by the small triangle drawn about Q. The total change in plate voltage is 20 volts, with the upper plate-voltage limit at 295 volts and the lower limit at 275 volts. The variations in plate voltage

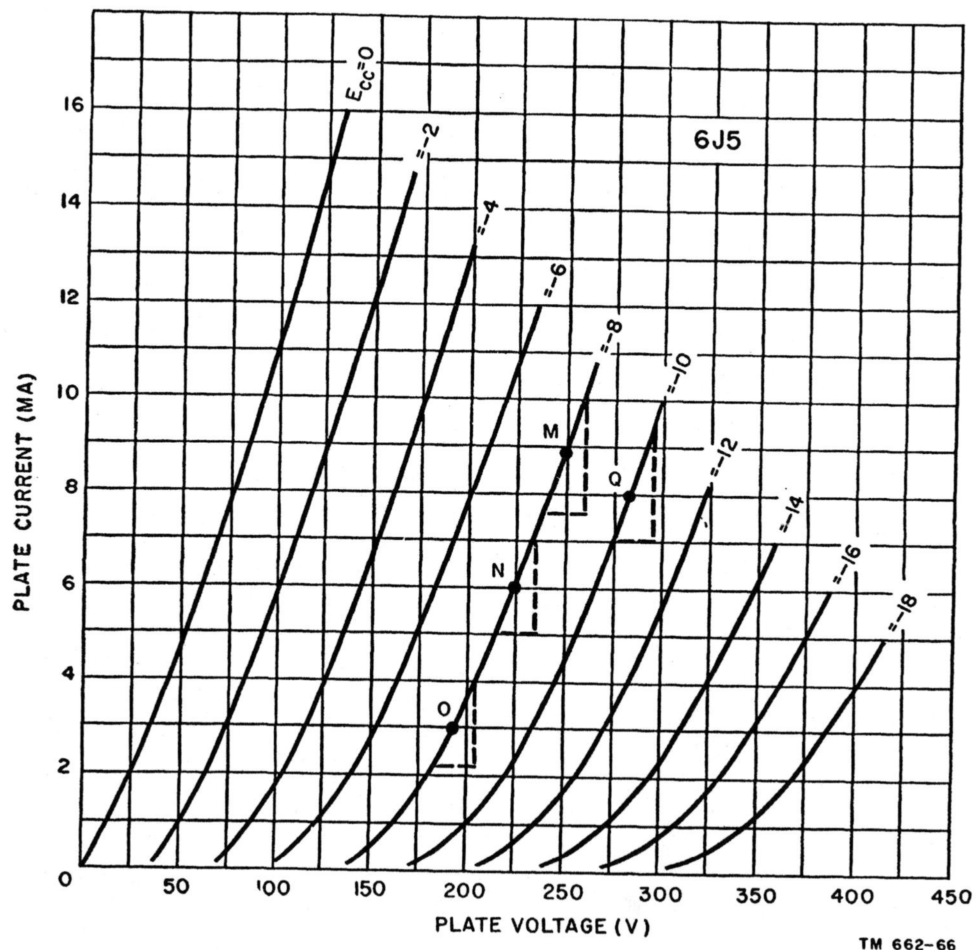

Figure 57. Plate family of curves used to determine a-c and plate resistance of 6J5 triode.

are arbitrary. The intersections with the appropriate grid-voltage curve are determined by projecting the voltage lines upward from the plate-voltage axis until they intersect the grid-voltage curve.

(2) The change in plate current, $\triangle i_b$, is determined by projecting the points of intersection between the —10-volt grid-voltage curve and the 295-volt and 275-volt plate-voltage lines, toward the left, to the plate-current axis. The upper and lower plate-current values are 9.4 ma and 7 ma, respectively. Consequently, the equation for the a-c plate resistance at the initial point Q is

$$r_p = \frac{\triangle e_b}{\triangle i_b} = \frac{295 - 275}{.0094 - .007} = \frac{20}{.0024} = 8{,}333 \text{ ohms.}$$

(3) This value does not agree exactly with the manufacturer's rating of 7,700 ohms, but it is sufficiently close for all practical purposes. As a matter of fact, the a-c plate resistance of a tube is a function of the operating point along the characteristic. This is shown by triangles M, N, and O in figure 57. Each represents a different operating point along the —8-volt grid-voltage curve. In each case, the plate voltage is changed by 10 volts each side of the operating point. For M, the $r_p = 8{,}000$ ohms, for N, $r_p = 10{,}000$ ohms, and for C, $r_p = 11{,}100$ ohms.

d. CONCLUSIONS. A comparison of the operating points discloses that the higher the applied plate voltage, the lower is the a-c plate resistance. Likewise, a change of the negative grid voltage in the positive direction causes a lower plate resistance. This is indicated by a comparison of the operating points M, N, O, and Q. In practice, one operating point is given, and the suitability of the tube is determined by the constants prevailing under the single set of conditions. As has been shown, each tube is capable of operation over a number of operating voltages, but only one of these is selected as being typical.

e. OTHER NAMES FOR R_P. Reference to *inter-nal plate resistance* invariably means a-c rather than d-c resistance. Sometimes the term *plate impedance* is used to express a-c plate resistance. The two have the same meaning. It is to be noted that the a-c term involves *a small change* in plate voltage and *a small change* in plate current, which are stated in an equation as

$$r_p = \frac{\triangle e_b}{\triangle i_b} \ (e_c \text{ constant}).$$

The d-c resistance value, on the other hand, is predicated on steady values of plate voltage and plate current.

49. Transconductance (Mutual Conductance)

a. GENERAL. Changing the plate voltage or the grid voltage or both causes changes in plate current. There is a tube constant which expresses the specific change in plate current for a unit change in grid voltage, with the plate voltage held constant. This constant is known as the *transconductance* or *mutual conductance*. The former term has become the common standard.

b. DEFINITION.

(1) By definition, transconductance is the ratio of the change in plate current to the change in grid voltage which produced it, with the plate voltage held constant. In equation form, it is

$$g_m = \frac{\triangle i_b}{\triangle e_c} \ (e_b \text{ constant}).$$

(2) When related to the plate and grid circuits, the precise notation for transconductance is g_{pg}, but it has become common practice to refer to it by the letters g_m. When given a quantitative interpretation, transconductance is the milliampere change in plate-current-per-volt change in grid voltage.

c. NUMERICAL EXAMPLE. The unit of transconductance is the *mho* (*ohm* spelled backward), but since its value generally is too large for common usage in connection with vacuum tubes, the *micromho*, or the millionth part of the mho, is the usual reference. A mho corresponds to 1 μa (microampere) change in plate current for a 1-volt change in grid voltage, or 1 mho = 1,000,000 umho (micromhos). Conversion from mho to micromhos is accomplished by multiplying the quantity expressed in mhos by 1,000,000. An electron tube which operates so that a 1-volt change in its grid voltage results in a 1-ma change in plate current is rated at a transconductance of 1,000 umho. Because of the single meaning, it has become customary to drop the word micromhos and to say simply that a tube such as that stated above has a transconductance of 1,000.

d. DETERMINING G_M GRAPHICALLY (fig. 58). Plate families of characteristic curves generally are more available than grid families. Therefore, it is desirable to know how to determine the transconductance of a tube from them. Since by definition the transconductance requires that the grid voltage be varied and the plate voltage remain constant, first select a point on a grid-bias curve which, when projected upward or downward vertically along a constant plate-voltage line, will intersect the adjacent grid-voltage curve. This is the initial point A, repre-

senting 235 volts on the plate and —8 volts on the grid. The plate current is 7.1 ma. A vertical projection from point A to the adjacent grid-voltage curve intersects the —6-volt curve and produces point B. The corresponding plate current is 12 ma and the plate voltage remains at 235 volts. The change in grid voltage occurs with the plate voltage constant. Substituting the plate-current and grid-voltage values in the equation, the transconductance is found to be

$$g_m = \frac{\triangle i_b}{\triangle e_c} = \frac{.012 - .0071}{8 - 6} = \frac{.0049}{2} = .00245 \text{ mho}$$

or 2,450 umho

Translated into plate-current change per 1-volt change on the grid, it amounts to 2.45 ma per volt. The abbreviation ma/v commonly is used in place of the words milliamperes per volt to express transconductance. Note how near this g_m value is to 2,600 which is given by the tube manufacturer. The slight discrepancy is due to the nonlinearity of the E_{cc} curves.

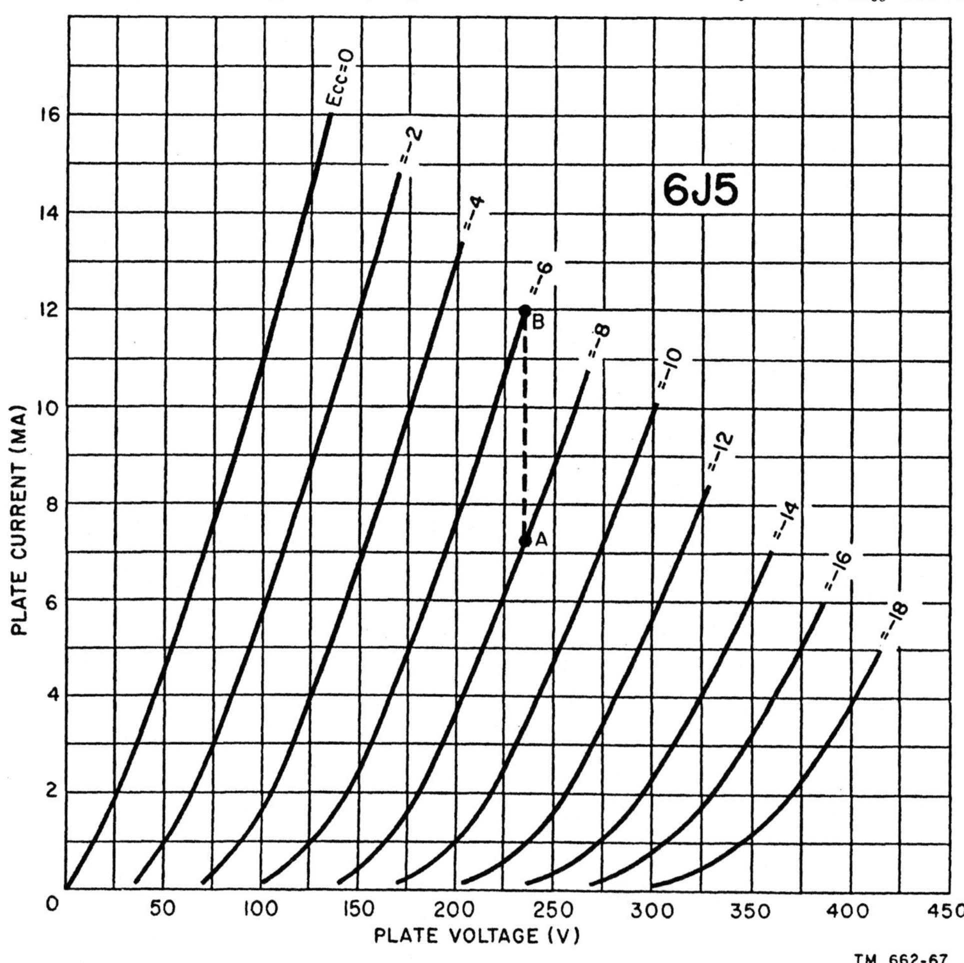

Figure 58. Plate family of curves used to determine transconductance of 6J5 triode.

e. EVALUATION OF G_M. The transconductance of a tube is an important tube constant. It is the most commonly used of all the constants when comparing tubes of like kind. A tube with a transconductance of 2,000 is a better tube than one rated at a transconductance of 1,000. The tube with the higher transconductance is capable of furnishing greater signal output than the tube with the lower transconductance, assuming like levels of signal voltage applied to the grid and like arrangements in the plate circuit.

50. Relation between μ, R_P, and g_M

a. GENERAL. The three triode tube constants—namely, amplification constant, μ, the a-c plate resistance, r_p, and the transconductance, g_m—are interrelated. The operating voltages applied to each type of tube determine the exact value of each constant, and for any one set of operating potentials to the electrodes each tube bears these three ratings. They vary in magnitude relative to each other in a definite manner. These data are illustrated in figure 59 for the 6J5.

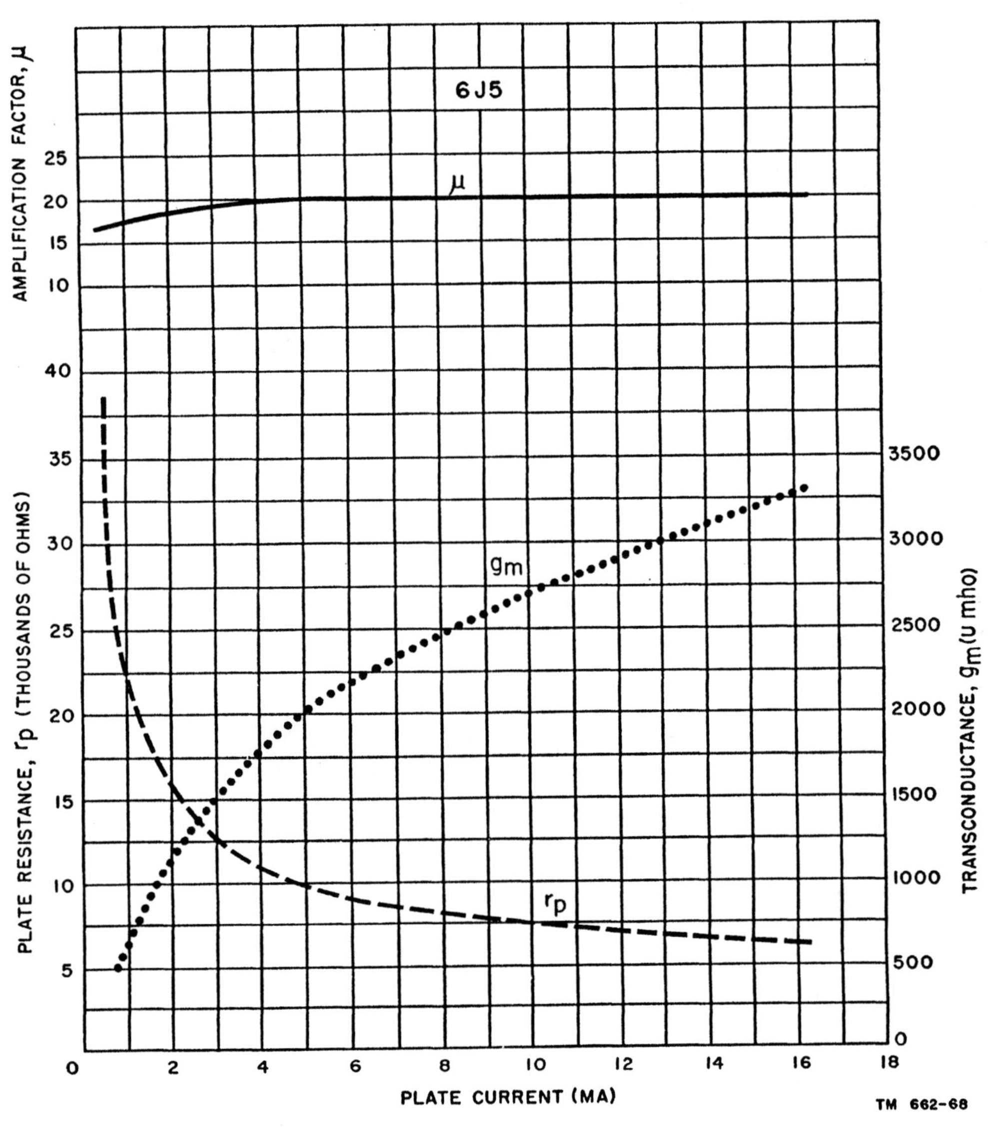

Figure 59. Curves showing relationships between μ, g_m, and r_p.

b. PLOTTING CONSTANTS. All three constants have individual Y-axis scales. As a reference, the plate current axis (X-axis) is common to all. When reading the values indicated upon the graph, the plate-resistance, r_p scale is shown on the lower left-hand Y-axis; the transconductance, g_m, appears on the right-hand Y-axis, and the amplification constant, μ, is on the upper vertical left-hand Y-axis.

c. GENERAL RELATIONSHIPS.

(1) Concerning the relationship between the constants, the amplification factor remains substantially constant under virtually all conditions. Being a property of the physical geometry of the tube, it varies little with changes in the operating voltages applied to the grid and the plate.

(2) Plate resistance, r_p, on the other hand, varies greatly with operating voltages, especially through the region of potentials which result in low values of plate current. This occurs at high negative grid bias or low plate voltage, or both. It is significant that increasing the physical separation between grid and plate in a triode results in a high amplification factor, also in a high a-c plate resistance. Usually, therefore, although not invariably, the higher the amplification constant of a triode, the higher the a-c plate resistance of that tube.

(3) As to the transconductance, it is seen to vary oppositely to a-c plate resistance; given any tube, the higher the plate resistance, the lower the g_m, and vice versa. All triodes with high transconductance ratings have low a-c plate resistance.

d. MATHEMATICAL RELATIONSHIPS.

(1) Following is the mathematical relationship that exists between the three constants. Since,

$$\text{amplification constant } (\mu) = \frac{\triangle e_b}{\triangle e_c}$$

$$\text{a-c plate resistance } (r_p) = \frac{\triangle e_b}{\triangle i_b}$$

$$\text{transconductance } (g_m) = \frac{\triangle i_b}{\triangle e_c}$$

and the term $\triangle i_b$ appears in both the r_p and g_m formulas, they can be made equal to each other; that is,

$$\triangle i_b = g_m \times e_c = \frac{\triangle e_b}{r_p},$$

resulting in

$$r_p \times g_m = \frac{\triangle e_b}{\triangle e_c}.$$

But $\mu = \dfrac{\triangle e_b}{\triangle e_c}$ also; therefore, the product of the a-c plate resistance and the transconductance is equal to the amplification constant. That is, $\mu = r_p$ times g_m or, by simple transformation, $r_p = \mu/g_m$ and $g_m = \mu/r_p$.

(2) Substituting the values for r_p and g_m that were previously obtained, by using $r_p = 8{,}333$ ohms and $g_m = 2{,}450$ umho, μ can be calculated as follows:

$$\mu = r_p \times g_m$$
$$= 8{,}333 \times .00245 = 20.4$$

which is in substantial agreement with the original value of 20.5. The accuracy of the constants varies because they are not read to the same decimals on each of the graphs, nor at exactly the same point. The a-c plate resistance is

$$r_p = \frac{\mu}{g_m}$$
$$= \frac{20.5}{.00245}$$
$$= 8{,}360 \text{ ohms}$$

and the transconductance is

$$g_m = \frac{\mu}{r_p}$$
$$= \frac{20.5}{8{,}340}$$
$$= .00246 \text{ mho.}$$

e. DISCREPANCIES PRESENT. The discrepancies present in the r_p and g_m values stem from the degree of accuracy present on the graphs. Errors of a few percent are unimportant and do not cause any complication. Moreover, the purpose of these calculations is to show how information contained in tube literature can be used for the development of unknowns, rather than to attempt to verify the information obtained from the graphs.

f. PURPOSE OF CONSTANTS. With reference to the constants of the triode, it is timely to say that these coefficients of tube behavior are not restricted to the triode only. They apply to all electron tubes which contain three or more electrodes. The 6J5 tube discussed in this chapter is just one example. Other triodes can have the same values of μ, r_p, and g_m, whereas some tubes may differ substantially. The important points to remember are that tube constants are a means of comparing tubes of like type; also, that they serve in selecting tubes to fit specific needs.

51. Dynamic Characteristics of Triode

a. GENERAL.

(1) The static characteristic of an electron tube provides an understanding of how it operates. For the information to have practical value, it must be translated into data which are related directly to an application.

(2) This requirement alters the fundamental circuit of the triode system, and in doing so, gives rise to a new set of characteristics. These are called *dynamic* characteristics. In the final analysis, they are the important ones rather than the static characteristics. The latter are used as a means for creating the familiarity necessary for comprehension of the behavior of the triode (and other electron tubes) as used in actual practice. All other exploration of how an electron tube works is done by means of the dynamic characteristics.

b. LOAD RESISTOR.

(1) The difference between an electron-tube system which allows determination of the static characteristics and a system which allows determination of the dynamic characteristics is not too great. In fact, it is a relatively minor change physically, although it is a major change relative to the actions occurring in its circuit. The dynamic characteristics can be described as a graphic portrayal of tube behavior under *load* (fig. 60). A shows the

basic triode circuit containing a load and, for comparison, the basic triode circuit without a load is shown in B.

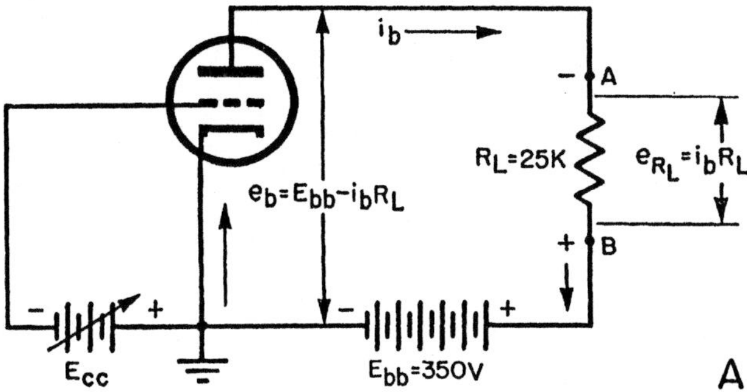

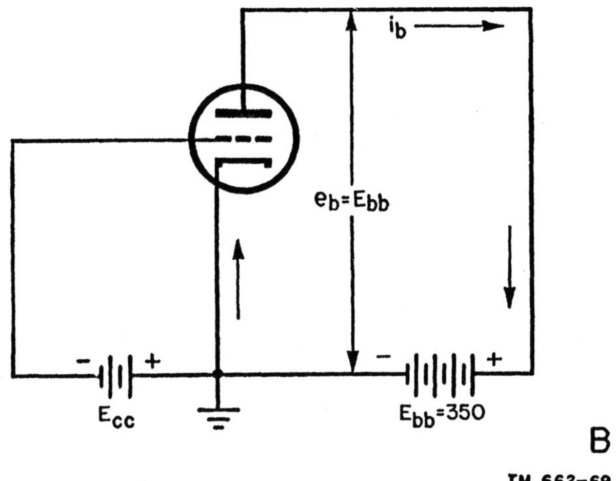

TM 662-69

Figure 60. Basic circuit of triode with and without load.

(2) One difference will be noted in these two basic circuits. In A, the presence of the resistor, R_L, is designated as the *load resistance.* It is located in the plate circuit external of the tube. It is in series electrically with the plate-voltage supply, E_{bb}, and the plate of the tube; consequently, the plate current must pass through R_L. This action greatly modifies the operating characteristics of the tube. It does not alter the three basic constants of the tube, but it does affect the plate-current grid-voltage characteristics.

c. USE OF LOAD RESISTOR. From a practical viewpoint, the circuit in A contains the essential elements for practical use of the triode. For an electron tube to be usable for any purpose whatsoever, it requires a load. The variations in plate current caused by a varying voltage applied to the control grid appear across the load as a varying voltage drop. Under proper conditions of use, the changing voltage across the load is the *output* signal. It is a replica, smaller or greater, as the case may be, of the varying voltage applied to the control grid.

d. LOAD CURRENT. The plate circuit contains several new terms, since a new action is described. To understand these new terms, it is imperative to note the plate resistor and its location. Load resistor R_L is directly in the path of plate current i_b. The movement of electrons comprising the plate current must be through this resistor, because the complete plate-current circuit is from cathode to the plate inside the tube, and, on the outside of the tube, through resistor R_L to the positive terminal of the battery, E_{bb}, and from the negative terminal of E_{bb} to the cathode.

e. CIRCUIT VOLTAGES.

(1) Under these circumstances, two voltage drops occur in the plate circuit. One of these is across the internal resistance of the tube, or $i_b r_p$, and the other is the drop across R_L, or $i_b R_L$. This leads to three voltages being present in the plate circuit: E_{bb} = the voltage available from the plate-voltage supply; e_b = the instantaneous voltage available from plate to cathode. This voltage is the difference between supply voltage E_{bb} and the voltage drop across load R_L, or $e_b = E_{bb}$ minus $i_b R_L$. Whether e_b is instantaneous or steady is determined by the character of the plate current, which relates back to the grid voltage; e_{RL} (the voltage drop across the load resistor R_L) = $i_b R_L$.

(2) In A of figure 60, assume that the grid voltage, E_{cc}, is adjusted to such a high negative value as to cut off plate current i_b. In that event, voltage e_b across the internal resistance of the tube (between plate and cathode) is the same as the plate-supply voltage, or $e_b = E_{bb} = 350$ volts, shown in B. This is true because with the plate current cut off there is no voltage drop

across R_L. The internal plate resistance is infinite and the finite value of 25,000 ohms in series with an infinite resistance has no significance when this action occurs.

(3) Now assume that the control-grid voltage is changed so that 1 ma of current flows in the plate circuit, or i_b = .001 ampere. Supply voltage E_{bb} is constant at 350 volts, at which value it remains regardless of the amount of plate current that flows. With .001 ampere of plate current, the voltage drop across R_L is

$$i_b R_L = .001 \times 25,000$$
$$= 25 \text{ volts.}$$

Since the voltage between the plate and cathode or the voltage drop across the internal resistance of tube r_p is the difference between the supply voltage and the voltage drop across R_L,

$$e_b = E_{bb} - i_b R_L$$
$$= 350 - 25$$
$$= 325 \text{ volts.}$$

Apparently most of the voltage drop takes place across the internal plate resistance, which now is finite since plate current flows. The voltage from plate to ground is high and the voltage across the load is relatively low.

(4) Now assume a change in grid voltage which causes the flow of 12 ma of plate current. By Ohm's law the voltage drop across the load is

$$e_{RL} = i_b R_L = .012 \times 25,000$$
$$= 300 \text{ volts}$$

and voltage e_b at the plate, or across the internal resistance of the electron tube, is

$$e_b = E_{bb} - i_b R_L$$
$$= 350 - 300$$
$$= 50 \text{ volts.}$$

During such operating conditions, most of the voltage drop occurs across load R_L and very little drop appears across the internal plate resistance of the tube; that is, the voltage from plate to cathode is relatively low.

(5) Bearing in mind that the plate-supply voltage, E_{bb}, is always constant at 350 volts, and that the change in plate cur-

rent is a function of the control-grid voltage, it is possible to imagine such an increase in plate current that all of the voltage drop in the plate circuit occurs across load R_L. If this occurs, the distribution of voltages in the plate circuit would be such that voltage e_b would be 0. This is an extreme condition and cannot occur in actual practice. The plate current may rise sufficiently that the effective voltage between plate and cathode is very low but never 0, since a drop always appears across r_p, even though it may be considered negligible.

f. EFFECT OF LOAD RESISTOR. All of these changes in effective plate voltage e_b result from the presence of load resistor R_L. The moment plate current flows, regardless of its value, the load resistance comes into play and affects the effective plate voltage. The load resistance reduces the effective plate voltage under all conditions of plate current except 0. This means that with a load present in the plate circuit, the actual change in plate current for a change in grid voltage is much less than it would be if load R_L were not present. This becomes evident upon consideration of the reduction in effective plate voltage simply because of the presence of a voltage drop across load R_L as soon as current flows.

g. DIFFERENCE BETWEEN CHARACTERISTICS. The relationship between grid voltages and plate current with the load present is called the *dynamic* characteristic. It differs from the static characteristic because the effective plate voltage, e_b, is not equal to the plate-supply voltage, E_{bb}, when plate current flows, as with static characteristics.

52. Load Line

a. GENERAL.

(1) The effect of the load applied to a triode (and other electron tubes) can be predicted in advance. This usually is accomplished by adding to the static plate family of characteristic curves (fig. 54) a graphical representation of the load, known as the *load line*. This shows the distribution of the output of plate-voltage supply E_{bb} between the load and the internal resistance of the tube under different conditions of plate current.

(2) For convenience, the plate family of characteristic curves is used for the dynamic characteristics. From it is developed the final transfer characteristic curve which correlates the plate-current grid-voltage relationship with the load present in the circuit. This curve is called the dynamic transfer characteristic. It is to be noted, however, that equal information can be obtained from the dynamic plate characteristic and from the dynamic transfer characteristic.

b. SELECTING LOAD LINE.

(1) Figure 61, showing a typical static plate family of curves, is the same as figure 54, except for the diagonal line XY, the load line. The load line corresponds to a load R_L of 25,000 ohms (A of fig. 60). The selection of 25,000 ohms is arbitrary. It can be a higher or a lower value, except that there are limits to the ohmic values of loads relative to the internal plate resistance of a tube, and also because of the fixed value of the applied plate voltage. These statements will be clarified later in discussions of voltage and power amplification.

(2) No connection exists between load line XY (fig. 61) and the static plate family. The load line has no fixed association with tube constants or characteristics. It is simply a means of developing information concerning the behavior of any tube that is used with a certain load. The load may be correct, or it may not be right for the tube. This information appears on the graph when the effects of the load are studied by means of the load line.

c. DISCUSSION OF LOAD LINE.

(1) The full length of the line represents one extreme condition — namely, a voltage drop across R_L equal to the full plate-supply voltage. This means

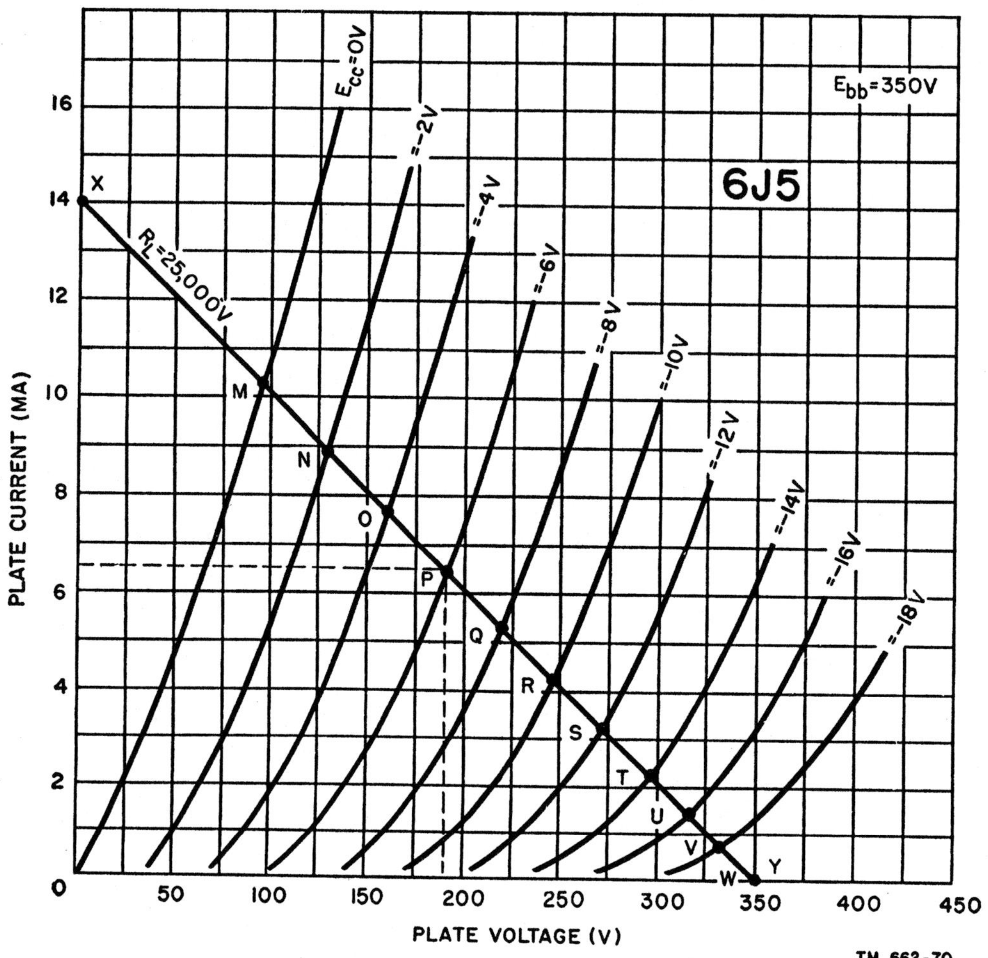

Figure 61. Plate family of characteristic curves for 6J5 triode, including load line.

that $e_b = 0$ and the plate current equals the full plate-supply voltage divided by the load resistor or

$$I_b = \frac{E_{bb}}{R_L} = \frac{350}{25,000}$$
$$= .014 \text{ ampere}$$
$$= 14 \text{ ma.}$$

This figure establishes one terminal of the load line—14 ma on the plate-current axis, point X. The other terminal of the load line corresponds to the other extreme condition: that is, zero plate current and the full plate-supply voltage of 350 volts on the plate of the tube. This is point Y on the plate-voltage axis.

(2) Further reference to figure 61 discloses that the load line intersects the plate-current curves at different points. To establish the distribution of voltages between the tube and the load, an operating point must be selected. Assume that the grid voltage chosen for examination is —6 volts, or $E_{cc} = $ —6. The load line crosses the plate-current curve for this bias at point P. A horizontal line drawn to the plate-current scale shows that the corresponding plate current is 6.35 ma. A vertical line dropped from point P to the plate-voltage axis intersects it at 190 volts. Therefore, point P can be described as

$$E_{cc} = \text{—6 volts}$$
$$E_{bb} = 190 \text{ volts}$$
$$i_b = 6.35 \text{ ma.}$$

The horizontal projection to the plate-voltage axis indicates the division of the plate-supply voltage across the load and across the internal resistance of the tube. The point of intersection between the horizontal projection to the plate-voltage scale and the axis indicates the voltage—190 volts—effective at the plate. The difference between this value and the total plate-supply voltage is the voltage dropped across the load, or 350 minus 190 = 160 volts. Further to verify this last statement, the voltage drop across the load resistor can be computed as follows

$$e_{RL} = i_b R_L = .00635 \times 25,000 = 160 \text{ volts.}$$

(3) The voltage on the plate and the voltage dropped across the load, for various values of grid voltage, can be determined in similar fashion. For example, point M, corresponding to 0 voltage on the control grid or $E_{cc} = 0$, can be described as

$$E_{cc} = -0 \text{ volt}$$
$$i_b = 10.1 \text{ ma}$$
$$e_b = 97 \text{ volts}$$
$$e_{RL} = 253 \text{ volts}$$
$$E_{bb} = 350 \text{ volts.}$$

Having selected $E_{cc} = -6$ volts as the operating point and $E_{cc} = 0$ volts as one extreme, the other extreme can be 6 volts in the other direction, or $E_{cc} = -12$ volts. This is point S and can be described as

$$E_{cc} = -12 \text{ volts}$$
$$i_b = 3.1 \text{ ma}$$
$$e_b = 272 \text{ volts}$$
$$e_{RL} = 78 \text{ volts}$$
$$E_{bb} = 350 \text{ volts.}$$

d. SUMMARY AND CONCLUSIONS.

(1) The three points of intersection, M, P, and S, establish a number of operating factors. For instance, $E_{cc} = -6$ was selected as the reference operating point. Then an extreme condition, $E_{cc} = 0$ volts, was selected as one limit of change in grid voltage. The second limit of change in grid voltage was set as $E_{cc} = -12$ volts. How have these changes affected the plate circuit? The answer is given in the tabulation which follows.

$E_{cc} = 0$ v	$E_{cc} = -6$ v	$E_{cc} = -12$ v
$i_b = 10.1$ ma $e_b = 97$ v $e_{RL} = 253$ v	$i_b = 6.35$ ma $e_b = 190$ v $e_{RL} = 160$ v	$i_b = 3.1$ ma $e_b = 272$ v $e_{RL} = 78$ v

(2) From the data available in the preceding tabulation, other factors can be determined as shown below:

Change from $E_{cc} = 0$ to $E_{cc} = -6$ v	Change from $E_{cc} = -6$ to $E_{cc} = -12$ v
At plate from 190 — 97 = 93 v Across load from 253 — 160 = 93 v	At plate from 272 — 190 = 82 v Across load from 160 — 78 = 82 v

As shown in the preceding tabulation, the change in plate voltage e_b is not the same for equal changes in grid voltage. That is, making the grid less negative by 6 volts (going from —6 to 0) causes a fall in the effective plate voltage by 93 volts. However, when the grid was made more negative by 6 volts (going from —6 to —12), the effective plate voltage increases by only 82 volts. Under ideal conditions (such as having perfectly linear characteristic curves), the rise and fall in voltage at the plate would be the same in both directions. It is not the same in this case because of the curvature in the plate-current characteristics over the range of grid voltages between —6 and —12 volts. The curvature reduces the change in plate current per increment of increase in negative grid voltage beyond approximately —8 volts, where the curvature is more predominant. This situation introduces another subject for discussion—namely, linearity of the resultant characteristic — but this is set aside for later discussion.

(3) Another pertinent fact is that the total change in voltage at the plate always

is equal to the change in voltage across the load. This is shown in the tabulation in (2) above. A 93-volt change in e_b is equaled by a 93-volt change in e_{RL} when the grid voltage is varied from —6 volts to 0 volt. An 82-volt change in e_b is equaled by an 82-volt change in e_{RL} when the grid voltage is increased from —6 to —12 volts.

e. CHANGES IN GRID VOLTAGE.

(1) This is an appropriate time to inject another view of the change in voltage drop across the load. Assuming a slow manual means for changing the grid voltage, the changes in plate voltage as well as the changes in voltage drop across the load R_L would occur at the same rate. Regardless of how slowly or rapidly this takes place, the fact remains that it is a swing in voltage. In one direction the peak amplitude is 93 volts and in the other direction it is 82 volts. This change in voltage frequently is described as the swing in voltage, and is referred to as the peak value expressed in volts.

(2) A single expression of the peak value does not suffice for the example under discussion because of the nonlinear changes above and below the reference —6-volt grid voltage. The expression *swing in voltage* is usable, nevertheless, except that two peak values must be mentioned, a positive and a negative value. A single peak reference is usable when the changes are in only one direction.

53. Dynamic Transfer Characteristic

a. The load line shown on a static plate family tells a great deal, but not as conveniently as does another organization of characteristics. This is the static family of plate-current grid-voltage curves to which is added the effects of the load. The resultant plate-current grid-voltage curve is known as the dynamic transfer characteristic (fig. 62). It has become common practice, when studying the behavior of the plate current under the influence of a signal voltage applied to the grid, to show the dynamic transfer characteristic and to plot the input signal and the resultant plate current along this characteristic.

b. In figure 62, A repeats the static grid family illustrated in figure 53, and B shows the 25,000-ohm load line on the static plate family. The latter will be projected on the former to show the plate-current grid-voltage curve which represents the plate current corresponding to certain grid voltages and the effect of the load in creating the effective plate voltages.

c. The two families of curves have three attributes in common: a common plate-current axis, like values of control-grid voltage, and like values of plate-supply voltage, although the last two named are illustrated differently. In order to project the effect of the load line in B on the static grid family in A of figure 62, it is necessary to plot, on the latter graph, the plate-current values for each value of grid voltage shown on the plate family.

d. This is done in the following manner. Point M, in B of figure 62, corresponds to $E_{cc} =$ 0 and 10.1 ma. This locates point M′ on the $e_c = 0$ grid-voltage line at the 10.1-ma point. Point N on the plate-family graph corresponds to $E_{cc} = $ —2 volts and 8.9 ma and this locates point N′ on the intersection of the appropriate plate-current and grid-voltage projections on the grid-family graph. In similar fashion, points O, P, Q, R, S, T, U, and V on the plate family are transferred to the grid family. When these points are formed, the result is the characteristic curve M′, N′, O′, P′, Q′, R′, S′, T′, U′, V′. This is the *dynamic transfer characteristic.*

e. The dynamic transfer characteristic is seen to differ substantially from the static characteristic. It is much less steep and shows that a change in grid voltage results in a smaller change in plate current than does any static characteristic. This is not a disadvantage, although it may appear so at first glance, because the change in plate current for a given change in grid voltage on both sides of a reference point is much more linear with the load present than with the load absent.

f. For example, if the reference point is O′, which corresponds to —4 volts on the grid and 7.6 ma plate current, a change in grid voltage to —2 volts and then to —6 volts results in 8.9 and 6.4 ma of plate current. This is an

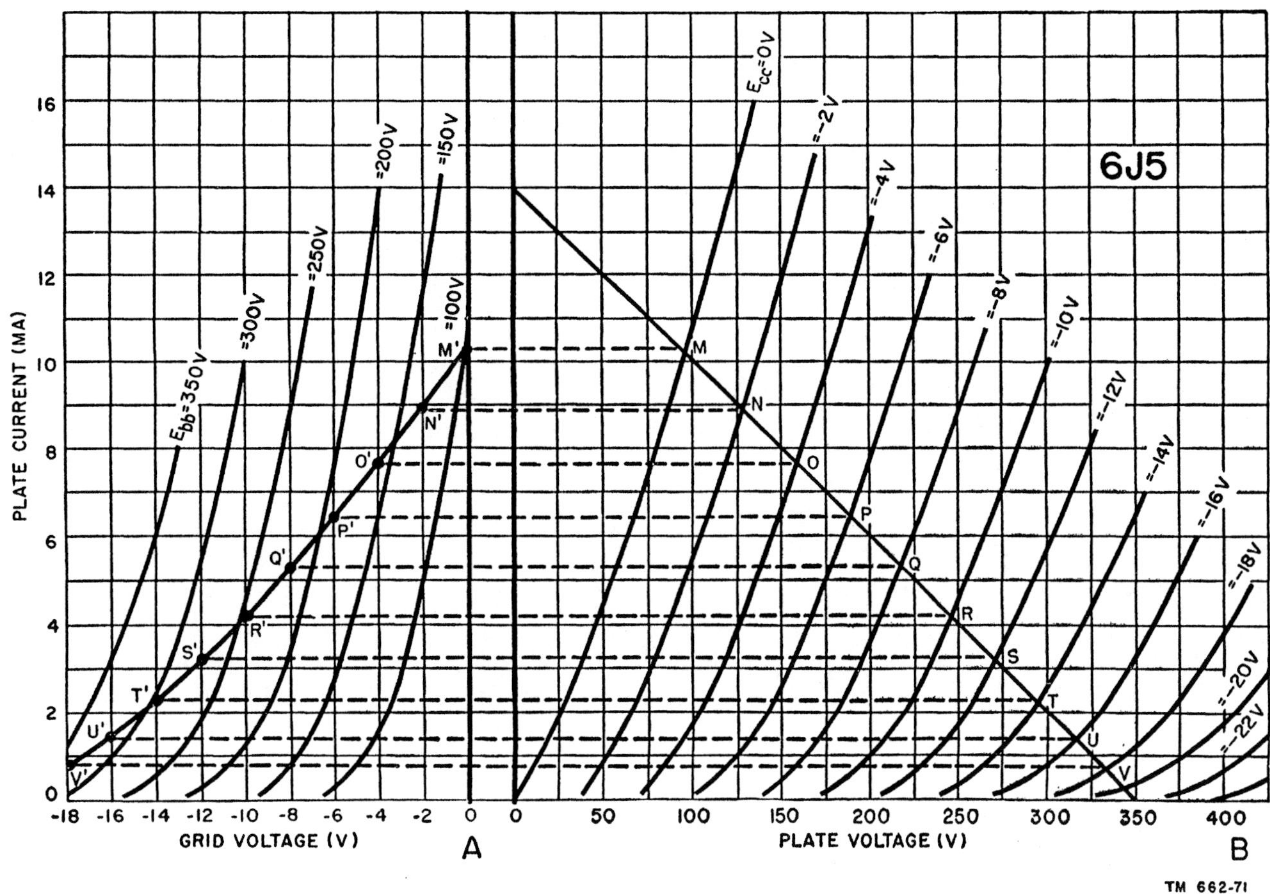

Figure 62. Plate and grid families of 6J5 which are used to construct dynamic tranfer characteristic.

increase of 1.3 ma in the positive direction and 1.2 ma in the negative direction. The swing in plate current is not linear, but the degree of nonlinearity is less than 8 percent even with the limited accuracy which is possible in graphs of this kind.

g. A corresponding point of operation along a static characteristic without any load is selected on the 150-volt curve. With —4 volts applied to the grid, the plate current is 6.4 ma. With the grid reduced to —2 volts, the plate current rises to 12 ma, a change of 5.6 ma. With the grid voltage changed to —6 volts, the plate current drops to 2.6 ma, a change of 3.8 ma. Nonlinearity between the positive and negative changes amounts to approximately 30 percent. Nonlinearity in electron tube behavior is a source of distortion and highly undesirable. Some value of distortion can be tolerated, but it always must be kept within the bounds permitted by the function of the system wherein the tube is being used. Too much distortion can destroy the usefulness of a complete system.

h. The dynamic transfer characteristic in A of figure 62 is seen to have some curvature. It is greater between approximately $e_c = 10.5$ volts and $e_c = -18$ volts than between $e_c = 0$ volts and $e_c = -10.5$ volts. Since linearity of plate-current changes is a general requirement, examination of the dynamic transfer characteristic discloses the range of grid voltages over which linearity in plate-current changes is available.

54. Effect of Different Loads

a. Before showing how the dynamic transfer characteristic is used with a simple triode amplifier system, it is necessary to dwell briefly on the subject of different loads. It was said that the selection of 25,000 ohms for 6J5 triode operated at a supply voltage of 350 volts was arbitrary when used in the circuit of figure 62.

It was seen also that the presence of a finite load produced a dynamic transfer characteristic which had some curvature or nonlinearity, although it was an improvement over the static plate-current grid-voltage without any load.

b. As apparent from the statements above and shown in figure 63, the higher the ohmic value of the load, the straighter should be the dynamic transfer characteristic. As the ohmic value of the load is increased, the over-all characteristic approaches a straight line; that is, it becomes more and more linear throughout its length. This can be seen by comparing characteristics A, B, C, and D. The curve for $R_L =$ 100,000 ohms is substantially straight for almost its full length, whereas the curve for $R_L =$ 25,000 ohms is straight for only a small portion, and that for $R_L =$ 15,000 ohms is straight for even a smaller segment of the characteristic. The reason for this behavior is that the higher the load resistance (within a reasonable value

of ohms), the greater the effect of the load in determining the plate current per unit of change in grid voltage and plate voltage rather than the change in internal plate resistance. With infinite load, the characteristic would be along the 0 plate-current line, because changes in grid voltage would have no effect on the circuit.

55. Simple Triode Amplifier

a. GRID SIGNAL. Previous discussions in this chapter have led to the explanation of the simple triode amplifier, shown schematically in figure 64. This amplifier is similar to the triode systems shown previously when they were used for the construction of static and dynamic characteristics, but it contains one important difference. This is the addition of a sine-wave signal source which generates the signal voltage e_g that is to be amplified. Signal e_g appears in the input circuit as a sinusoidal variation of voltage between the control grid and the E_{cc} supply. An amplified version of the signal re-

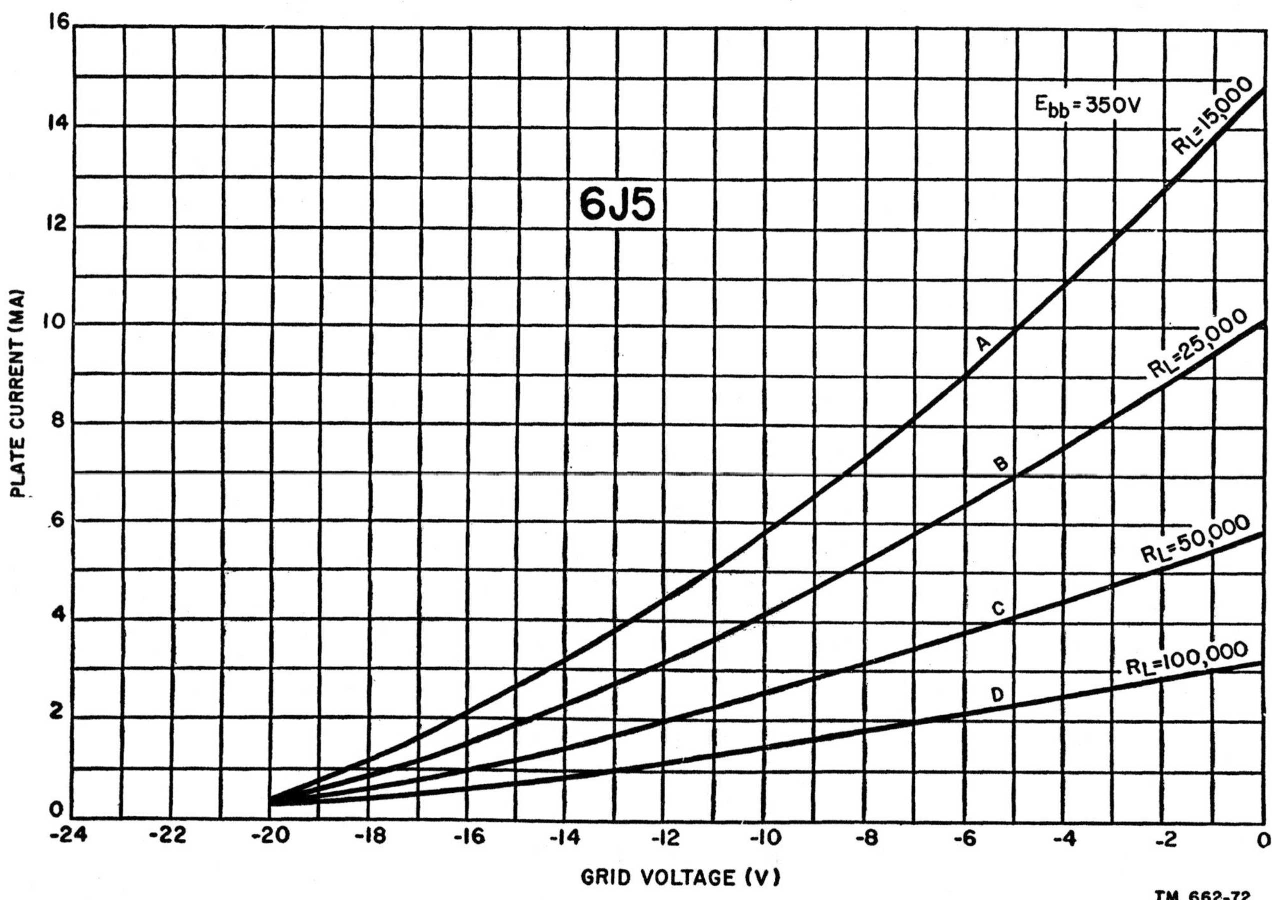

Figure 63. Dynamic transfer characteristic curves for different loads.

appears in the plate circuit as the varying drop, e_{RL}, across the load resistance. As described previously, this voltage drop developed across the load is the result of the varying plate current, i_b, that flows through the load.

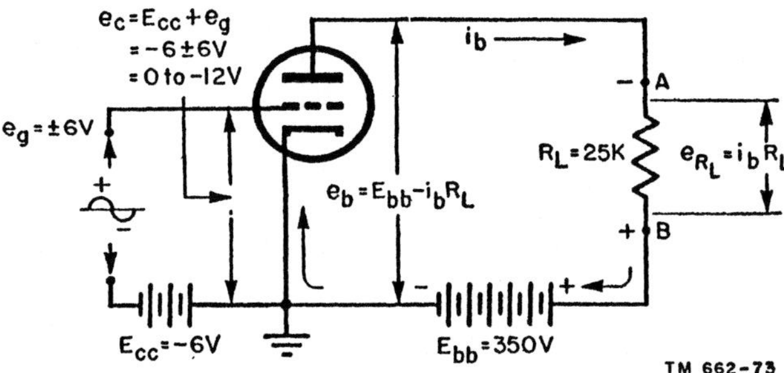

Figure 64. Basic circuit of triode, including input signal and load.

b. GRID VOLTAGES.

    (1) By examination, three significant voltages now are present in the control grid-cathode circuit:

$E_{cc} =$ the bias voltage
$e_g =$ the instantaneous value of signal voltage
$e_c =$ the instantaneous voltage between grid and cathode.

    (2) The relative values of grid bias E_{cc} and signal voltage e_g warrant only brief comment at this time because the operation of this amplifier is explained in a later chapter. If the assumed requirement is that the triode amplifier must operate without grid current, the amount of the fixed grid bias, E_{cc}, must be at least equal to the peak value of the positive half of the sine-wave signal, e_g. In that way, the positive alternation of the signal voltage will not drive the grid into the positive voltage zone; consequently, no grid current is drawn.

c. VOLTAGE COMPUTATIONS.

    (1) If the signal, e_g, is established at 6 volts peak, then the fixed grid bias can be set at the same figure, or $E_{cc} = -6$ volts. This amount of bias will be just offset by the 6-volt peak of the positive alternation of the signal. This is one extreme of voltage conditions in the input circuit and can be shown numerically as follows:

$e_c =$ signal voltage plus grid bias
$= e_g + E_{cc}$
$e_c = (+6) + (-6) = 0$ volts.

    (2) The other extreme occurs at the instant when the peak value of the negative alternation of the signal voltage appears in the grid circuit. At that moment, the signal is -6 volts. With the fixed bias of -6 volts active in the circuit, the instantaneous voltage between the grid and the cathode is the arithmetical sum of the two voltages, or -12 volts. This can be shown numerically as

$e_c =$ signal voltage plus bias voltage
$= e_g + E_{cc}$
$e_c = (-6) + (-6) = -12$ volts.

d. GENERAL.

    (1) Between these extremes of grid voltage, the instantaneous voltage between grid and cathode is a function of the instantaneous amplitude and polarity of the signal voltage plus the fixed negative bias. At no time does the control grid become positive.

    (2) Another important item is the association between the operating conditions created by the voltages applied and the so-called *operating point* on the dynamic transfer characteristic, illustrated in figure 65, where the load resistor equals 25,000 ohms. This is the same dynamic transfer characteristic developed in A of figure 62 with the static curves omitted. The dynamic transfer characteristic alone is sufficient for a graphical display of the way in which a signal voltage applied to the grid reappears amplified in the plate circuit via the plate current, and also to show the meaning of the operating point and possible conditions of operation. These curves furnish the means for predicting the behavior of the plate current for a signal input.

56. Operating Point

a. QUIESCENT CONDITIONS. To demonstrate the method of predicting the plate current behavior from the dynamic transfer characteristic, it is necessary first to establish the oper-

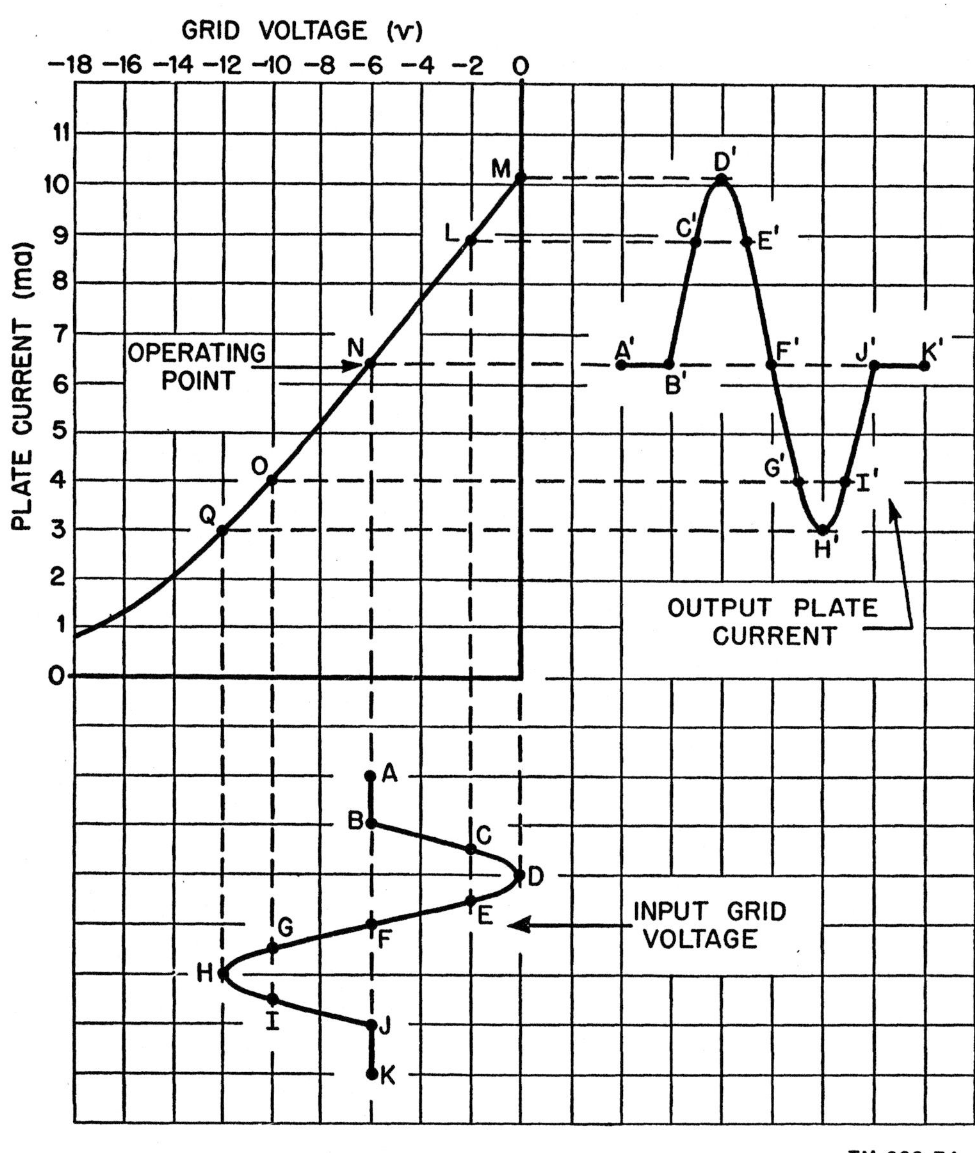

Figure 65. Operating on linear portion of 6J5 dynamic transfer characteristic (R_L equal to 25,000 ohms).

ating point on the characteristic curves. This is set by the amount of fixed grid bias applied to the tube. It establishes a steady value of plate current which prevails for a 0-input signal voltage and is generally referred to as the *quiescent* value of plate current. Based upon the data in figure 64, where a 6-volt peak signal and a —6-volt fixed bias are used, the operating point is N in figure 65. This is seen to equal 6.4 ma and is the quiescent value of plate current.

b. CHOICE OF OPERATING POINTS.

(1) The location of the operating point determines how much change occurs for input signals of different magnitudes. Since it is always desirable to develop a maximum change in plate current for a unit change in grid voltage, the location of the operating point is determined by the maximum amount of signal voltage anticipated in the grid circuit. The reason for this is obvious. Establishing the operating point for a 6-volt peak signal is satisfactory for all signals less than 6 volts peak, but it is not adequate for signals

78

exceeding 6 volts peak. Such signals will drive the grid positive and cause grid current to flow.

(2) Another factor related to the location of the operating point is the desired behavior of the plate current. Inasmuch as the signal to be amplified is a sine wave, it is necessary for the plate current to rise as much as it falls during the alternate positive and negative half-cycles of the signal. This requirement is described simply by saying that linear changes are required in the plate current. It can be satisfied only by locating the operating point along the linear part of the plate-current curve (N of fig. 65). For instance, selecting the operating point at P (fig. 66) by the application of —12-volt bias permits a 6-volt upward and downward swing in grid voltage, but the change in plate current is nonlinear. The reason will be explained shortly.

c. AMPLIFICATION.

(1) Examples of the plate current result-

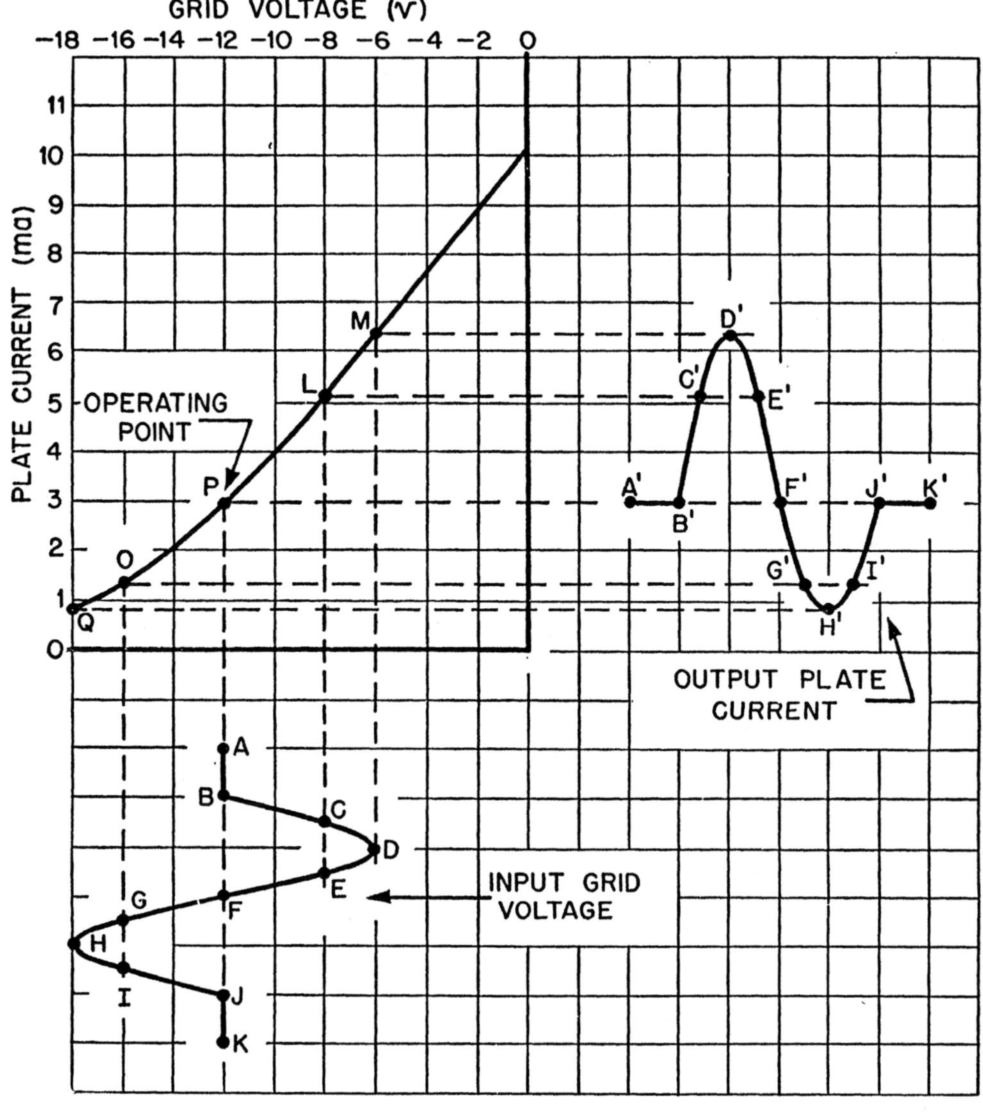

Figure 66. Operating on nonlinear portion of 6J5 dynamic transfer characteristic (R_L equal to 25,000 ohms).

ing from a 6-volt peak signal applied to a simple triode are illustrated in figures 65 and 66. Each graph shows the plate-current results for a different operating point. These correspond to N and P, previously discussed. In each case, the input signal is a sine wave of voltage. The plate-current variations resulting from the applied signal are shown in similar fashion, that is, as graphical representations of a current wave. Faithful amplification of the signal is indicated by similar shapes for the plate-current wave and the input signal-voltage wave. For proper correlation between input signal voltage and output plate current, the same time scale must be used for both. This is accomplished by displaying the cycle of the signal voltage and the plate current along zero axes of similar physical length. Faithful amplification of the input signal is indicated by sine-wave shapes for the plate-current wave, since the input signal voltage in each case is a sine wave. If the plate-current wave-shape departs from the original grid sine wave, it is the equivalent of distortion.

(2) The development of the plate-current changes for instantaneous changes in signal voltage requires a point-by-point plot of the instantaneous resultant of the signal and bias voltages on the transfer characteristic. Projections of these plots form the plate-current wave, as accomplished step-by-step in figure 65, where the signal voltage and the developed plate current bear corresponding identifying letters at designated points.

(3) Referring to line AB in figure 65, the input signal-voltage wave is drawn so that its 0-volt point coincides with the fixed bias voltage. This is standard procedure and in this instance coincides with $E_{cc} = -6$ volts. All other instantaneous amplitudes of the signal then are projected into the characteristic and indicate the instant-

aneous plate currents. Horizontal projections of these intersections become the instantaneous amplitudes of the plate-current wave.

(4) For instance, line AB is a period during which the signal voltage is 0 (point N), producing the period A'B' wherein the quiescent plate current is 6.5 ma. This is established by the no-signal voltage conditions—that is, the 350-volt supply, the load resistor of 25,000 ohms, and the fixed bias of —6 volts. The signal voltage starts increasing in the positive direction and at point C equals +4 volts. It offsets 4 volts of the negative bias; therefore, the effective grid voltage is —2 volts at this time. A vertical projection from point C on the input signal wave to the characteristic passes through the —2-volt mark on the grid-voltage scale. It intersects the characteristic at L and equals 8.9 ma. A horizontal projection on the plate-current wave scale produces the point C'.

(5) The peak of the input signal labeled D has a voltage magnitude of +6 volts. The effective grid voltage, therefore, equals 0 and the projection on the characteristic produces point M and equals 10.5 ma. Projecting this point horizontally on the plate-current time-base produces the peak amplitude D'. Point G is projected to produce point O; point H is projected to produce point Q. Treatment of every other point on the signal-voltage wave in similar fashion develops the plate-current waveshape of figure 65.

d. ANALYSIS OF ILLUSTRATIONS.

(1) An analysis of the graphical display of the signal voltage, dynamic transfer characteristic, and resultant plate-current waves affords pertinent information. A casual inspection of the signal-input and plate-current waves discloses a substantial similarity, which signifies relatively little distortion. A more critical examination shows that the positive alternation of the plate current has a

slightly higher peak amplitude than the negative alternation. Nonlinearity of this kind is a form of distortion, but in this case it is not too great. Even this small amount of distortion can be eliminated by increasing the load resistance, reducing the signal voltage, and operating entirely along the straight part of the characteristic.

(2) An aggravated example of distortion appears in figure 66. It results from the choice of the operating point on the dynamic transfer characteristic relative to the swing in grid voltage caused by the signal. The plate-current wave is developed as shown in figure 65, but the steps have been omitted, although corresponding parts of the signal and plate-current waves are labeled similarly.

(3) The nonuniformity in plate-current changes on both sides of the quiescent value is evident in the plate-current wave of figure 66. The curvature in the dynamic transfer characteristic for relatively high values of grid bias results in *flattening* of the peak of the negative alternation. This is serious distortion when the intended function of the triode amplifier is distortionless amplification of a sine wave. It can be said in passing that wave flattening of this kind is the equivalent of introducing frequencies which were not present in the single-frequency input signal.

(4) The statements in the preceding paragraphs are a brief treatment of the subject of amplification. More elaborate and complete details are contained in chapter 6. In the meantime, it should be noted that the amplitude of the plate-current curves in figures 65 and 66 relative to the pictured amplitudes of the input signals is not an indication of the amount of amplification being obtained. This can be derived only from calculation (or measurement) of the voltage drop developed across the triode load by the changing plate current and comparison with the amplitude of the input signal.

e. GENERAL. Concerning the types of distortion that can be developed in an electron-tube amplifier, they are more numerous than the two examples illustrated in figures 65 and 66. Additional examples are given later in this manual.

57. Interelectrode Capacitances

a. An important electrical effect is associated with triode tubes. It is the capacitance presented by the facing metal surfaces which are the tube electrodes. Capacitance exists between the cathode and the control grid, between the cathode and the plate, and between the control grid and the plate.

b. The interelectrode capacitance between the control grid and the cathode bears the label C_{gk}, the subscripts indicating the tube electrodes which form the tiny capacitor. The capacitance between the control grid and plate is labeled C_{gp}, and between plate and cathode C_{pk}. These capacitors are illustrated in figure 67. These capacitances are relatively small, being possibly 2 to 10 $\mu\mu$f. Nevertheless, they are very important and cannot be ignored in some cases.

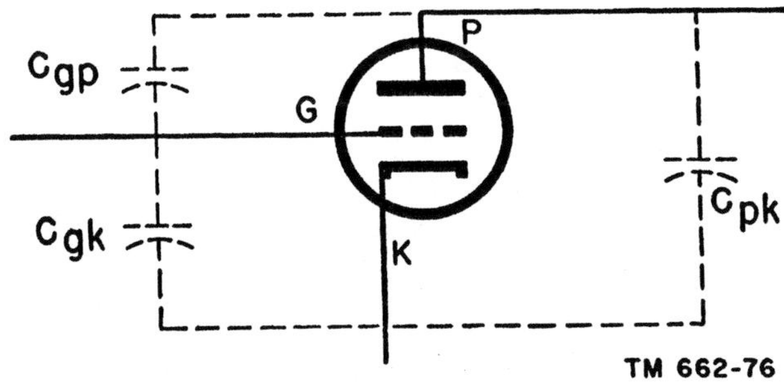

Figure 67. Location of interelectrode capacitances within triode.

c. The triode is used in many circuits that involve the phenomenon of resonance. The capacitance between the triode electrodes such as C_{gk} and C_{gp} is additive to any already there as a part of the resonant systems connected between these electrodes. Consequently, substitution of one tube for another is capable of impairing the condition of resonance because of changes in tube capacitance.

d. In all applications of the triode and other multielectrode tubes, a form of circuit isolation is attained between the circuit components which form the control-grid circuit and those which are connected to the plate circuit of the tube. Under ideal conditions, the only coupling between the control-grid and plate circuits is the stream of electrons inside the tube. The presence of capacitance inside the tube between the control-grid and plate electrodes tends to defeat the usefulness of the triode as a means of attaining circuit isolation. The interelectrode capacitance between the control-grid and plate circuits acts as a path by which energy present in the plate circuit is transferred back to the grid circuit. The result is interaction between the control-grid and plate circuits. This can be serious when the grid circuit and the plate circuit of the triode contain systems which are resonant to the same or approximately the same frequency. This action is called *feedback.* The higher the frequency of operation the worse is the situation, since the reactances of C_{gk} and C_{gp} become negligible and increase the feedback effect.

e. The results of uncontrolled feedback are bad. The circuit behaves as a very unstable amplifier. If the feedback is excessive, the system becomes useless as an amplifier; instead, it generates a signal of its own. This action is called *oscillation.* In order to minimize the tendency toward oscillation and so permit the triode to function as an amplifier, recourse is made to a circuit action known as neutralization, which will be discussed later. In neutralization, the interelectrode capacitance between the control grid and the plate is neutralized by means of a special capacitor located outside the tube and connected between the grid and plate circuits.

f. Freedom from uncontrolled feedback in amplifiers can be attained by use of electron tubes which contain more than three—sometimes four or five—electrodes. These *tetrodes* and *pentodes* are discussed fully in the following chapter.

58. Summary

a. Control grids, as used in triodes, are composed of such metals as molybdenum, nichrome, iron, and an alloy of nickel and iron.

b. The inside view of a glass triode is obscured by the opaque coating formed by a *getter.* The purpose of the getter is to absorb any gases that are initially liberated from the electrodes during operation.

c. The purpose of the control grid in triodes is to govern the movement of electrons from the space charge to the plate.

d. One distinguishing feature of the triode is that its plate current can be cut off despite the presence of a positive d-c voltage on the plate.

e. Amplification occurs in a triode when it delivers a stronger signal to the output circuit than it receives.

f. D-c power sources for a triode consist of an A-battery, which supplies the filament voltage; a B-battery, which supplies the plate voltage; and a C-battery, which supplies the control-grid voltage.

g. Operating potentials of a triode are the fixed d-c operating voltages of the filament, control grid, and plate.

h. The reference point of the triode usually is the cathode (or the common junction point of the voltage sources) and it usually is grounded.

i. An electrostatic field exists between the cathode (space charge) and the control grid in a triode. The direction of this field is such as to pull electrons to the plate when the grid is made positive. The reverse is true when it is negative.

j. Grid current is a condition that prevails in a triode that exists when electrons attach themselves to the control grid and complete a path to it. It is negligible when the grid is at zero or negative potential and is more predominant as the grid is made more and more positive.

k. Placing a voltage on the control grid sets up two electrostatic fields, one between cathode and control grid and the other from control grid to plate. The direction of these fields is dependent on the polarity of the voltages applied to the electrodes.

l. A low voltage applied to the control grid can exert on the space-charge electrons as much influence as (or even more than) a very much higher voltage applied to the plate.

m. When the potential on the control grid is such that it completely stops the movement of electrons through a tube, the tube is said to be *cut off*.

n. When the potential on the control grid is positive, it aids the plate by pulling more electrons to it because more electrons pass through the grid openings.

o. In some tubes, a grid-voltage change of a few volts can have the same effect on plate current as changing the plate voltage by 10, 20, or even 50 volts. (In some instances, the effect can be less and in others more.)

p. When the maximum and minimum points of one waveshape occur at the same instant as the maximum and minimum of another waveshape, it is said that the two waveshapes are *in phase*.

q. The plate current of a triode always remains unidirectional even though it may vary above and below a reference level.

r. The polarity of the plate (with reference to the cathode) in a triode always remains positive.

s. The control-grid voltage can change very slowly or very rapidly and the plate voltage varies accordingly.

t. In most cases, the presence of grid current is undesirable because it represents a consumption of power.

u. Bias can be defined as the d-c voltage existing between the grid and cathode of a triode.

v. Characteristic curves display the behavior of electron tubes easily and specifically.

w. In the plate-current grid-voltage characteristics, the grid voltage is made variable (causing changes in plate current) and the plate voltage is made constant. In the plate-current plate-voltage characteristics, the plate voltage is made variable (also causing changes in plate current) and the grid voltage is made constant.

x. Static characteristics are curves that indicate tube behavior when the voltages applied to its electrodes are determined by voltage sources E_{bb} and E_{cc}.

y. Plate-current saturation is reached when no further increase in plate current occurs as the grid is made more positive.

z. In general, grid-family and plate-family curves furnish the same information. The grid family displays the plate current for small increments of grid-voltage change and substantial fixed differences in plate voltage; the plate family displays the effects of small increments of plate-voltage change and larger fixed increments of grid-voltage change.

aa. The plate current corresponding to a fixed change in grid voltage is a function of the *operating region* on the plate-current curve. The *operating point* determines the linearity and nonlinearity of the output waveshape of a triode.

ab. The tube *constants* are:

$$\text{Amplification factor } (\mu) = \frac{\triangle e_b}{\triangle e_c} (i_p \text{ constant})$$

$$\text{Transconductance } (g_m) = \frac{\triangle i_b}{\triangle e_c} (e_b \text{ constant})$$

$$\text{A-c plate resistance } (r_p) = \frac{\triangle e_b}{\triangle i_b} (e_c \text{ constant}).$$

ac. Triodes generally are classified as low-μ (μ is less than 10), medium-μ (μ is between 10 and 30), and high-μ (μ is 100 or more).

ad. R_p is defined as the d-c internal resistance of a tube when steady values of voltage are applied to its electrodes. The term r_p is defined as the a-c internal resistance of a tube when varying values of voltage are applied to its electrodes.

ae. The higher the applied plate voltage (or the lower the negative grid voltage), the lower is the a-c plate resistance.

af. When comparing μ, g_m, and r_p, μ remains relatively constant under virtually all conditions. The value of r_p varies greatly at high values of negative grid bias or low plate voltage. G_m varies oppositely to r_p.

ag. The tube constants are a means of comparing tubes of a similar type and permit the selection of a tube to fit specific needs.

ah. The dynamic characteristics of a tube can be described as a graphic portrayal of tube behavior under *load*.

ai. When a signal source and a load are used, the following mathematical relationships hold true:

$$e_o = E_{cc} + eg; \quad eb = E_{bb} - i_b R_L; \quad e_{RL} = i_b R_L.$$

aj. A load line displays the way in which the output of the plate-supply voltage is distributed between the load and the internal resistance of the tube under different conditions of plate current.

ak. The dynamic transfer characteristic curve correlates the plate-current grid-voltage relationship with the load present in the circuit.

al. The total change in plate voltage always is equal to the change in voltage across the load.

am. Nonlinearity in electron-tube behavior is a source of distortion and is highly undesirable.

an. If a triode amplifier is to operate without grid current, the amount of fixed negative bias must be at least equal to the peak value of the positive half of the input grid signal.

ao. The plate current in a triode amplifier is said to be *quiescent* when no signal is applied to its input circuit.

ap. The interelectrode capacitances of a triode, C_{gp}, C_{pk}, and C_{gk}, are in the order of 2 to 10 $\mu\mu$f.

aq. The effect of interelectrode capacitance can lead to feedback and oscillation.

59. Review Questions

a. Name some metals that are used to manufacture control grids in triodes.

b. What famous person inserted the *third element* into the diode?

c. Why are the elements of triodes generally larger in transmitters than in receivers?

d. What substances are used as *getters* in triodes?

e. Why are getters necessary in triodes?

f. What is the electrical difference between a diode and a triode?

g. Is it possible for the control grid to overcome the influence of a very high plate voltage in the order of a few thousand volts? Why?

h. Why is the control grid in a triode sometimes compared to a *valve*?

i. Comparatively speaking, what are the approximate voltage differences between the A, B, and C supplies in a triode?

j. Are the polarities of B- and C-batteries the same for all triodes?

k. What is meant by the operating potentials of a triode?

l. What point usually is grounded in a triode circuit?

m. In physical construction, why is the control grid placed closer to the cathode than to the plate?

n. What influence does the electrostatic field existing in a triode have on the electron flow through the tube?

o. Will grid current flow when the control grid is at zero potential?

p. Explain the differences in direction of the electrostatic fields in a triode when the grid is positive and negative.

q. What is meant by *cut-off* in an electron tube?

r. Upon what factors does the cut-off voltage of an electron tube depend?

s. Name the effects that exist in making the control grid positive in a triode.

t. What is meant when two voltages are said to be *in phase*?

u. Can the plate voltage of a triode ever go negative? Why?

v. Does the plate current have the same shape as the control-grid voltage in a triode?

w. Why is a negative control grid desired in the majority of cases?

x. Define bias, grid voltage, and cathode voltage.

y. What is the main purpose of characteristic curves?

z. What is meant by the notations E_{bb}, i_b, I_b, e_c, and E_b?

aa. What are the dependent and independent variables in both the plate-current grid-voltage and plate-current plate-voltage characteristics?

ab. What are the differences between static and dynamic characteristic curves?

ac. Why is plate current reduced when the control grid is made *too* positive?

ad. What are the advantages and disadvantages of using oxide-coated emitters?

ae. Why are grid-family characteristic curves used instead of a single characteristic curve?

af. Define μ, g_m and r_p.

ag. What is the differencee between R_p and r_p?

ah. What is meant by a variable-μ tube?

ai. How does r_p vary with the plate-supply voltage? Grid-supply voltage?

aj. What is the mathematical relationship between μ, g_m, and r_p?

ak. Is the change in plate current for a change in grid voltage more or less when a load is used in a triode circuit? Why?

al. Explain what a dynamic transfer characteristic curve is.

am. How would you go about drawing a load line for a given set of grid-family curves?

an. Why does nonlinearity of characteristic curves cause distortion? How can it be minimized?

ao. What three voltages are present in the control grid-cathode circuit of a triode amplifier?

ap. When can a quiescent value of plate current be obtained?

aq. What is the approximate value of interelectrode capacitance existing between the electrodes of a triode tube?

ar. Is interelectrode capacitance more noticeable at higher or lower frequencies? Why?

as. Why is *neutralization* used in some triode amplifiers?

CHAPTER 5

MULTIELECTRODE TUBES

Section I. TETRODES

60. General

a. Although the triode is an important device in communications, its use in amplifying systems is limited in some respects. The principal reason is the interelectrode capacitance between its electrodes, especially the capacitance between the control grid and the plate. As the frequency of operation is increased, the grid-plate capacitance affords an easier path for the transfer of energy back from the output to the input circuit. This action is most pronounced in a resonant system in which the grid and plate circuits are tuned to the same or similar frequencies. The result is that triode tubes seldom are used as amplifiers without recourse to special neutralizing systems to counterbalance the undesired feedback.

b. Neutralization is not wholly satisfactory. The higher the frequency of operation, the more critical is the adjustment. Sometimes it is impossible to neutralize properly over the entire band. Above all, neutralization is a critical adjustment and, frequently, a bothersome one. It still is used in some equipment, but in general its use has been reduced by the development of new tube types which obviate the necessity for neutralization by greatly reducing the interelectrode grid-plate capacitance as a feedback path.

c. In addition to the problems of feedback, the triode does not satisfy all the amplifying needs encountered in receivers, transmitters, and related apparatus. The physical relationship between the electrodes of the triode is such as to set unsatisfactory limits on the degree of amplification that can be achieved in a practical tube. At one time this posed a serious problem for design engineers because progress in communication depended on increasing the amplifying capabilities of equipment.

d. Today the answer is found in two types of tubes. Both are based on the triode but represent modifications of the original three-element electron tube. One of the versions is the *tetrode,* or four-electrode electron tube, which contains an electron emitter, two grids, and a plate. The other is the *pentode,* which contains five electrodes: an electron emitter, three grids, and a plate.

e. The development of electron tubes since the early 1920's has resulted in more than just the multielectrode tetrode and pentode. Other tubes which have been developed are various combinations of diodes, triodes, tetrodes, and pentodes in the same envelope. These are identified by the general name of *multiunit* tubes.

61. Tetrode

a. SCREEN GRID. The four-electrode tube contains all the electrodes of the triode (with generally similar functions), and in addition a fourth electrode. This is the *screen grid.* As a rule, the four-electrode tube is called a tetrode, although upon occasion it is referred to as a screen-grid tube.

b. PHYSICAL CONSTRUCTION.

(1) The physical organization of the tetrode (fig. 68) does not differ too much from that of the triode. In A, the screen grid of the tetrode (type 48) is rectangular in shape. The plate is fin-shaped for heat dissipation purposes. In B, the outer screen grid is a perforated-metal structure

of circular shape (type 32), located between the glass envelope and the plate electrode. The inner screen grid is oval and is located between the control grid and plate. There is variation in the shape of the grids; in some instances a helical form is used. The functioning of the electrodes is fundamentally the same regardless of the shape.

to the base are called *double-ended* tubes. If *all* the electrodes appear as pins or prongs in the base of the tube, they are called *single-ended* tubes. In some tetrodes designed for use in receivers and similar low-power equipment, the cap illustrated in C affords electrical connection to the control grid. The plate, screen-grid, and heater (or filament) junc-

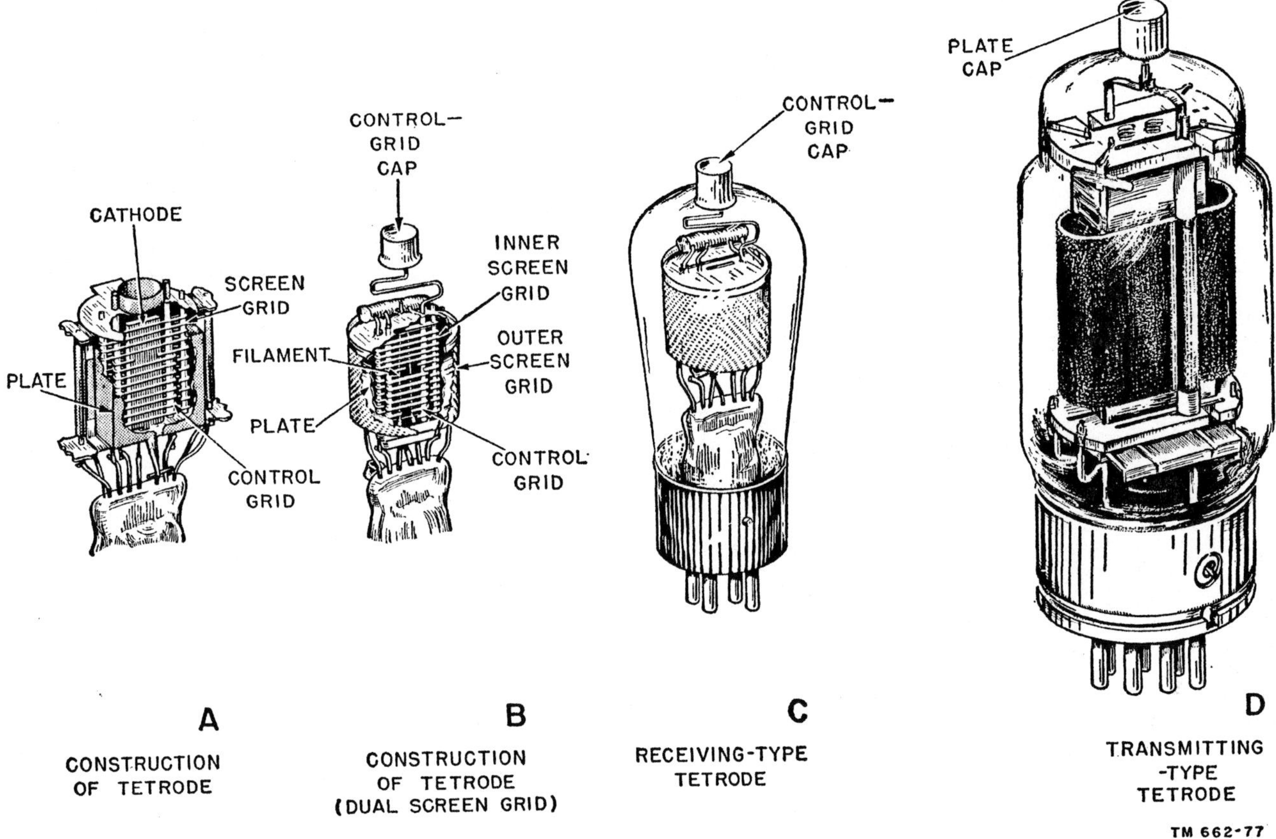

Figure 68. Physical construction of tetrodes.

(2) C and D show external views of two tetrodes. The tubes are substantially alike in appearance, but differ in one respect. The practice of using pins in the tube base as connecting points to the electrodes inside the envelope is followed in the tetrode, but an exception is the use of a metal connecting cap on the top of the tube. Tubes which have electrodes connected to portions of the envelope in addition

tions are made through the tube base pins. In some higher-power tubes, such as are used in transmitters, the cap furnishes electrical contact with the plate electrode, as in D. The remaining tube electrodes terminate at the base pins. The reason for cap connections to the control grid or to the plate, as the case may be, is the desire to reduce the capacitance between the connecting pin terminations of the control grid and the plate.

c. TETRODE CIRCUIT. It is common practice to use the full names, such as control grid and screen grid, but for tube schematics it is advantageous to designate these electrodes in abbreviated form as G1 and G2. The smaller number is assigned to the grid which is closer to the cathode, this being the control grid. Therefore, G2 represents the screen grid. These notations are included in two schematic representations (fig. 69). The old tube symbol, in A, finds occasional use in modern literature, and therefore, is shown here. B shows the modern tube symbol. It also is interesting to note that abridgement of the words *screen grid* to the single word *screen,* as meaning the same thing, is practiced regularly. When illustrated schematically with operating voltages applied to all the electrodes, the tetrode appears as in figure 70. The main difference between the tetrode and the triode is the screen circuit and the source of its operating voltage. In normal use the screen is made positive relative to the cathode by receiving a voltage from source E_{bb}, which also supplies the plate voltage. In the majority of instances, the d-c screen voltage is appreciably less than the d-c plate voltage.

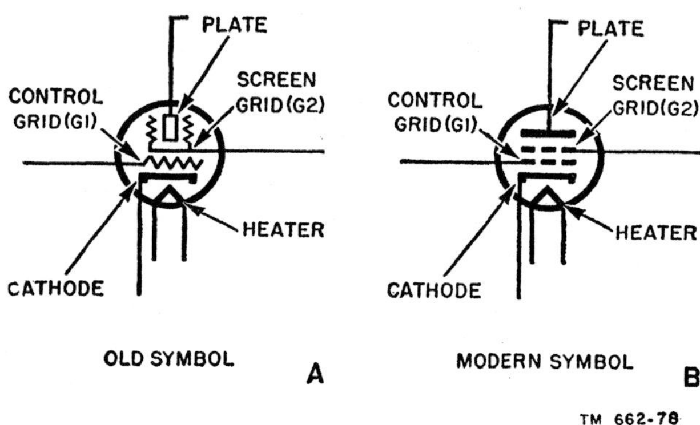

Figure 69. Tube symbols for tetrode.

d. ANALYSIS OF CIRCUIT.

(1) The organization of the tetrode circuit relative to signal transfer (fig. 70) is the same as that of the triode. The control-grid-to-cathode circuit is the *input* part of the tube system (including e_g and E_{cc}), and the plate-to-ground circuit is the *output* part of the tube system (including e_b and

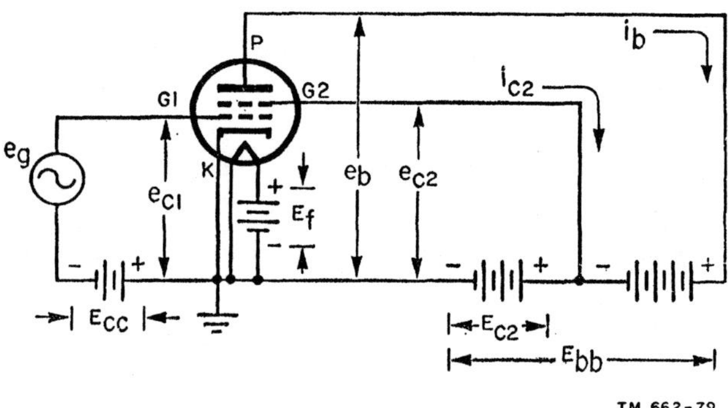

Figure 70. Circuit representation of tetrode.

E_{bb}). The circuity between the emitter and the common junction point, shown grounded, is the *cathode* circuit. This, too, conforms with the breakdown of the triode.

(2) The presence of the screen electrode introduces new voltage and current notations. The voltage between the control grid and the cathode bears association with G1, and is, therefore, identified as e_{c1}. On the other hand, the screen-grid voltage source is a part of E_{bb}, but the voltage applied to the screen electrode is identified apart from the plate voltage. Being related to G2, it bears the notation e_{c2}, and its portion of E_{bb} is labeled E_{c2}.

(3) The current label i_{c2} identifies the screen-grid current. The screen is located in the space between the control grid and the plate, and therefore, it is in the path of the electrons flowing through the tube. Since it is subjected to a positive voltage, it does more than just accelerate electrons toward itself and the plate. It attracts some of the electrons, which comprise the screen current. Again, the current notation is seen to bear an association with the subscript letter *c* which is used with all electrodes identified as grids. The screen current usually is much smaller than the plate current.

e. FUNCTION OF SCREEN GRID.

(1) The main purpose of the screen-grid electrode is to reduce the plate-to-grid capacitance, which is a path for feed-

back in high-frequency amplifiers. Feedback may lead to self-oscillation of the amplifier, which is undesirable. With an average of about 2 $\mu\mu$f grid-plate capacitance in a triode, this value is reduced to approximately .01 $\mu\mu$f or less in a tetrode. The screen-grid accomplishes this by acting as an electrostatic shield between the control grid and the plate, consequently reducing the grid-plate capacitance. When an outer screen-grid is used, it serves to isolate the plate of the tube from the external circuit.

(2) The screen current, i_{c2}, that flows as a result of E_{c2} performs no useful purpose. Most of the emitted electrons from the cathode flow to the plate through the openings in the screen-grid mesh. Consequently, the screen reinforces the action of the plate by helping it to attract more electrons to the plate. At the same time, the screen has another effect; it makes the plate current practically independent of plate voltage because of its shielding action. Inasmuch as the screen is between the control grid and the plate, changes in the value of the voltage on the plate have little effect on the space charge. As a matter of fact, it is possible to view the plate-voltage field as terminating on the screen. This does not imply that the plate voltage is unimportant. It accounts for the advance of electrons to it and the plate circuit is still the output circuit. However, the attracting force responsible for the movement of electrons beyond the confines of the control grid is the voltage on the screen rather than the voltage on the plate.

62. Plate-current Plate-voltage Characteristic Curves

a. The behavior of a tetrode under various conditions can be analyzed from its plate-current plate-voltage characteristic curves (figs. 78 and 71, with e_g excluded) as was done in the analysis of the triode. The now obsolete UY–224 tetrode was used to obtain the curves illustrated because it exhibits certain properties of fundamental interest which later types of tetrodes do not show equally well. Figure 71 shows curves for only one control-grid and screen-grid voltage setting. Since e_g is not introduced in the circuit the control-grid voltage, e_{c1} or E_{c1}, is held constant at —1.5 volts and the screen-grid voltage, E_{c2}, is held constant at +75 volts. The filament voltage, E_f, is a d-c voltage of 2.5 volts.

b. Referring to figure 71, curve i_b is the plate current and curve i_{c2} is the screen current (Y-axis) for the stipulated changes in plate voltage (X-axis). It can be seen that great changes occur in these two curves below a certain value of plate voltage, that is, from E_{bb} values of 0 to about 90 volts. It is in this region that great interest lies, although the operating range of the two curves, when used for amplifier design, is substantially above the 90-volt value. Curves i_b and i_{c2} are added to produce curve $i_b + i_{c2}$. This illustrates that changes in plate voltage E_{bb} do not appreciably affect the total current through the tetrode.

c. The previous statement that the plate current is not affected too much by changes in plate voltage in a tetrode appears to be contradicted in an examination of the curves. This is because, with voltage applied to the plate and +75 volts to the screen, an electrostatic field exists between the screen and the cathode and not between the cathode and plate (cathode is at 0 potential). This field attracts electrons to the screen grid and a small screen current (4 milliamperes) flows in the external screen circuit. Some electrons do reach the plate, however, as evidenced by the flow of a very small plate current (.5 milliampere) at 0 plate voltage.

d. Raising the plate voltage from 0 to +2 volts causes an increase in plate current and a decrease in screen current. This is indicated by the upward slope in the plate-current characteristic and the downward slope of the screen-current characteristic. Note that as the plate voltage is increased to +5 volts and slightly higher, the plate-current curve slopes downward and continues so for an appreciable distance. This appears to be the contrary of what

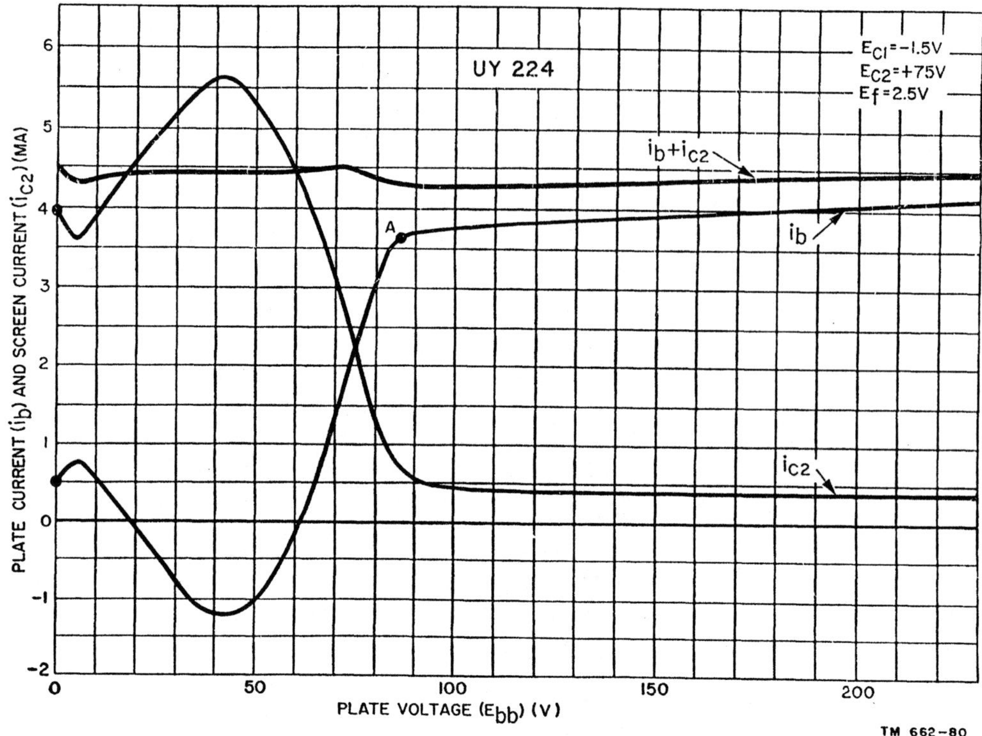

Figure 71. Plate-current plate-voltage characteristic curve for UY–224 tetrode.

should happen. By ordinary tube behavior the plate current should increase rather than decrease as the plate voltage increases. The decrease is caused by a phenomenon known as *secondary emission.*

e. Secondary emission occurs when electrons from the space charge strike the plate with sufficient force to dislodge electrons from the plate material itself. These dislodged electrons are known as *secondary* electrons. With the screen voltage fixed, the velocity with which the space-charge electrons (known as *primary* electrons) arrive at the plate increases as the plate voltage increases. At low plate voltages, secondary electrons dislodged from the plate under primary electron bombardment are attracted to the higher-potential screen. The movement of these secondary electrons is in the direction *opposite* to the regular flow of electrons from cathode to plate. The result of this dual-directional current flow is that the current in the plate circuit decreases as more and more secondary electrons are dislodged from the plate. These secondary electrons are attracted to the screen and the screen current increases.

The decrease in plate current and increase in screen current resulting from secondary emission are represented by the curves in figure 71. The plate current, therefore, continues to decrease as the screen current increases. When the plate voltage is increased and approaches the screen potential of +75 volts, some of the secondary electrons are attracted back to the plate, causing the plate current curve to turn in an upward direction. When the plate voltage reaches the screen voltage, the plate field is now strong enough to prevent the secondary electrons from leaving the vicinity of the plate because they are nearer to the plate than they are to the screen.

f. As the plate voltage exceeds the screen voltage, most of the electrons which pass through the screen openings arrive at the plate. This condition represents the relatively flat portion of the i_b curve (beyond $E_{bb} = 90$ volts). Increasing the plate voltage beyond 90 volts does not cause an appreciable increase in plate current, because the increased attracting force caused by the higher positive voltage on the plate does not affect the space charge as much as the screen does; therefore, its action on the space-charge electrons is negligible.

g. Normal operation of a tetrode as a distortionless amplifier is in the zone where the plate current curve i_b is not subject to radical flunctuations caused by secondary emission effects, to the right of point A.

h. In the plate family of characteristics for the more modern 24A tetrode (fig. 72), the control-grid voltage is varied from 0 volts to —6 volts in steps of .5 volt; the screen-grid voltage, E_{c2}, is held constant at +90 volts and the plate voltage is varied over a wide range, from 0 volt to +500 volts. Several significant differences can be observed in figures 71 and 72.

volts. It is simply that the extent of the negative resistance effect is not as great.

j. This improvement is attributable to the kind of material used for the plate. The plate of the 24A tetrode is treated chemically so that fewer secondary electrons are dislodged by the primary electrons. This reduces the number of secondary electrons that are available to be attracted to the screen when its voltage exceeds that of the plate.

k. As to the limitations of its use, the modern tetrode is subject to secondary emission also, although it performs as a somewhat better amplifier than the old-type tetrode.

63. Constants of Tetrodes

a. GENERAL. Like the triode, the tetrode is identified with three basic tube constants: a-c

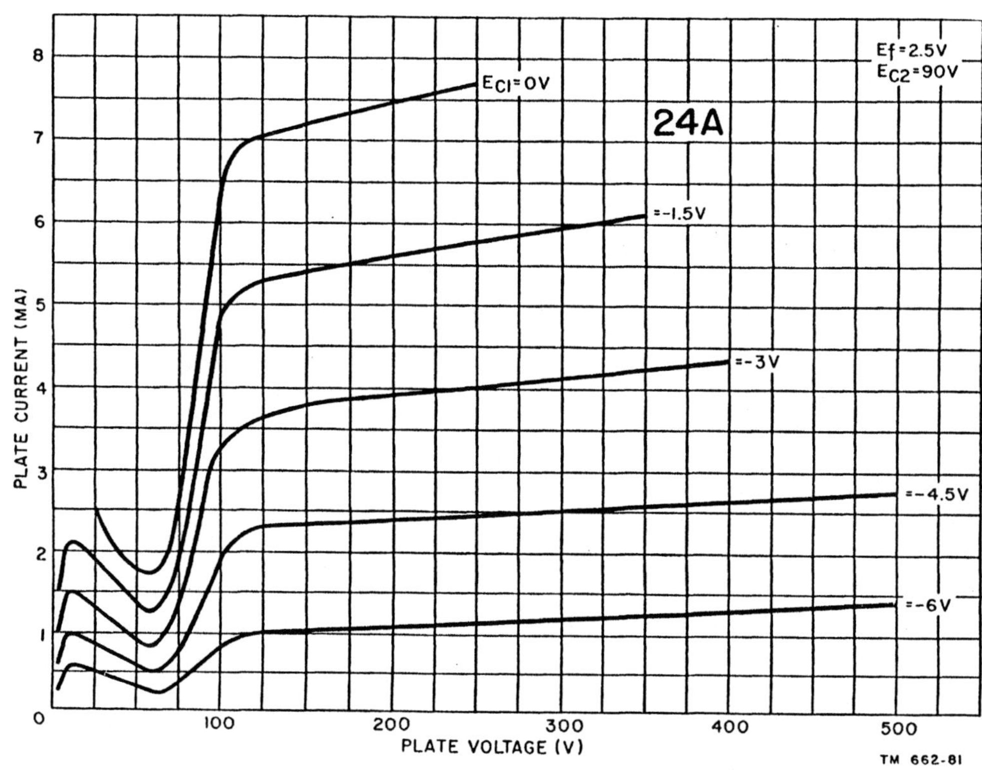

Figure 72. Plate family of characteristic curves for 24A tetrode.

i. It is to be noted that none of the characteristic curves for the modern 24A drop below the 0 value of plate current, as they do for the obsolete UY–224. This does not mean that the modern tetrode is free from negative resistance effects, for, as can be seen, the plate current does decrease with increasing plate voltage over the region where E_{bb} is approximately 0 to 70

plate resistance r_p, amplication factor, μ, and transconductance g_m. The meaning of each is the same as when they are used for the triode, although the different organization of the tetrode establishes a contrasting range of values for these constants.

b. A-C PLATE RESISTANCE. The a-c plate resistance of the tetrode is very high in comparison with that of the triode. The reason for this is the action of the screen grid; it reduces very substantially the effect of a change in plate voltage on the plate current. The a-c plate resistance of the few receiving-type (low-power) tetrodes averages about 400,000 to 500,000 ohms, although in the receiving-type power tetrodes it is much lower—approximately 70,000 to 100,000 ohms. High-power tetrodes used in transmitters display much lower values of plate resistance, usually about 6,000 to 10,000 ohms.

c. AMPLIFICATION FACTOR. The amplification factor of tetrodes is many times higher than in triodes. In contrast to triodes, which possess μ values of from about 5 to perhaps 50, low-power tetrodes are rated at from about 400 to perhaps 600. High-power tetrodes used in transmitters have μ values approximating 160.

d. TRANSCONDUCTANCE. The transconductance is not too high in tetrodes, despite the high μ rating. The reason for this can be explained by the equation $g_m = \mu/r_p$. The value of r_p, in the denominator, increases to a greater proportion than μ, in the numerator, does in a tetrode. Therefore, the ratio of μ/r_p decreases in proportion and consequently g_m decreases. In general, the g_m of low-power tetrodes is about 1,000 to 1,500 micromhos; special power types average between 4,000 and 4,500 micromhos.

64. Advantages and Disadvantages of Tetrode

a. ADVANTAGES. The advantages of the ordinary tetrode are relatively few. It does reduce the capacitance between plate and control grid, and therefore the amount of feedback. Also, it affords greater amplification than is available with a triode. These are the background items which explain its use as an amplifier in a number of communication receivers. However, it is necessary to understand that the receiver and other similar equipments which utilize the low-power tetrode are not the most modern devices. Present-day apparatus generally uses the pentode, which is an elaboration created to overcome the disadvantages of the tetrode.

b. DISADVANTAGES.

(1) The disadvantages of the tetrode outnumber the advantages. As stated previously, it is limited to a portion of the over-all plate-current characteristic. Only in this zone, where the plate voltage exceeds the screen voltage, are secondary emission effects negligible. In view of the effects of the plate load on the plate voltage when a signal voltage is applied to the control grid, the use of a very high value of plate-supply voltage, E_{bb}, is warranted. Only then can the required swing in plate voltage take place and still have the lowest value of plate voltage, e_b, exceed the applied screen voltage. Under such conditions the tetrode functions as a linear amplifier. However, if the plate-supply voltage is insufficient and causes the effective voltage at the plate momentarily to be less than the screen voltage, secondary emission occurs, with consequent impairment of the performance of the amplifier. Excessive distortion is the result.

(2) The necessity for a very high plate-supply voltage, E_{bb}, to assure that e_b always exceeds, or at least never goes below E_{c2}, may be inconvenient. Moreover, the fulfillment of the voltage requirements at the plate and at the screen introduces a limitation on the amount of signal, e_g, that can be fed to the input circuit of the tube. The changes in plate current must be kept to a reasonable minimum in order that the drop across the load resistor never makes e_b less than E_{c2}. This limits the tetrode to the handling of relatively weak signals.

c. GENERAL. The fact remains that few receiving-type tetrodes are manufactured today. For all practical purposes they can be considered obsolete. This is not true, however, of transmitting-type tetrodes. Many communication needs demand high amplification, relatively low plate-supply voltages, and the handling of substantial amounts of power. These are accomplished by means of the *pentode* and the *beam power tetrode*.

65. Physical Construction

a. The *pentode* is a five-electrode electron tube. It contains an emitter, three grids, and a plate. The grid closest to the cathode, G1, is the control grid; next is the screen grid, G2, and the third, located between the screen grid and the plate, is the new *suppressor grid*, G3. The construction of a metal-type pentode is shown in A of figure 73; symbolized, the pentode is shown in B.

from these because of the action of the suppressor grid.

c. In external appearance some pentodes resemble the tetrode. This similarity is so great, even to the use of control-grid or plate caps on the top of the tube envelope, that identification by visual inspection is difficult. Three receiving-type pentodes are shown in figure 74. The envelope of a pentode may be glass, as in A, or metal. A departure from the tetrode appear-

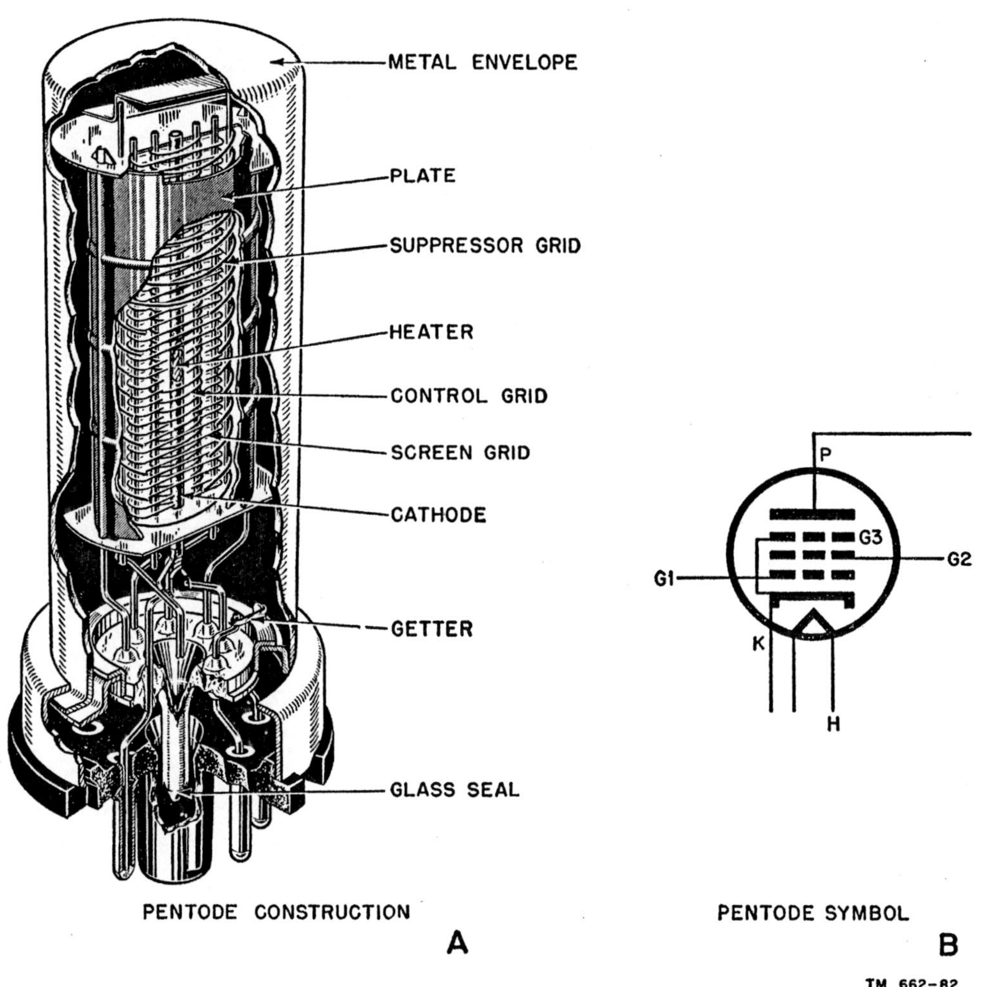

Figure 73. Physical construction (metal-type) and symbol of pentode.

b. Functionally speaking, the action of the emitter, control grid, screen grid, and plate in the pentode are the same as in the tetrode, excepting that, whereas the tetrode suffers from negative resistance effects, the pentode is free

ance is the acorn-type tube shown in B. This is a comparatively small tube requiring a special socket and having wire extensions serving as the tube pins. Another physical feature of the acorn pentode is the location of the plate

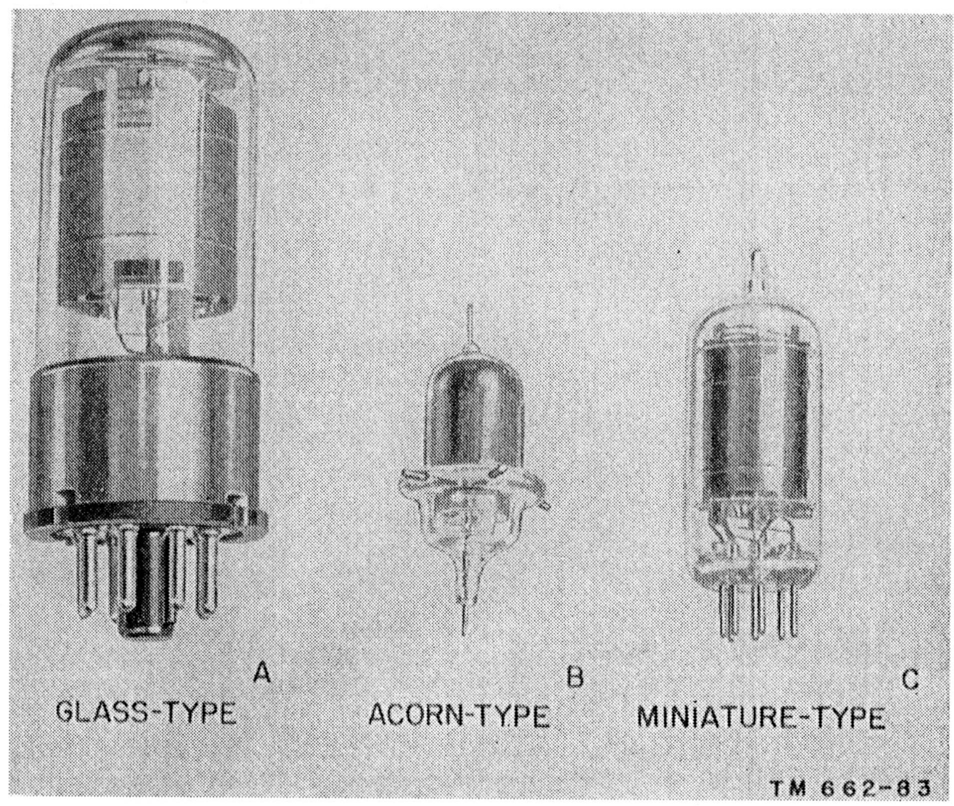

Figure 74. Different types of receiving pentodes.

connection at the top and the control-grid connection at the bottom of the envelope. These are stiff wires which protrude through the envelope. Another type of miniature tube is shown in C.

66. Pentode Circuit

a. When arranged as a basic amplifier, the pentode (fig. 75) is similar to the tetrode (fig. 70). One difference between the two circuits reflects the fundamental difference in behavior between the pentode and the tetrode. In some types of pentodes, as in figure 75, the suppressor is electrically connected to the cathode inside the tube; in other types, the connection is made to its own prong on the tube base. Such flexibility of connection between the suppressor and the cathode enables some pentodes to be arranged for action as triodes.

b. The electrode voltage and current notations for a pentode parallel those for the tetrode. The electrical connection between the suppressor and the cathode places the former at the same potential relative to the adjacent electrodes as the cathode. Under the circumstances, the suppressor bears no current or voltage notations of its own. It is viewed as being at 0 or ground potential.

c. Referring again to figure 75, primary electrons are accelerated toward the plate by suitable voltages applied to G1 and G2. They pass through the openings of the control grid and screen wires and also through the openings of the suppressor, G3. When the electrons

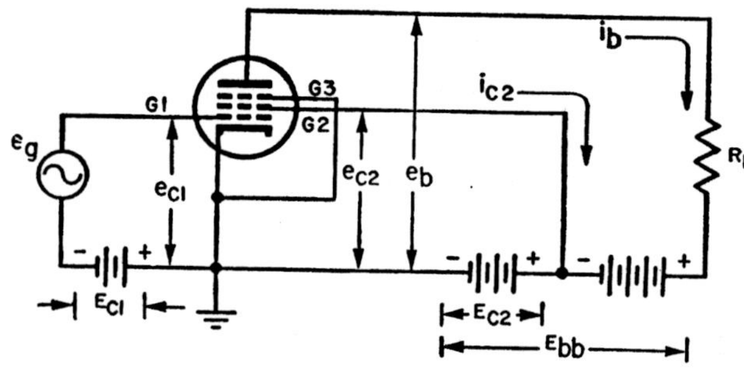

Figure 75. Circuit representation of pentode.

strike the plate they do so with sufficient velocity to cause secondary emission.

d. Since the plate is positive with relation to the cathode and the cathode is connected to the suppressor, the suppressor is negative with relation to the plate. Therefore, secondary electrons emitted by the plate are repelled by the suppressor and returned to the plate. Reverse current between plate and screen is therefore avoided, even if the screen voltage momentarily exceeds the plate voltage. The existence of the suppressor between the screen and plate has another beneficial effect. It reduces the control-grid-to-plate capacitance even more than in the tetrode; therefore, it greatly reduces the feedback problem. This is one reason that accounts for the wide application of the pentode.

67. Characteristic Curves

a. The plate family of characteristic curves for a typical pentode (6SJ7) is shown in figure 76. Plate-supply voltage E_{bb} is variable from 0 volts to 400 volts (X-axis). Since the curves are made without a load in the plate circuit, the plate potential $e_b = e_{bb}$. The screen voltage $E_{c2} = e_{c2}$ is held constant at 100 volts; the suppressor is at 0 potential and the control-grid voltage E_{c1} is varied between 0 volts and —5

volts in steps of 1 volt. The filament voltage, E_f, is equal to 6.3 volts.

b. A number of significant details are shown in figure 76. Dips in the curves (fig. 72) due to negative resistance are absent. At no time does the plate current fall with increasing plate voltage. A zone of critical plate potential (the range of plate-voltage changes over which the plate current rises rapidly) still prevails. However, this happens over a limited area only. The top of the knee of the plate-current curve is reached much sooner than in the tetrode.

c. As in the tetrode, the plate current is relatively independent of the plate voltage. The usable portion of each characteristic is to the right of the knee in each curve, where the curves are relatively flat. A significant condition is implied by the unequal spacings of the plate-current curves at the flat portions. Each plate-current curve corresponds to a specific value of control-grid voltage, each of which represents an equal increment of change. The plate current does not change equally for equal changes in control-grid voltage, with the screen and plate voltage fixed in value.

d. The tabulation below shows this plate-current control-grid voltage behavior. The plate voltage is held constant at 300 volts; the screen-

Figure 76. Plate family of characteristic curves for pentode.

grid voltage is constant at 100 volts and the suppressor is joined to the cathode and equals 0 volts.

Plate voltage (E_{bb})	Control grid voltage (E_{c1})	Plate current (i_b)	Plate current change
300	0	9.3 ma	
300	—1	7.0 ma	(9.3 —7.0) = 2.3 ma
300	—2	4.8 ma	(7.0 —4.8) = 2.2 ma
300	—3	3.0 ma	(4.8 —3.0) = 1.8 ma
300	—4	1.6 ma	(3.0 —1.6) = 1.4 ma
300	—5	.6 ma	(1.6 — .6) = 1.0 ma

e. These figures imply a nonlinear behavior between plate current changes and control grid voltage changes, and consequently, the appearance of distortion. Therefore, it can be said that pentode operation is productive of more distortion than is triode operation. By proper choice of operating constants the distortion is reduced to tolerable proportions. The advantages gained from the use of pentodes in certain sections of communications equipment exceed the disadvantages caused by distortion, and therefore, the tube enjoys wide popularity.

68. Constants of Pentodes

a. AMPLIFICATION FACTOR. The amplification factor of a pentode is great, being approximately 1,500 for receiving tube types. This is about 100 times more than the amplification obtainable with triodes, and two to three times as great as the amplification factor of tetrodes. The reason for the unusual amplifying ability of the pentode can be found by graphically locating μ in the plate family of curves in figure 76.

b. PLATE RESISTANCE.

(1) The a-c plate resistance of the pentode is very high. For example, the 6SJ7 pentode has a plate resistance of about 1 megohm, or nearly 200 times greater than the plate resistance of a triode and several times that of a tetrode. The high plate resistance of the tetrode and the pentode sets them apart from the triode in the manner of use with different devices. It is significant to note that the higher the negative grid bias, the higher the plate resistance of the pentode, as can be seen in the characteristic curves.

(2) Some types of pentodes, known as *power pentodes,* manage to keep the plate resistance lower than it is in the usual type of pentode, but even in these varieties r_p is many times the value found in triodes intended for similar uses in communications equipment. The transconductance ratings, however, do not differ greatly from those of triodes designed for similar applications.

c. TRANSCONDUCTANCE.

(1) Despite the high amplification factor of pentodes, the transconductance ratings are comparable to the ratings of triodes and tetrodes because of the extremely high plate resistance found in the pentode. It is obvious, therefore, that the specific change in plate current per unit change in control-grid is very low.

(2) Figure 77 illustrates the variation in transconductance of a typical pentode with changes in control-grid voltage, screen voltage, and plate voltage. The highest values of transconductance prevail for low values of control-grid bias, E_{c1}, high values of screen voltage, E_{c2}, and plate voltage, E_{bb}. With fixed plate voltage, the greater the screen voltage the greater the transconductance. This is understandable because of the great effect that is displayed on the plate current by the screen voltage. Also, in figure 77, a lower value of plate voltage (dotted lines at $E_{bb} = 100$) introduces a lower value of g_m than a higher value of plate voltage ($E_{bb} = 200$), since raising the plate voltage lowers the value of r_p, which in turn raises the value of g_m.

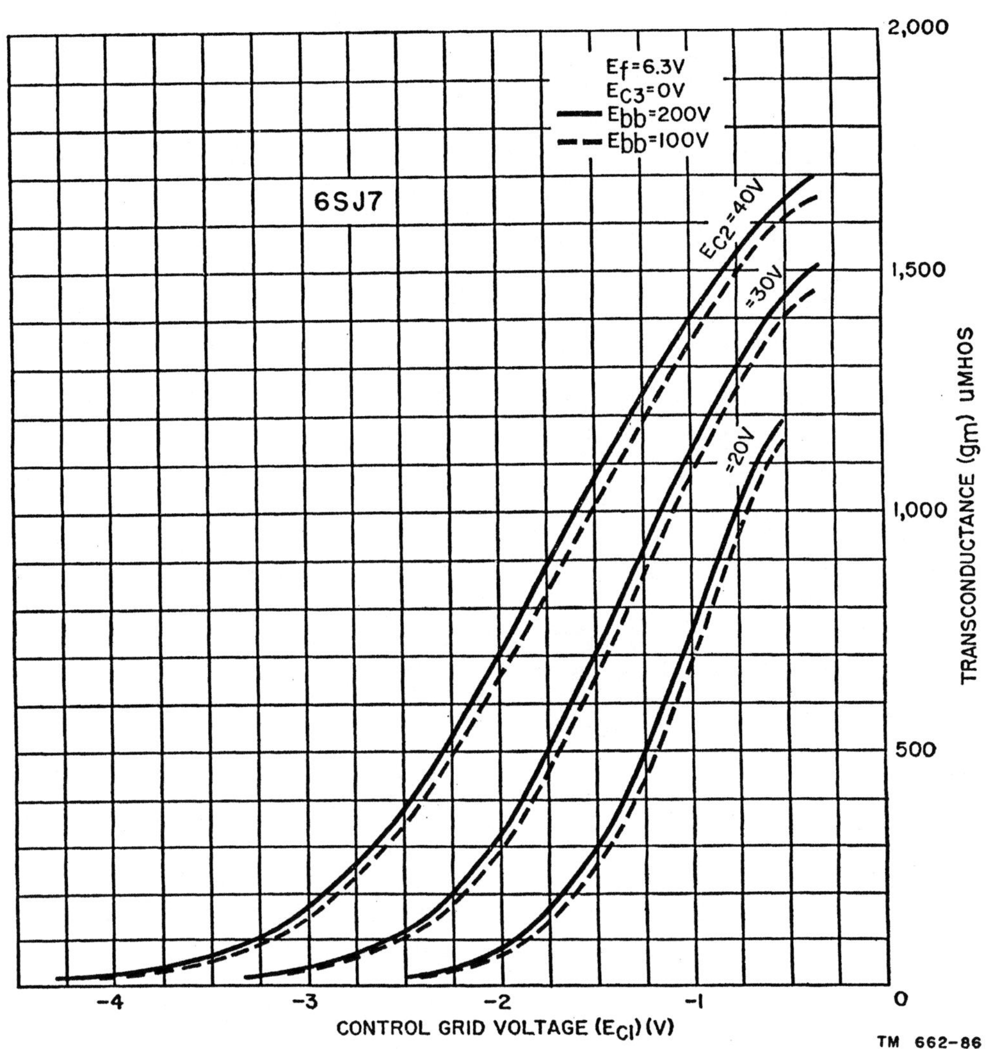

Figure 77. Family of curves showing variation of transconductance of 6SJ7 pentode.

69. Dynamic Transfer Characteristic

a. The development of the dynamic transfer characteristic of a pentode follows the method described for the triode. The effects developed in the pentode as compared with those of the triode are shown in the dynamic transfer characteristics in figure 78. These curves are drawn for a 6SJ7 pentode with $E_{bb} = 300$ volts, $E_{c2} = 100$ volts, and $E_{c3} = 0$ volts. The load lines are for $R_L = 0$ ohms, 30,000 ohms, 50,000 ohms, and 100,000 ohms.

b. The discussion concerning the dynamic transfer characteristics of the triode (pars. 54 and 55) brought out the advantages of linear curves and the desirability of operating over

the linear portions of the transfer characteristic curve. This was the condition established for distortionless amplification. It was pointed out, moreover, that the higher the ohmic value of the load, R_L, the straighter was the characteristic, and the more faithful the pattern of changes in plate current relative to the pattern of changes in control-grid voltage.

c. The contrasting conditions encountered in the pentode operated as stipulated are shown in figure 78. The characteristic curves are not straight when R_L equals a finite value. The curve labeled $R_L = 30,000$ ohms appears to be the nearest approach to a usable condition, but even this curve is not linear throughout its entire length. As the value of the load R_L is in-

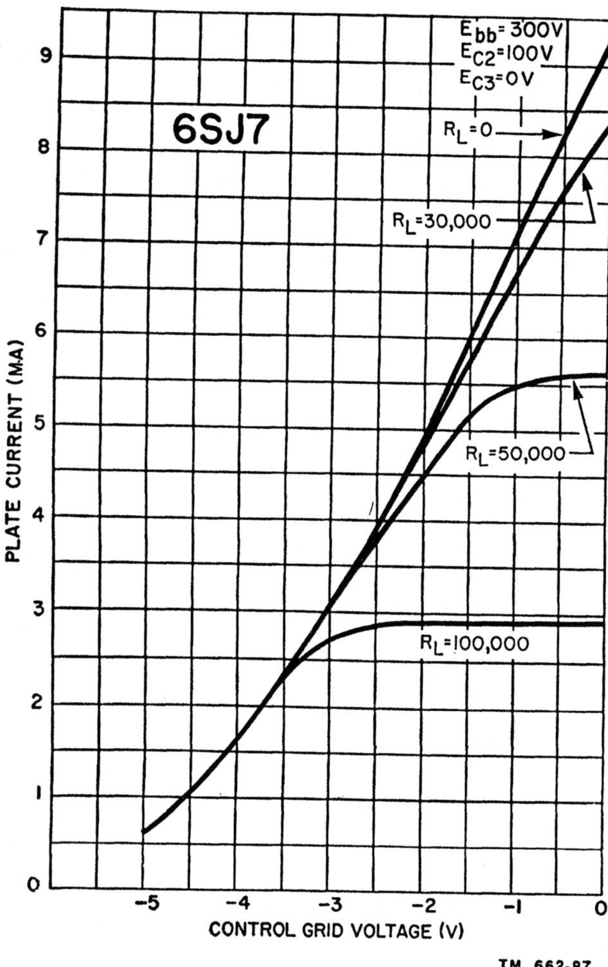

Figure 78. Dynamic transfer characteristic curves of 6SJ7 pentode.

tics by connecting two points; one point is on the Y-axis (plate current) and the other on the X-axis (plate voltage). On the Y-axis, the point is obtained in the following manner. By Ohm's law, $i_b = E_{bb}/R_L$, or the current through the load resistor, i_b, equals the voltage across it, E_{bb}, over the value of its resistance, R_L. Consequently, for different values of R_L different load lines are obtained. On the X-axis, the point is obtained by applying the value of E_{bb} that is used in the circuitry. Under this condition, no plate current is flowing through the circuit and, therefore, no voltage drop appears across the load resistor. The load lines in figure 79 intersect the plate current curves below the knee of each curve, resulting in very little change in plate current as the grid bias is reduced.

e. By reducing the screen voltage, E_{c2}, to perhaps one-seventh or one-tenth of the plate-supply voltage, the slope of the plate-current rise for low values of plate voltage is made very steep. For instance, with a screen voltage of 100 volts ($E_{c2} = 100$) and a control voltage of 0 volts, the knee of the plate-current curve is reached with the plate-supply voltage equal to approximately 40 volts (fig. 79). When the screen voltage is reduced to 40 volts, the same point is reached with a plate voltage of only about 16 volts. A similar improvement is attained for each value of control-grid voltage. Of course, the reduced screen voltage reduces the plate current, but this is secondary in importance to the improvement in the characteristics and utility.

f. The benefits derived by reducing the value of E_{c2} are shown in the dynamic transfer characteristics illustrated in figure 80. Compare these curves with those in figure 78, especially the curve for $R_L = 100,000$ ohms. Note that while all curves are nonlinear, the 100,000-ohm load curve approaches $R_L = 0$ and is without the flat portion shown in figure 78. As a matter of fact, a load somewhat in excess of 100,000 ohms is tolerable. The $R_L = 100,000$-ohms curve has been straightened because its load

creased, the nonlinearity becomes more pronounced. In fact, the characteristic actually flattens over a range of control-grid voltage changes. Note how flat curve $R_L = 100,000$ becomes. These flat portions represent very little change in plate current. Dynamic transfer characteristics of this kind are usable under special circumstances, but not when distortionless amplification is desired.

d. The reason underlying the flattening of the dynamic transfer characteristics (fig. 78) is brought out in figure 79. This is the static plate family of the 6SJ7 with load lines drawn on them. Load lines may be drawn very simply on the plate-current plate-voltage characteris-

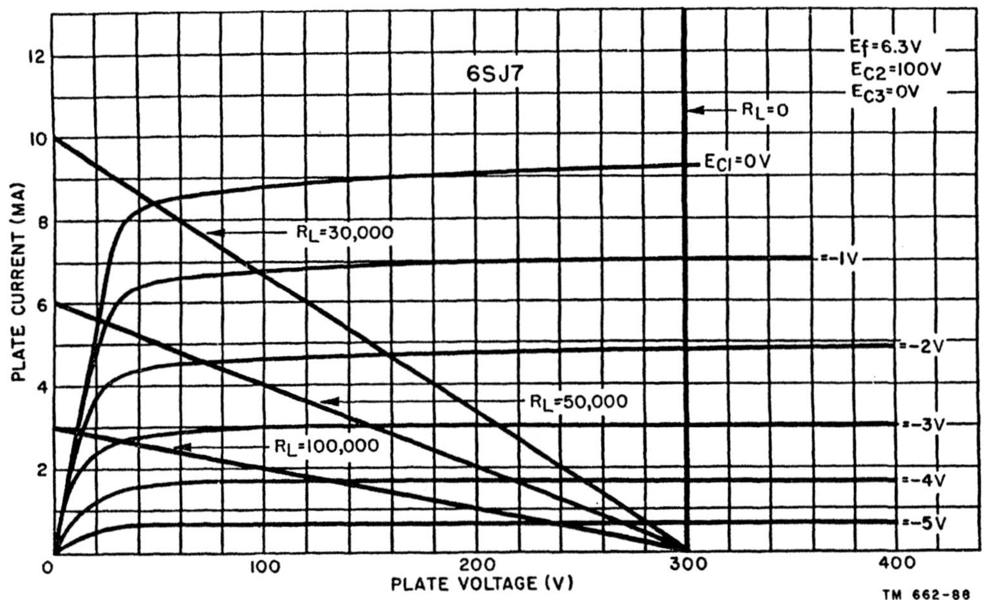

Figure 79. Load lines of 0 ohms, 30,000, 50,000, and 100,000 ohms drawn on plate family of 6SJ7 pentode.

line intersects the static plate-current curves over their relatively flat portions. Curves for $R_L = 30,000$ and $50,000$ are not included in figure 80 because they closely resemble the $R_L = 10,000$ curve.

g. From the transfer characteristics for $R_L = 250,000$ ohms and $R_L = 500,000$ ohms (fig. 80), it is evident that loads in excess of perhaps 150,000 ohms again cause considerable distortion. Even $R_L = 200,000$ ohms begins to show some reverse curvature.

h. Another significant fact present in figure 80 relates to the usable range of control-grid voltages. Obviously, the control-grid voltage cannot exceed a maximum of slightly more than 2.5 volts, or plate-current cut-off is reached. This sets the operating point at about —1.5 volts. Even then, the curvature in $R_L = 0$ ohms and $R_L = 100,000$ ohms indicates unavoidable distortion.

i. Relatively low grid-voltage swing or signal input is a requirement in pentodes. This is a standard compromise. For example, in figure 80, distortion is minimized by selecting —1.25 volts as the operating point and limiting the control-grid voltage swing to about .5 volt peak-to-peak. In this way, the increased curvature of $R_L = 100,000$ ohms below —1.5 volts does little harm.

70. Beam Power Tube

a. The beam power tube has the advantages of both the tetrode and the pentode tube. This tube is capable of handling relatively high levels of electrical power for application in the output stages of receivers and amplifiers, and in different parts of transmitters. The power-handling capacity stems from the concentration of the plate-current electrons into beams or sheets of moving charges. In the usual type of electron tube the plate-current electrons advance in a predetermined direction but without being confined into beams (fig. 81).

b. The external appearance of these tubes is like that of other receiving-type tetrodes or

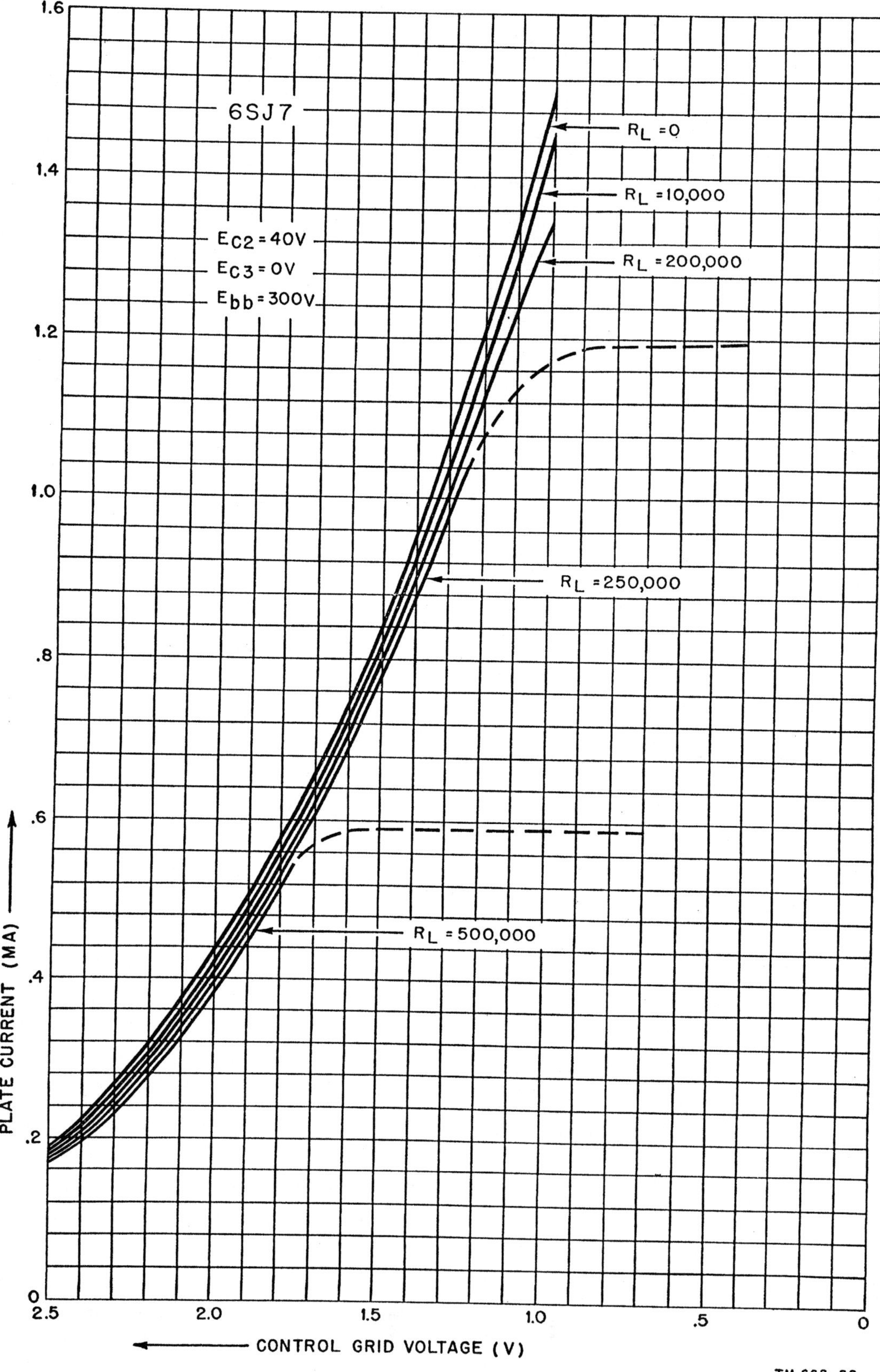

Figure 80. Dynamic transfer characteristic curves of 6SJ7 pentode under improved conditions.

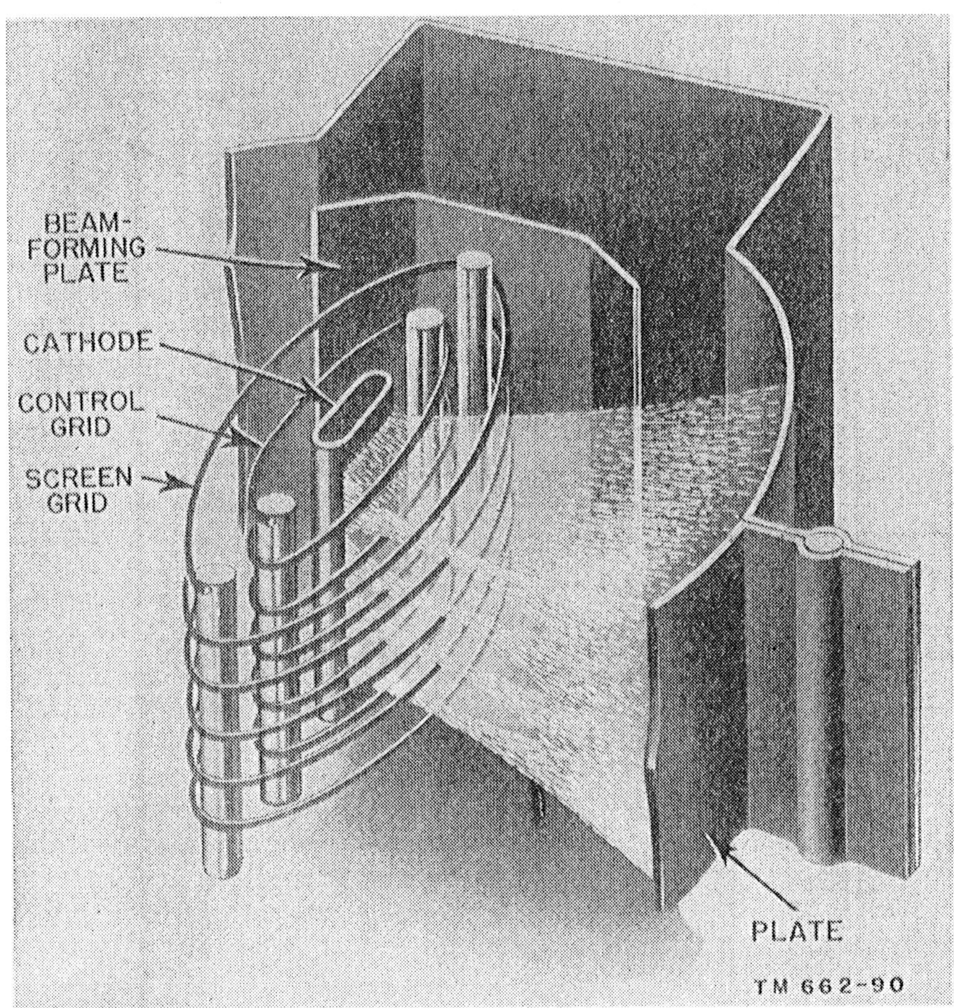

Figure 81. Electrons concentrated into beams in beam power tube.

pentodes. They are slightly larger in dimension because they are called upon to handle somewhat more power, but they have no distinctive external identifying features.

c. The constructional details of a beam power tube which contains beam-forming plates are shown in figure 81. It is not evident, however, that the control-grid and screen-grid electrode windings are of the same pitch, and that the wires of these electrodes are physically in line with each other relative to the paths of the plate-current electrons (fig. 82).

d. A of figure 82 illustrates how the screen-grid wires and control-grid wires in the ordinary tetrode determine the electron paths. The wires are out of alinement; therefore, electrons which pass through the control-grid wires are partly deflected from their paths and many strike the screen wires. This produces a screen current and thus limits the value of plate current.

e. In B the results are different. Because of the arrangement of the control and screen grids, the screen intercepts fewer electrons; therefore, the relative screen current is less in the beam power tube than in the ordinary tetrode, or pentode. In turn, more electrons reach the plate, thereby making the plate current higher.

f. The cumulative effect of the arrangement in B is a tube in which the plate and the control grid are electrically isolated; the plate current is high, the plate resistance is relatively low, and a substantial amount of electrical

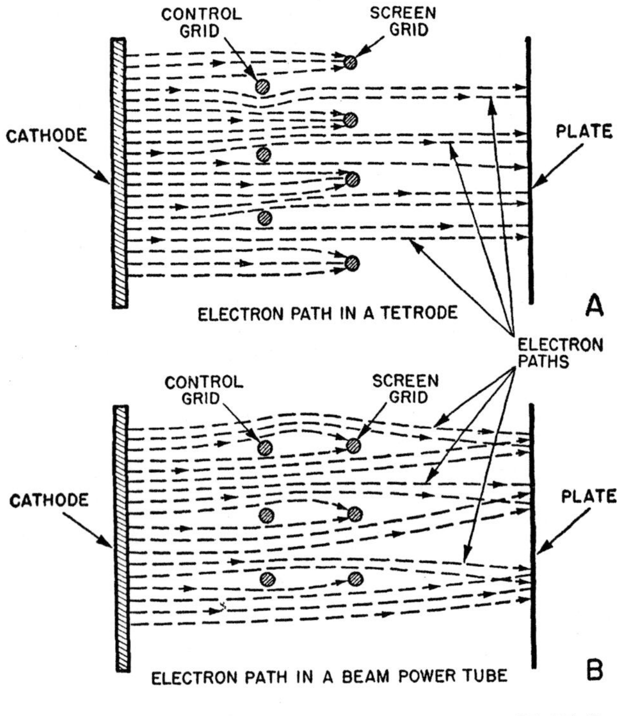

Figure 82. Electron paths in tetrode and beam-power tube.

power can be handled with reduced distortion.

g. The beam-forming plates (fig. 81) further influence the movement of the plate-current electrons from the time they pass the screen electrode and strike the plate. The beam electrodes are connected internally to the cathode and consequently they are at the same potential as the latter.

h. Because of this potential of the beam-forming plates, an effect equivalent to a space charge is developed in the space between the screen and the plate. The effect is as if a surface (dashed lines joining the ends of the beam-forming plates in fig. 83) existed in the screen-plate space. This is identified as the *virtual cathode.* The presence of this electric plane repels secondary electrons liberated by the plate and prevents them from moving to the screen. Figure 83 represents a beam power tube as seen from the top of the envelope.

i. In some tubes, the effect of a virtual cathode is achieved by the use of a third grid in place of the beam-forming plates. The results are identical in both versions. In order to satisfy the two types of construction, two references are made to the symbol of the pentode (fig. 84). A symbolizes the beam power tube with the beam-forming plates and B symbolizes the version in which a grid replaces the beam-forming plate. There is no difference between the symbol shown in B and that of the ordinary pentode (B of fig. 73).

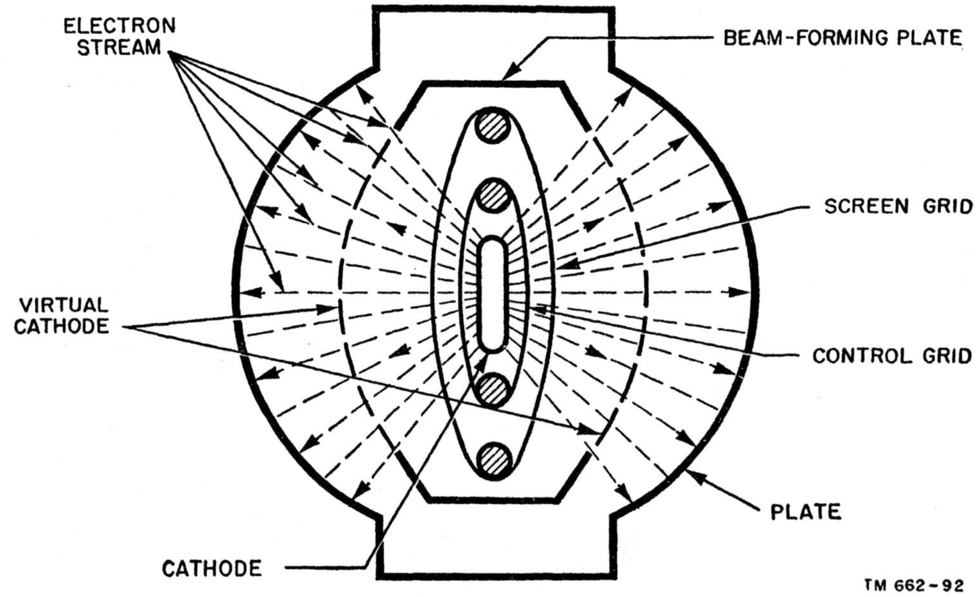

Figure 83. Effect of beam-forming plates on electron flow through beam power tube.

negative voltage on the grid, all parts of the grid electrode winding act to cut off the plate current, but the negative grid voltage required to attain this is perhaps three to four times as much as, if not more than for the conventional tube operated at like screen and plate voltages. The plate-current grid-voltage characteristic of a variable-μ tube, 6SK7, is shown in B of figure 87. A standard pentode curve, 6SJ7, is shown for comparison.

g. Variable-μ tetrodes and pentodes are used in locations in communications equipment where high bias voltages may be necessary to provide control of the signal level.

Section III. MULTIGRID AND MULTIUNIT TUBES

73. Multigrid Tubes

a. INTRODUCTION AND SYMBOLS. Tubes which have more than three grids commonly are referred to as *multigrid* tubes. For instance, if a grid is added to a pentode, a six-electrode multigrid tube results. This tube is known as a *hexode*. The schematic symbol for an indirectly heated hexode is shown in A of figure 88. Other multigrid tubes are the *heptode,* shown in B, which contains five grids, and the *octode,* in C, which has six grids. In these tubes, as in the basic types, the grids are designated in numerical order starting with the control grid, as shown.

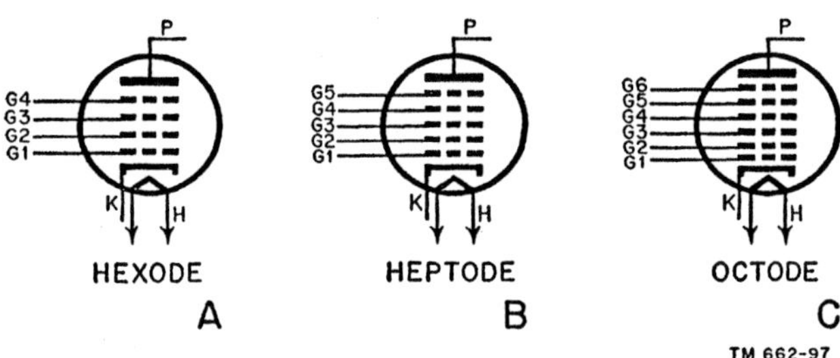

Figure 88. Schematic symbols of multigrid tubes.

b. USES OF MULTIGRID TUBES. The hexode is an experimental tube and is never manufactured commercially. Heptodes, also known as *pentagrid* tubes because of their five grids, are used mostly in frequency converter or mixer circuits. When a heptode is used as a frequency converter, two voltages having different frequencies are each impressed on a separate grid of the tube. Heptodes also are used as volume compressors and expanders. In these applications the gain of an amplifier is controlled automatically. The octode, just as the heptode, also is used as a frequency converter.

c. SUPERHETERODYNE ACTION.

(1) Multigrid frequency-converter tubes are used in superheterodyne receivers. In this application, two different frequencies beat or heterodyne with each other to form additional frequencies called beat frequencies. One of these beat frequencies is equal numerically to the sum of the two applied frequencies; the other beat frequency is equal numerically to their difference. It is possible to obtain the sum and difference frequencies because the tube has a nonlinear dynamic characteristic. In superheterodyne receivers only one frequency output from the converter system is desired. Whether the sum or the difference frequency is chosen, it is known as the intermediate frequency of the receiver. It is common practice to use the difference frequency (fig. 89). The r-f (radio-frequency) input signal from the transmitter and the local oscillator signal are injected simultaneously into the frequency converter. Suppose that the desired intermediate frequency is 465 kc (kilocycles) and that the transmitting station frequency is 1,000 kc. The oscillator then can operate at a frequency of 1,465 kc (1,000 plus 465) to produce the desired intermediate frequency.

(2) Two systems can be used to obtain the desired intermediate frequency. In one system the oscillator is a separate

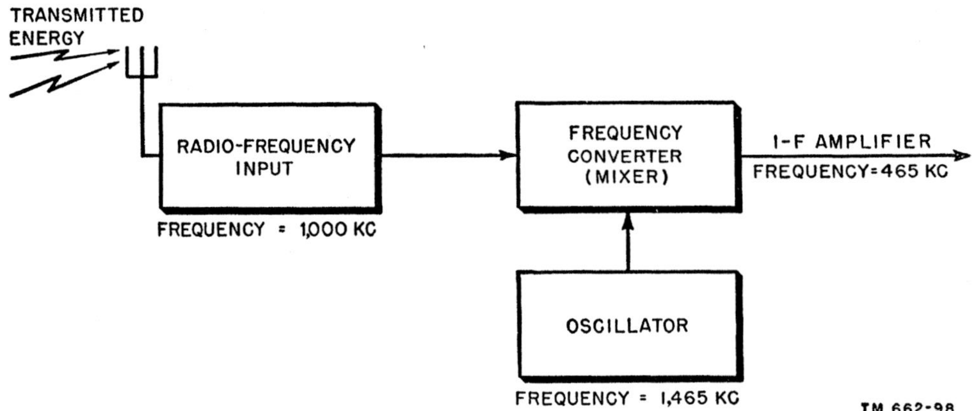

Figure 89. Superheterodyne action.

tube, usually a triode, whose output signal is coupled to the frequency converter tube, usually a pentode or a heptode. In this system the frequency-converter tube commonly is known as a *mixer*. The other system uses a single tube as both the frequency converter and the oscillator. In the latter arrangement, a heptode is used most often. The tube then is known as a *pentagrid converter*. Both systems commonly are found in superheterodyne receivers.

d. PENTAGRID CONVERTER.

(1) The circuit arrangement of a heptode used as a pentagrid converter is shown in figure 90. The purpose of the pentagrid converter is threefold. It acts as an oscillator, as a mixer, to obtain the intermediate frequency, and as an amplifier. The r-f input signal originating at a transmitter is designated as e_g and the oscillator circuit is shown as a block. An L-C (inductance capacitance) tuned circuit is the load. Capacitor C is made variable to tune the circuit to the intermediate frequency. The d-c operating potentials are indicated by E_{cc}, E_{bb}, and E_{c2}.

(2) The pentagrid converter tube can be considered as made up of two parts, a triode and a tetrode. In the triode section, K is the cathode, G1 is the control grid, and G2 operates as the plate. Physically, G2 consists only of a pair of vertical rods across which no grid wires are strung. For the tetrode section, G4 is the control grid, G3 and G5 are the screen grids, and P is the plate. Grids G3 and G5 are connected

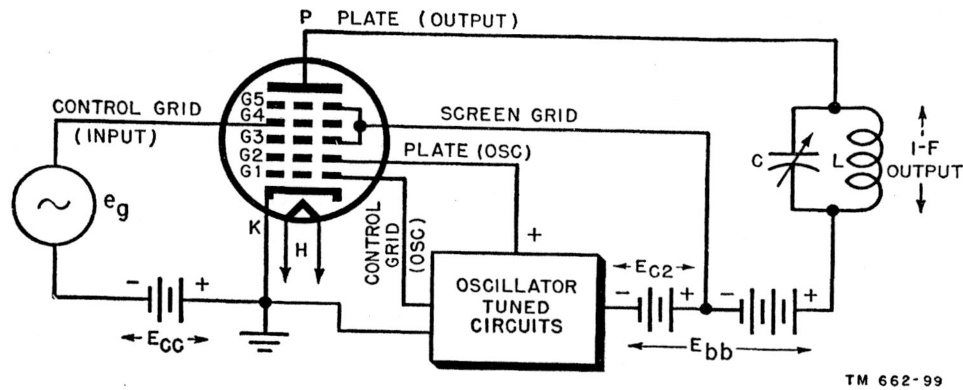

Figure 90. Pentagrid converter circuit.

internally as shown. G3 prevents interaction between the oscillator circuits and the r-f input signal circuits. G4 prevents interaction between the i-f (intermediate-frequency) output on the plate and e_g. The cathode for the tetrode section exists virtually between G3 and G4.

(3) Electrons emitted from the surface of the cathode, K, are influenced by the various elements within the tube. In practice, G1 receives 3 to 6 percent of the electrons emitted, G2 receives about 40 percent, G3 and G5 receive about 25 to 30 percent, and the plate receives the remaining electrons, about 30 percent. The tube is biased so that G4 does not draw current.

(4) Oscillation is produced because of feedback from G2 to G1 in the oscillator circuit. This causes G1 to become alternately positive and negative. Consequently, the current flow through the tube is alternately increased and decreased. Many of the electrons which approach G2 flow beyond it, because of its open structure, even though it is highly positive. Electrons passing G2 are further accelerated by the high positive potential of G3. Grid G4 is always negative, and therefore tends to retard the electrons approaching it. This action produces a virtual cathode between G3 and G4. The electron flow from this virtual cathode is varied or modulated by the r-f signal, e_g, that is injected on G4. The electrons then are drawn to the plate, which is highly positive.

(5) If G1 is slightly negative, or even somewhat positive, the virtual cathode is able to supply the plate of the tetrode section with sufficient electrons. On the other hand, if G1 is made very negative, a deficiency of electrons exists at the virtual cathode which may even cut the tube off. As a result of this variation in the virtual cathode, a current is produced at plate P which varies at the oscillator fre-

quency. However, the electron stream is further modulated by the signal injected on G4. In this way, both the oscillator signal and the r-f input signal on G4 modulate the electron stream, with the result that i-f signals are produced in the plate circuit. The L-C tuned circuit selects the desired intermediate frequency.

(6) Several tube types can be used in the circuit shown in figure 90. Examples are the 1A7, 6A7, 6A8, and 7B8. These tubes operate satisfactorily at medium and low frequencies. However, because of reduced oscillator output and undesirable interaction effects at higher frequencies, other pentagrid converters sometimes are used. In this type no electrode operates solely as the oscillator plate (fig. 91).

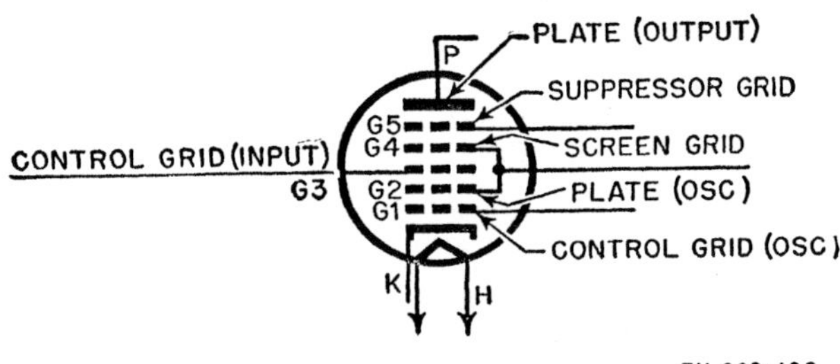

Figure 91. Another form of pentagrid converter.

(7) This form of pentagrid converter can be considered as a triode and a pentode. The triode section forms the oscillator and consists of cathode K, control grid G1, and grid G2. The r-f signal input is injected into G3. Grid G4 is connected internally to G2. The two grids serve as the screen grid of the pentode section in addition to the plate of the triode (oscillator) section. They also minimize any interaction between the oscillator and input signal circuits by shielding G3. Grid G5 is the suppressor of the pentode section; in some tubes it is connected internally to the cathode; in other heptodes it has an external pin connection. Ex-

amples of heptodes with a suppressor grid are the IR5, 6SA7, 7G7, 12SA7, and 14Q7.

74. Multiunit or Dual-purpose Tubes

a. A multiunit or dual-purpose tube is one in which two or more individual tube-element structures are combined within a single envelope. As a result, compactness, economy, and more satisfactory operation for certain purposes are achieved. The most commonly used multiunit tubes, known as duo-diodes and duo-triodes, combine two diode or two triode elements. Frequently, a single common cathode is used which supplies electrons to both sets of elements in the multiunit tube. Occasionally, an electrode of one set of elements is connected internally to an electrode of another set of elements.

b. There are many types of multiunit tubes, used for a wide variety of purposes. The schematic symbols of the most common ones are shown in figure 92. The tube whose symbol is shown in A can be used as a full-wave rectifier, an f-m (frequency-modulated) discriminator, or a combination detector and avc (automatic volume control) rectifier. The diode-triode, in B, can be used as a diode detector and a triode amplifier. The duo-triode, in E, is used as a push-pull amplifier, two amplifiers in cascade, or as a special type of complex-wave generator. The triode-hexode, in L, and the triode-heptode, in M, are used as mixer-oscillators in superheterodyne receivers.

c. Table I lists typical examples of multipurpose tubes to be found in receiving-type equipment. The letters in the first column refer to the symbols used in figure 92.

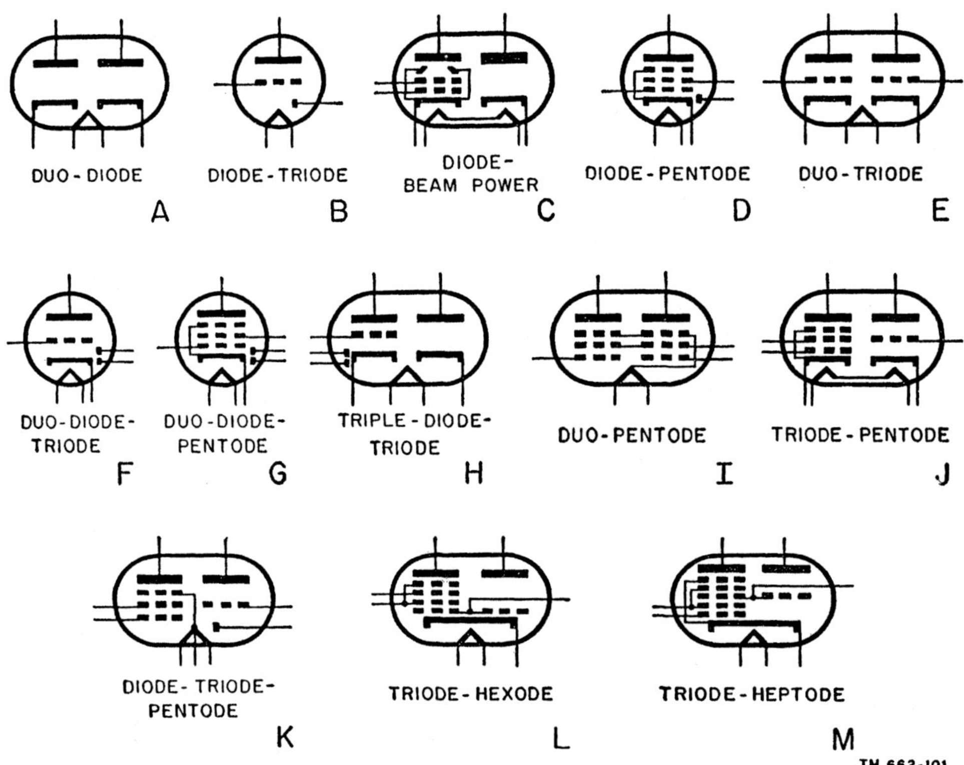

Figure 92. Tube symbols for multiunit tubes.

Table I. Examples of Multipurpose Tubes in Receiving Equipment

Symbol designation	Multipurpose tube	Examples of tube types
A	Duo-diode	5U4, 5V4, 5Y3, 6AL5, 6X5, 12AL5
B	Diode-triode	1H5, 1LH4
C	Diode-beam power	70L7, 117L7, 117N7, 117P7
D	Diode-pentode	1N6, 1T6, 6SF7, 12A7, 12SF7
E	Duo-triode	1J6, 6J6, 6N7, 6NS7, 6SL7, 12AT7, 12AU7
F	Duo-diode-triode	1H5, 6AQ7, 6SQ7, 6V7, 7K7, 7X7, 12SQ7
G	Duo-diode-pentode	1F6, 7R7, 14E7, 14R7
H	Triple-diode-triode	6S8, 6T8, 12S8
I	Duo-pentode	1E7
J	Triode-pentode	6AD7, 6P7, 25B8
K	Diode-triode-pentode	3A8
L	Triode-hexode	6K8, 12K8
M	Triode-heptode	7J7, 7S7, 14J7

Section IV. SUMMARY AND REVIEW QUESTIONS

75. Summary

a. The use of a triode is unsatisfactory in high-frequency systems because of feedback which is caused by interelectrode capacitance.

b. A tetrode contains an electron emitter, a control grid, a screen grid, and a plate.

c. In receiving-type tetrodes, the cap on the top of the tube is connected to the control grid. In transmitting-type tetrodes, the cap is connected to the plate to reduce the capacitance between control grid and plate.

d. In a tetrode, the screen grid is made positive with relation to the cathode. In most cases the d-c screen voltage is appreciably less than the d-c plate voltage.

e. The main purpose of the screen grid is to reduce the interelectrode capacitance between the plate and control grid, thereby preventing self-oscillation.

f. The plate current in a tetrode is not greatly affected by changes in plate voltage.

g. Secondary emission occurs in a vacuum tube when electrons from the space charge strike the plate with sufficient force to dislodge secondary electrons from the plate itself.

h. The movement of secondary electrons in the opposite direction to the regular current flow results in a dip in the plate-current plate-voltage characteristic curve of a tetrode.

i. In a tetrode, r_p and μ are high and g_m remains about the same as in a triode.

j. The main advantage of a tetrode compared to a triode is that it has lower interelectrode capacitance between control grid and plate and it affords greater amplification.

k. The pentode uses a suppressor grid to reduce the effects of secondary emission.

l. The pentode produces more distortion than the triode but it is more widely used because of its other advantages.

m. The beam power tube is capable of handling high power levels. It is used in the output stages of receivers and amplifiers and in various parts of transmitters.

n. To minimize the undesirable effects of distortion, variable-μ tubes are used. In these tubes, the amplification factor is determined by a nonuniform grid structure.

o. A six-electrode tube is a hexode; a seven-electrode tube is a heptode; and an eight-electrode tube is an octode. These are called *multigrid tubes*.

p. A tube which contains five grids and is used to operate as a frequency converter in a superheterodyne receiver is known as a pentagrid converter.

q. Multiunit tubes contain more than one set of elements within a single envelope.

76. Review Questions

a. Why are triodes undesirable for amplification at high frequencies?

b. What are the fundamental differences between the triode, the tetrode, and the pentode?

c. To what electrode is the metal cap on top of a receiving tube usually connected?

d. What is the difference between single-ended and double-ended tubes?

e. What are the major differences between the circuits for a triode and a tetrode?

f. What is the approximate value of inter-electrode capacitance in a triode? In a tetrode?

g. Do changes in E_{bb} affect the total current, $i_b + i_{c2}$, flowing through a tetrode? Why?

h. Explain what is meant by secondary emission.

i. Why do *dips* appear in plate-current plate-voltage characteristic curves of tetrodes?

j. How do the plate-current plate-voltage characteristic curves of modern tetrodes differ from those of the older types?

k. Generally speaking, how do μ, g_m and r_p of a tetrode compare with the same characteristics of a triode?

l. How does a suppressor grid reduce the effects of secondary emission?

m. In external appearance, what is the difference between the tetrode and the pentode?

o. Does the suppressor grid in a pentode reduce the control grid-to-plate capacitance more than the screen grid in a tetrode?

p. How do the plate-current plate-voltage characteristic curves of a pentode compare with those of a tetrode? A triode?

q. Why are pentodes used more often than triodes even though they produce more distortion?

r. Compare r_p, g_m, and μ of a pentode and a triode.

s. Why is the g_m of a pentode about the same as the g_m of a triode and a tetrode?

t. Why are the dynamic transfer characteristic curves of most pentodes not linear?

u. What effect is noted on the dynamic transfer characteristic if the screen voltage is reduced to about one-seventh the value of the plate supply voltage?

v. What are the electrical and physical differences between beam power tubes and ordinary tetrodes and pentodes?

w. How does the physical arrangement of the screen grid in a beam power tube help produce a greater concentration of electrons?

x. What are the physical and electrical differences between a variable-μ pentode and an ordinary pentode?

y. What determines the amplification factor of a tube?

z. What is the difference between a multigrid and a multiunit tube?

CHAPTER 6

AMPLIFICATION

---

77. General

a. INTRODUCTION. The most important use of an electron tube is to amplify or increase the amplitude of input signal voltages. For example, a minute amount of power at the input of a broadcast receiver is amplified by a number of amplifier stages in the receiver to the level necessary to operate a loudspeaker. The amount of amplification or *gain* that results is dependent primarily upon the number of amplifier stages used. Gain is defined as the ratio of output to input. The greater the number of stages, the greater the over-all gain will be. One stage may have a larger gain (gain per stage) than another.

b. CLASSIFICATION OF AMPLIFIERS.

  (1) Electron tubes may be classified according to the functions they perform. They operate as amplifiers, oscillators, detectors, and so on. Amplifiers, in turn, often are classified in terms of the frequency or range of frequencies at which they operate. An ideal circuit amplifies all signals it receives, ranging in frequency from a few cycles to several thousand megacycles. Such an ideal amplifier cannot be obtained in practice. Instead, amplifiers usually operate within a rather restricted frequency range. For example, an a-f (audio-frequency) circuit amplifies frequencies in the audio range. This is from about 16 to 16,000 cps.

  (2) If a circuit amplifies signals beyond the audio range, it often is called an r-f amplifier. In this type of amplifier, usually only one frequency is amplified at one time. However, r-f amplifiers frequently are tunable so that they can operate over a range of frequen-

cies. Very-high-frequency amplifiers, which operate at frequencies above 30 mc, and ultra-high-frequency amplifiers, which operate at frequencies above 300 mc, also are classed as r-f amplifiers.

  (3) An i-f amplifier operates at one fixed frequency, usually a radio frequency. Such amplifiers are used in superheterodyne receivers. The intermediate frequency of a standard broadcast superheterodyne receiver is frequently about 455 kc. In many radar and television receivers, the intermediate frequency may be as high as 45 mc.

  (4) The video amplifier is one that amplifies frequencies in both the audio-and radio-frequency ranges. Frequently, a video amplifier can amplify all signals from a few cycles to several megacycles. Video amplifiers are used commonly in conjunction with the cathode-ray tube displays and in such equipment as television sets, radar, and special test apparatus.

  (5) Still another type is the d-c amplifier in which the output of one amplifier stage is coupled directly to the input of the following amplifier stage. This amplifier is capable of amplifying direct and alternating voltages.

c. AMPLIFIER DISTORTION.

  (1) The amount of distortion that an amplifier introduces is partially dependent on the linearity of the dynamic characteristic of the electron tube used. This depends on several factors such as the d-c operating potentials, tube type, and load.

  (2) Distortion in amplifiers is dependent

also on the amplitude of the input grid signal. If the amplitude of the grid signal extends into the nonlinear region of the dynamic characteristic, severe distortion results. Also, if the grid swing extends into the plate-current cut-off or plate saturation regions, distortion occurs as a result of clipping or limiting of the input signal.

(3) Proper selection of an operating point on the dynamic characteristic curve helps minimize distortion. The selection of different operating points results in an amplifier having differing characteristics. Amplifiers can be classified further in accordance with their mode of operation. Such amplifiers are classified as A, B, AB, or C.

78. Class A Operation

a. The class A amplifier frequently used is so biased that plate current flows constantly. In figure 93, point B is the operating point and is determined by the bias voltage, E_{cc}. The grid signal voltage varies on both sides of the operating point. The output plate-current waveform is obtained in the figure by extending dotted lines from various points of the input grid voltage signal to the dynamic characteristic curve. Note that the output plate-current waveform is practically undistorted—that is, it closely resembles the input. Minimum distortion takes place because operation occurs along the linear portion of the curve. In addition, the peak-to-peak amplitude of the input grid signal is comparatively small. This prevents it from extending into the nonlinear portions of the curve.

b. It may seem desirable to have the grid signal extend along the entire length of the characteristic from point A to C. This produces a plate-current waveform of maximum amplitude which, in turn, produces maximum amplification. However, distortion results because the dynamic characteristic is not perfectly linear along this entire length. This is true of any tube. In most cases of class A operation, a distortionless output is preferred to large amplification.

c. If the positive half of the grid signal goes beyond point C, the control grid becomes positive in respect to the cathode and grid current flows. Part of the input grid signal is lost or clipped and the positive half of the output plate-current waveform is distorted. Similarly, if the negative half of the input grid signal goes beyond point A, plate current stops flowing and the negative half of the output plate-current waveform is clipped. Distortion results once again.

d. From this analysis, the following conclusions can be made concerning distortion in class A operation. The more linear the dynamic characteristic of a tube, the less distortion it introduces. Grid signal voltages should extend only into the most linear portion of the dynamic characteristic. The operating point usually is in the center of this linear portion. Class A amplifiers generally are biased to about one-half their plate-current cut-off value. Finally, as explained previously, increasing the load produces a more linear dynamic characteristic curve. If the load is increased excessively, however, a great deal of power is lost. Therefore, some intermediate value of load is selected. In this way, a minimum amount of distortion is obtained with a reasonable amount of power output.

e. The question often arises whether triodes or pentodes should be used for class A operation. The advantage of using triodes is that their dynamic transfer characteristics usually are more linear, and therefore, produce less distortion. Pentodes have the advantage of producing a greater power output.

f. Classes of amplifiers often are compared with each other in terms of *plate efficiency*. Plate efficiency is defined as the ratio of a-c power output that is developed across the load to the d-c power supplied to the plate. In class A amplifiers, the plate efficiency is about 20 percent or less. This low efficiency is due to the high average value of plate current, and consequently high plate dissipation.

79. Class B Operation

a. A class B amplifier is one that is biased at or near cut-off (fig. 94). Plate current flows during the positive half of the input grid signal and stops flowing during the negative half-cycle.

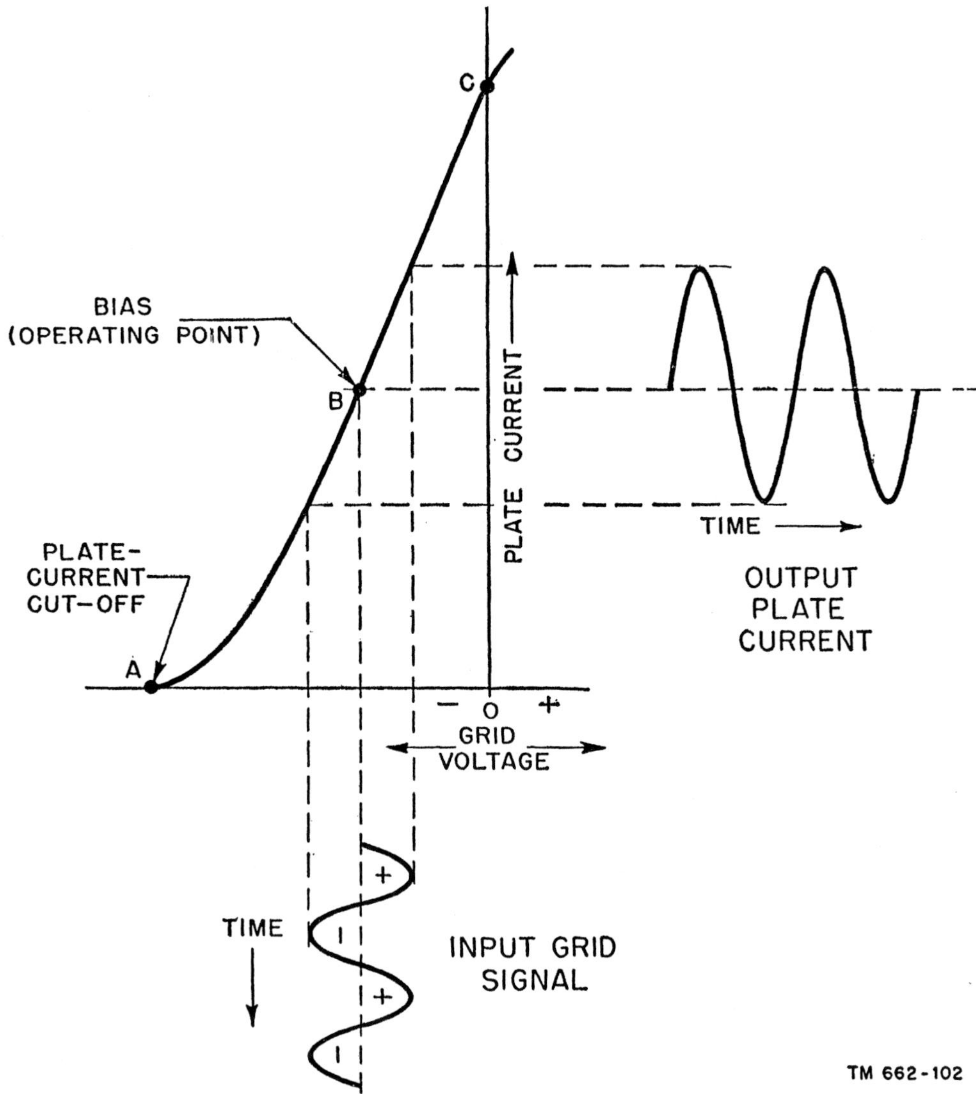

Figure 93. Class A operation on dynamic characteristic.

b. Operating point E equals the bias voltage, E_{cc}. Plate current starts to flow when the instantaneous value of grid voltage, e_c, rises above the plate-current cut-off point. Plate current continues to flow during the positive half of the input grid signal (the unshaded portion of the input grid signal). During the shaded portion of the input signal the plate current is 0. The shaded portion indicates the part of the input grid signal that has been clipped. This results in severe distortion of the output plate-current waveform. Note that the grid signal swing extends into the nonlinear portion of the dynamic characteristic; that is, from the plate-

current cut-off point to point F. This also causes distortion.

c. If the amplitude of the grid signal is increased so that it extends beyond point F, the amplitude of the plate-current waveform is increased, resulting in a greater power output. If the positive peak of the grid signal extends beyond point G, grid current flows, causing the positive peak of the plate-current waveform to be clipped. Therefore, distortion is increased.

d. A single tube that operates as a class B amplifier produces a great deal of distortion. Therefore, it is undesirable to use this class as an a-f amplifier where fidelity of reproduction

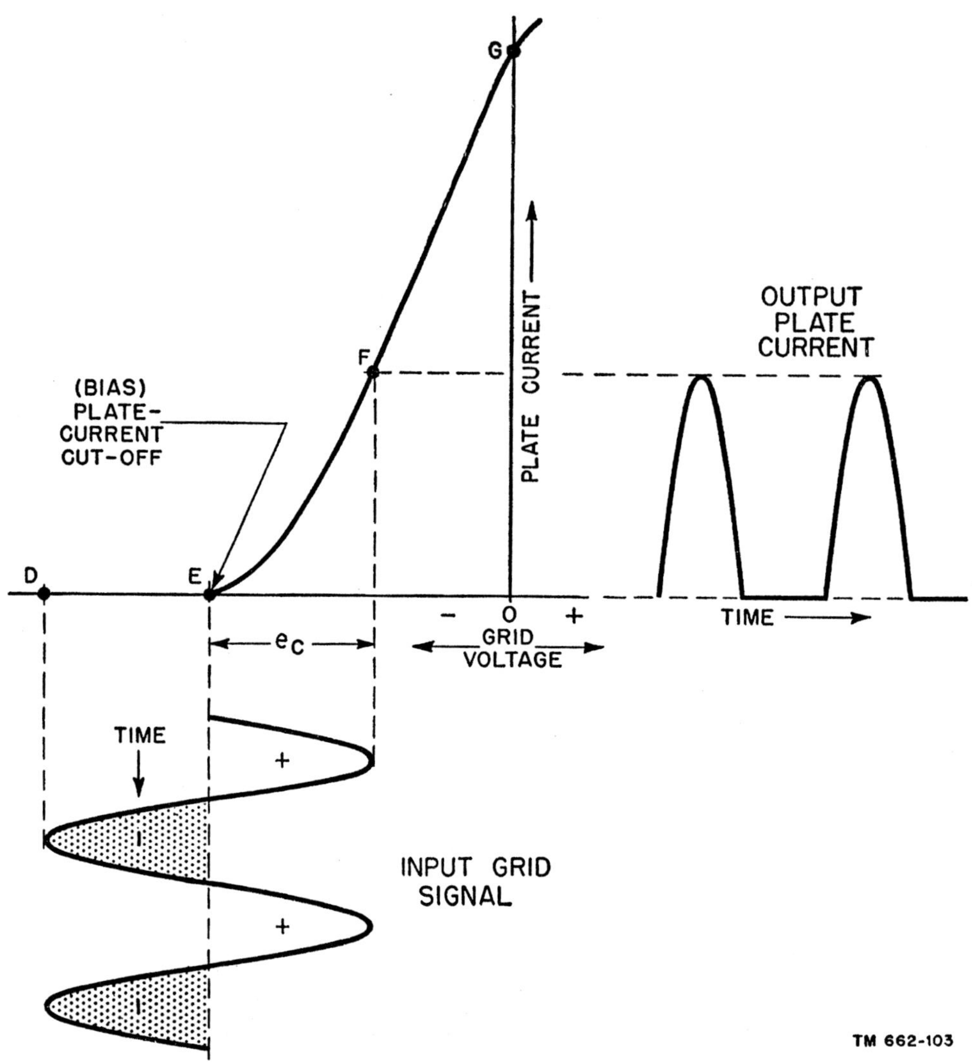

Figure 94. Class B operation on dynamic characteristic.

is necessary. However, two tubes operating in *push-pull* often are used in amplifying systems. A push-pull amplifier produces twice the output of a single tube. In addition, the out-put waveform contains a minimum of distortion, as will be shown later.

e. A single-tube class B amplifier delivers a greater amount of power than a single-tube class A amplifier. This is in part because of the greater voltage swing that is permissible in class B operation. In a class B push-pull amplifier, the power output is about five times as great as in a single-tube class A amplifier using the same tube type.

f. The plate efficiency of a class B amplifier is about 40 to 60 percent—about double that

obtained in a class A amplifier. The high plate efficiency permits the use of smaller power supplies for the d-c operating potentials. Sometimes the fluctuating output current of a class B amplifier is unintentionally coupled into the power source, thereby causing irregularities in the d-c operating potentials. When this happens, additional filter circuits are needed.

80. Class AB Operation

a. An amplifier that operates in the region between class A and class B is called a class AB amplifier. In a class A amplifier, plate current flows during the entire cycle of input grid signal; in a class B amplifier, plate current flows

114

during the positive half-cycle of input grid signal. Therefore, in a class AB amplifier, plate current flows for more than a half—but less than the entire cycle of input grid signal. Class AB operation may be subdivided into classes AB_1 and AB_2. Subnumber 1 indicates that grid current does not flow during any part of the input cycle. Subnumber 2 indicates that grid current does flow during some portion of the input cycle.

b. Class AB_1 operation is shown in A of figure 95. The operating point is located between plate-current cut-off and the linear portion of the dynamic characteristic. The positive peak of the grid signal extends into the linear region of the dynamic characteristic. To prevent the flow of grid current, the positive peak of the grid signal cannot exceed the fixed bias E_{cc}. The negative peak of the grid signal extends beyond plate-current cut-off. The shaded areas of the input grid signal indicate those portions of the input signal which are cut off. The output plate-current waveform is distorted because of the clipping action of the negative peaks.

c. Class AB_2 operation is illustrated in B of figure 95 with the same operating point H, shown in A. In class AB_2 operation, the peak value of grid signal exceeds the fixed bias of the tube. The positive peaks of input grid signal extend into the positive region of grid voltage. This causes grid current to flow. Just as in class AB_1 operation, the negative grid signal peaks go beyond the plate-current cut-off point.

d. Clipping is much greater in class AB_2 operation than in class AB_1. However, because of the greater grid voltage swing in class AB_2 operation, a greater output exists. The plate efficiency in class AB_1 operation is somewhat greater than in class AB_2. Compared with class A, class AB operation produces more distortion, more power output, and a greater plate efficiency.

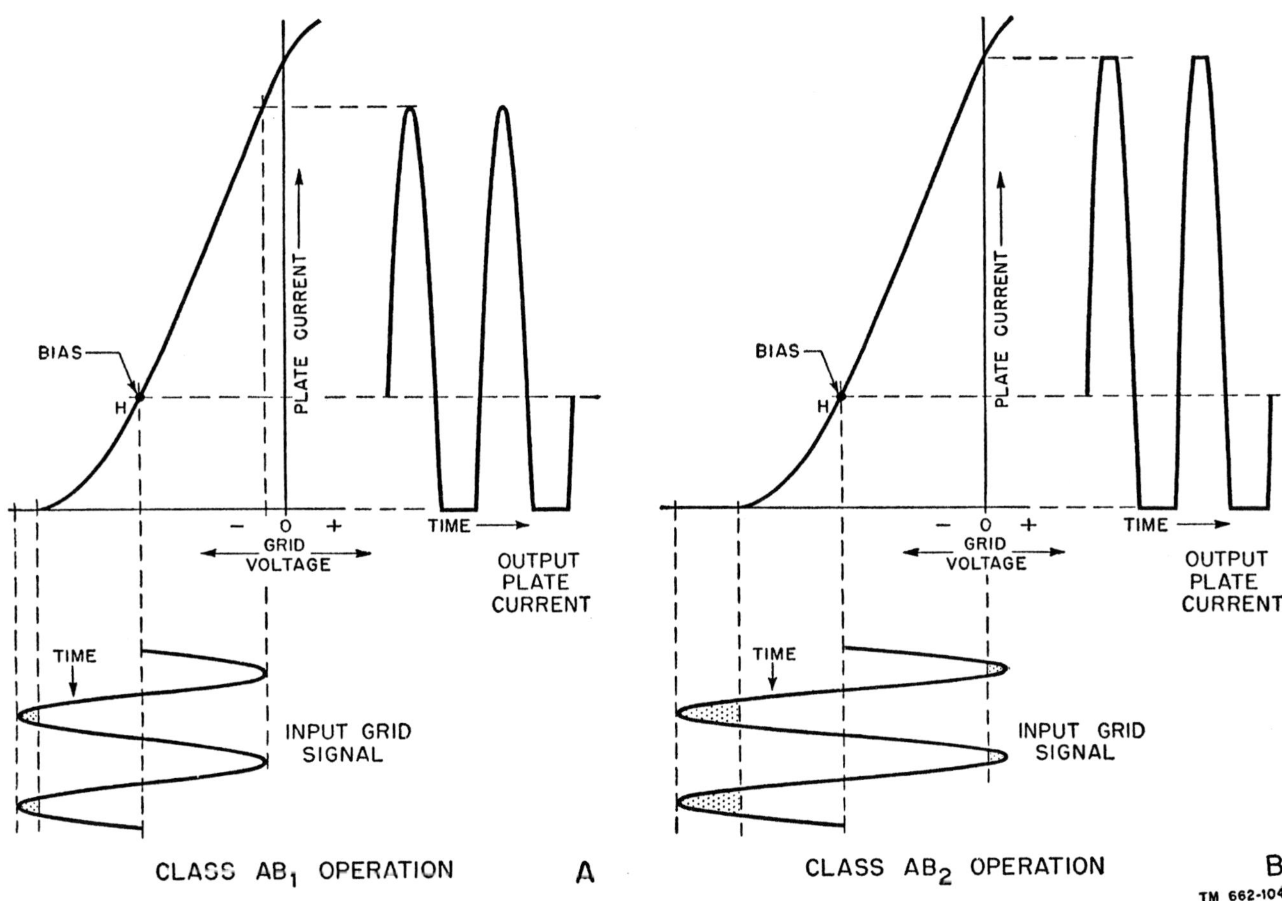

Figure 95. Class AB operation on dynamic characteristic.

81. Class C Operation

a. A class C amplifier is one whose operating point is located well beyond the plate-current cut-off point so that plate current flows for appreciably less than a half-cycle. Class C operation is shown in figure 96. The operating point for the input grid signal shown is point J. The voltage at this point is one and a half times the cut-off value. The operating point for class C operation usually is made from one and a half to four times cut-off. The positive peak of the input grid signal extends to point K on the dynamic characteristic and the negative peak extends to point I. The shaded areas indicate the portions of the input grid signal which are cut off. The output plate-current waveform represents a small part of the positive peaks of input grid signal.

b. If the positive peak of the grid signal extends beyond point L on the dynamic characteristic and into the positive region, grid current flows. The result is an output plate-current waveform that is greater in amplitude; however, it contains more distortion since its positive peaks are clipped. In a class C amplifier, the large grid signal swing produces a greater power output as compared to classes A and B.

c. Class C amplifiers generally are used as r-f amplifiers where a large power output and a high plate efficiency are desired. The plate efficiency of a class C amplifier is usually 60 to 80 percent. The high distortion of a class C stage is overcome by the flywheel effect of tuned circuits.

82. Push-pull Operation

a. GENERAL. A push-pull amplifier consists of two electron tubes whose grid and plate signals are 180° out of phase. The two tubes are operated class A, AB, or B. Push-pull amplifiers are used frequently in a-f amplifier circuits. Less distortion with a greater power output and plate efficiency is obtained in push-pull operation.

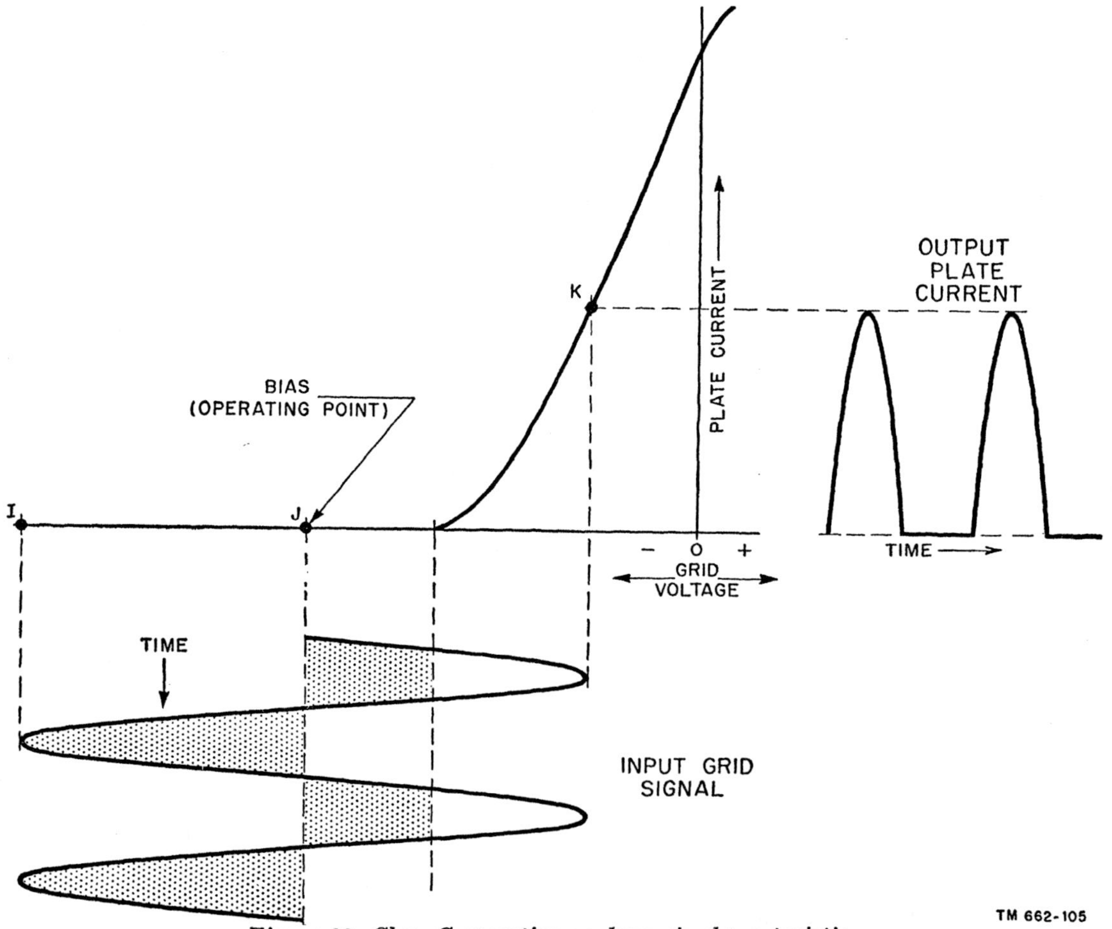

Figure 96. Class C operation on dynamic characteristic.

b. PUSH-PULL CIRCUIT.

(1) A push-pull triode amplifier circuit is shown in figure 97. The upper and lower sections of the circuit are similar. Triodes V1 and V2 are the same type of tube and, therefore, have similar characteristics. The two grid signals, e_{g1} and e_{g2}, have the same amplitude and frequency but are 180° out of phase with each other. A single bias battery and a single plate-supply battery are used to supply both tubes with their proper d-c operating potentials. Transformer T acts as the load for the circuit. The primary of transformer T is center-tapped (point A) so that the output voltages, e_{b1} and e_{b2}, are equal in magnitude. Plate currents i_{b1} and i_{b2} also are equal in magnitude.

(2) The symbols for the various voltages and currents are similar to those used in a simple triode circuit. Subnumber 1 is used to indicate a voltage or current associated with tube V1. Subnumber 2 is used to indicate a voltage or current associated with tube V2. For example, e_{g1} is the instantaneous value of grid signal voltage applied to V1; i_{b2} is the instantaneous value of plate current in the plate circuit of V2 and so on. The instantaneous voltage across the entire primary is e_p. The instantaneous current in the secondary is i_o.

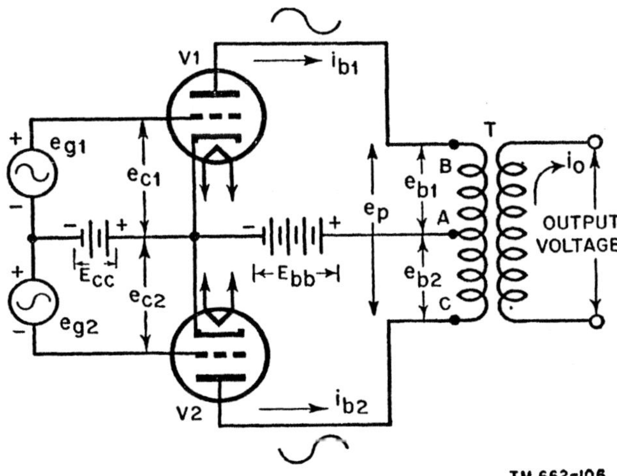

Figure 97. Push-pull triode amplifier circuit.

(3) With no input signals applied to the push-pull circuit, the secondary current i_o is 0. Since steady d-c plate current induces no voltage into the secondary, the secondary current is 0. The d-c plate current of V1 flows from cathode to plate, through the upper half of the primary winding (from points B to A), and back to the cathode. The d-c plate current of V2 flows from cathode to plate, through the lower half of the primary winding (from point C to A), and back to the cathode. Points B and C are equally negative with relation to point A since the magnitudes of the plate currents are equal. Thus the total magnetizing force is 0, and d-c saturation of the core does not result.

(4) When the sinusoidal signals, e_{g1} and e_{g2}, are applied to the respective grids, sinusoidal plate currents i_{b1} and i_{b2} flow in the primary of the transformer. Current i_{b1} is 180° out of phase with i_{b2} since the two grid signals are 180° out of phase with each other. During the positive swing of i_{b1}, point B on the primary becomes more negative with relation to point A. At the same time, the fall in i_{b2} causes point C to become less negative with respect to point A by an equal amount. Therefore, the voltage across the entire primary, e_p, is twice the value of either e_{b1} or e_{b2}. In other words, e_p is equal to e_{b1} plus e_{b2}. A half-cycle later, all the polarities reverse. Here again the voltage across the primary, e_p, is equal to e_{b1} plus e_{b2}. The relationship of e_p in terms of e_{b1} and e_{b2} holds true for all instantaneous values of plate current. Transformer T couples the output of the push-pull amplifier to another circuit.

(5) There are two common methods of obtaining a 180-degree phase reversal of the input grid signals. The first is the transformer method. The secondary of a center-tapped transformer is connected to the two control grids of V1 and V2. The center tap is con-

nected to the negative side of the bias battery. With a sinusoidal voltage impressed across the primary, two sine waves of voltage appear in the secondary. These sine waves are 180° out of phase with each other because of the tapped winding. In the other method, a paraphase amplifier is used.

c. DYNAMIC CHARACTERISTICS FOR PUSH-PULL OPERATION.

(1) The dynamic characteristic for two tubes operating in push-pull is constructed from the individual dynamic characteristics of both tubes. Figure 98 shows a dynamic characteristic of two tubes operating in class A push-pull. It is obtained in the following manner. The dotted curve labeled V1 is the dynamic characteristic of one tube and V2 is the dynamic characteristic of the other. These dynamic characteristics are identical and be-cause the plate currents of both tubes are 180° out of phase, they are placed 180° out of phase with each other so that their horizontal axes are common. They are then lined up so that the bias voltage of one tube meets the same value of bias voltage of the other tube. For example, if $E_{cc} = -5$ volts, this value of voltage occurs at the same place on both grid-voltage axes. The resultant dynamic characteristic is obtained by algebraically adding the instantaneous values of plate current for different values of grid signal voltage.

(2) In single-tube class A operation, little distortion exists in the output plate-current waveform, since the grid signal operates along the most linear portion of the dynamic characteristic. In class A push-pull operation, the distortion is even less because the dy-

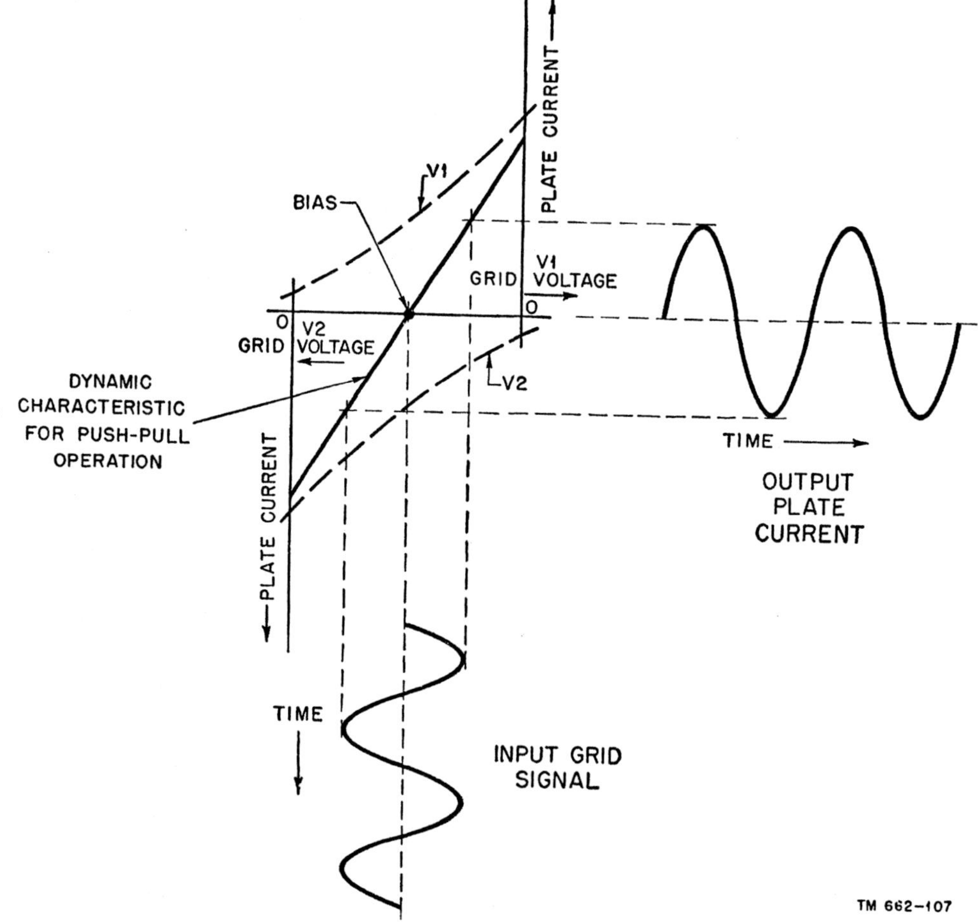

Figure 98. Dynamic characteristic of two tubes in push-pull class A operation.

namic characteristic is even more linear. By projecting various points of the input grid signal to the push-pull characteristic (shown by dashed lines in the figure), the output plate-current waveform is obtained. A greater grid voltage swing is possible in push-pull operation without resulting in noticeable distortion. The reason is that the dynamic characteristic is linear for a greater amount of voltage variation. If a greater grid signal swing is possible in push-pull operation, then a greater power output can be obtained. The efficiency of a class A push-pull amplifier can go as high as 30 percent; in class A single-tube operation, the limit is about 20 percent.

(3) Class AB push-pull operation is also possible and frequent. The resultant dynamic characteristic is obtained in the same manner as the resultant characteristic in figure 98. The main advantage of class AB push-pull oper-ation as compared to class A push-pull operation is that a greater grid signal swing is permissible because of the longer resultant characteristic. This results in a greater power output and plate efficiency. The plate efficiency of class AB push-pull amplifiers can go as high as 55 percent.

(4) The resultant characteristic of class B push-pull operation is shown in figure 99. The bias voltage is nearer to plate-current cut-off than in class AB push-pull operation. Here again, the output plate-current waveform is fairly undistorted. Note that the bias is not quite at the plate-current cut-off value. If exact cut-off bias were used, distortion would occur (fig. 100). Here, the resultant dynamic characteristic has an S shape which causes severe distortion of the output plate-current waveform. The plate efficiency of a class B push-pull ampli-fier is about 60 to 65 percent.

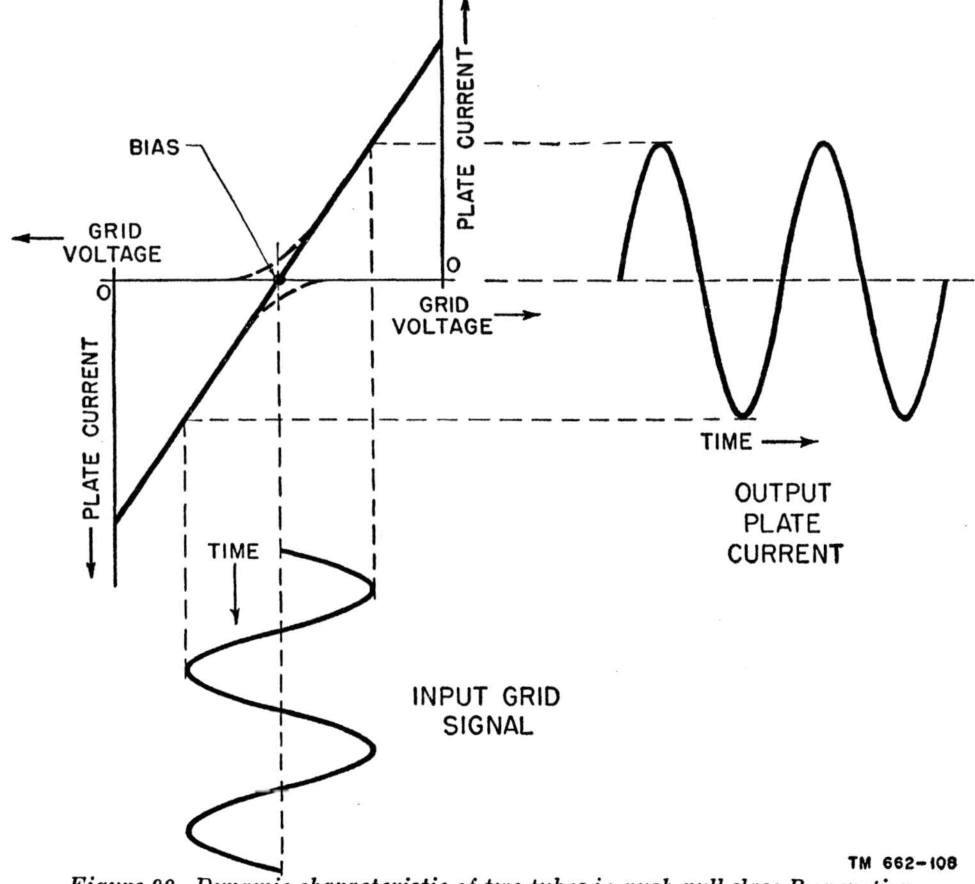

Figure 99. Dynamic characteristic of two tubes in push-pull class B operation.

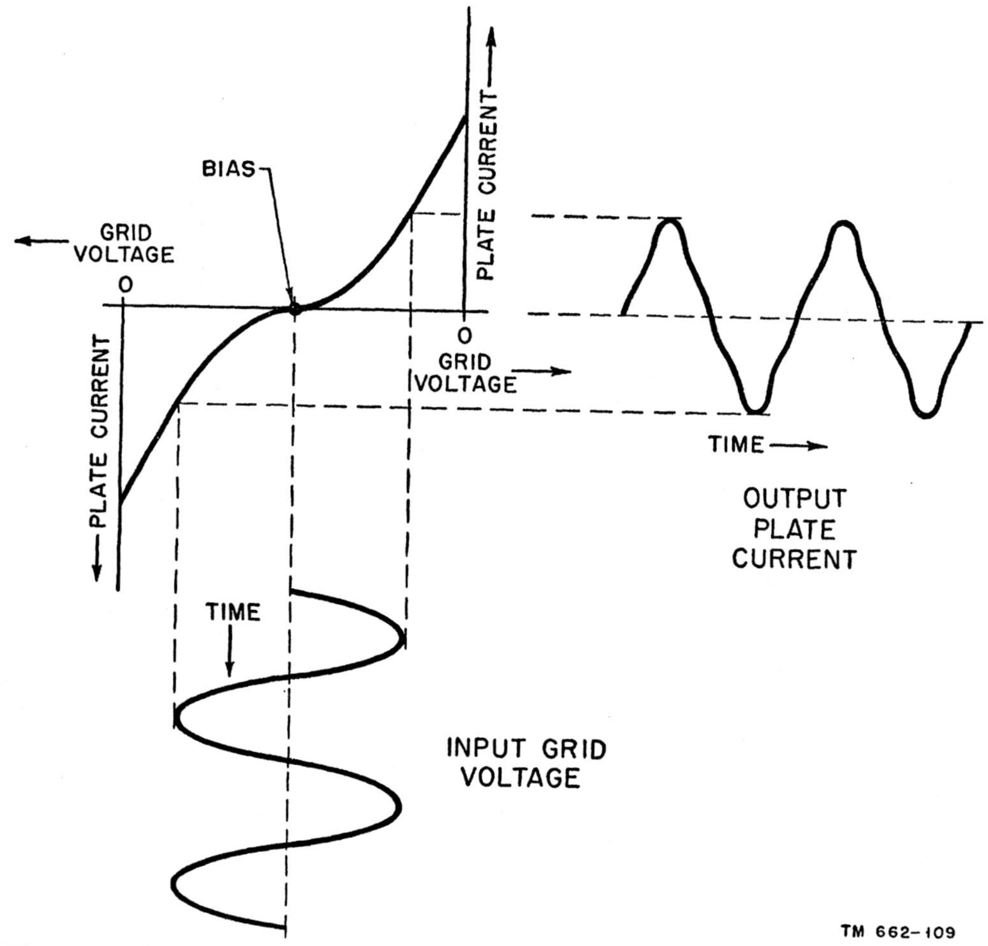

Figure 100. Operation of push-pull class B amplifier biased exactly at plate-current cut-off.

83. Table of Characteristics

Table II gives a resume of the operation points, distortion, power output, and plate efficiency of the different classes of amplifiers.

84. Phase Relationships in Amplifiers

a. Grid signal e_g is always *in phase* with plate current i_b. Two waveshapes of equal frequency are said to be in phase when they pass through

Table II. Amplifier Characteristics

Class	Location of operating point on dynamic characteristic	Relative distortion	Relative power output	Approximate percentage of plate efficiency
A single-tube		Low	Low	Under 20%
A push-pull	On linear portion	Very low	Moderate	20 to 30%
AB single-tube		Moderate	Moderate	40%
AB push-pull	Between linear portion and plate-current cut-off	Low	High	50 to 55%
B single-tube		High	High	40 to 60%
B push-pull	At vicinity of cut-off	Low	Very high	60 to 65%
C single-tube	About 1½ to 4 times plate-current cut-off	Very high	Very high	60 to 80%

corresponding points at the same time; that is, the curves reach their maximum positive and negative values at the same time and they pass through 0 at the same time. The phase relationship of the various a-c voltages and currents in a triode amplifier is indicated in figure 101. In this circuit, e_g, e_c, i_b, are in phase with each other, but they are 180° out of phase with e_{out}.

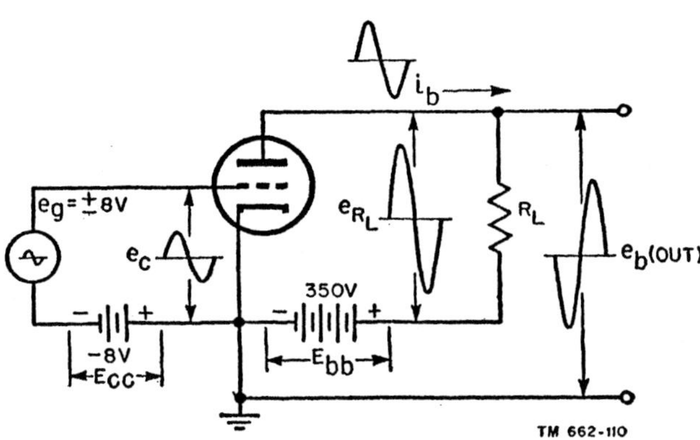

Figure 101. Current and voltage phase relations in triode amplifier circuit.

b. Assume that the triode in figure 101 is a 6J5 tube operating at a grid bias voltage of —8 volts and a plate-supply voltage of 350 volts. Under quiescent conditions, plate current I_{bo} is 5.2 ma, plate voltage E_{bo} is 220 volts, and R_L is 25,000 ohms. The voltage drop across the load resistor under quiescent conditions E_{Lo} is E_{bb} minus E_{bo} which equals 350 minus 220, or 130 volts. The various voltages and currents appearing in the triode circuit are shown graphically in figure 102.

c. These waveshapes are obtained in the following manner: Points A, A1, A2, A3, and A4, connected by a vertical dotted line, represent conditions which occur at quiescence. The signal voltage, in A, is 0 at point A and the total grid voltage in B equals —8 volts at point A1, which is the value of E_{cc}. At the same time, the total plate current, I_{bo}, in C, equals 5.2 ma at point A2. The d-c voltage drop across load resistor E_{Lo}, in D, equals 130 volts at point A3. The total plate voltage, E_{bo}, in E, is equal to 220 volts at point A4.

d. When the grid signal voltage reaches its most positive value at point B, total grid voltage e_c is 0 volt, at point B1. The maximum signal voltage on the grid causes the plate current to rise to a maximum value of 10.1 ma at point B2. This maximum current causes the voltage drop across the load resistor to be a maximum value of 252 volts at point B3. R_L, r_p, and E_{bb} comprise a series circuit. The voltage drops across R_L and r_p equal the supply source E_{bb}. If the voltage drop across R_L increases, then the voltage drop across r_p must decrease. The output voltage, e_b, from plate to cathode, is the voltage drop across r_p. Since the voltage drop across R_L is at a maximum, at point B3, then voltage e_b is a minimum value. This is shown at point B4, which equals 98 volts.

e. At quiescence, the same conditions prevail at points C, C1, C2, C3, and C4 as at points A, A1, A2, A3, and A4. When e_g reaches its most negative value at point D, total grid voltage e_c is —16 volts at point D1. This minimum input voltage causes i_b to be a minimum (1.3 ma at point D2). In turn, the minimum plate current causes e_{RL} to be a minimum value of 33 volts at point D3. Since e_{RL} is a minimum, the voltage drop across r_p is a maximum value and e_b is a maximum (317 volts at D4). Points F, F1, F2, F3, and F4 occur at quiescence and have the same numerical values as the corresponding points C, C1, C2, C3, and C4.

f. From the foregoing analysis, it is noted that the waveshapes shown in parts A, B, C, and D are in phase with each other but are 180° out of phase with waveshape E. Consequently, the following conclusion can be made: The signal on the control grid of an electron tube is always in phase with the plate current but is 180° out of phase with the output plate voltage. This statement holds true for all types of electron tubes, whether they are triodes, tetrodes, or pentodes.

85. Methods of Biasing

a. TYPES OF BIAS. In all amplifier circuits previously discussed, bias has been supplied by a battery, E_{cc}. This type of bias, known as *fixed bias*, obtained from a separate voltage source. Power supplies, to be discussed later, and d-c generators are other examples of fixed-bias supplies. The type of bias most commonly used is *self-bias* in which the bias voltage is developed across a resistance by the tube itself. The volt-

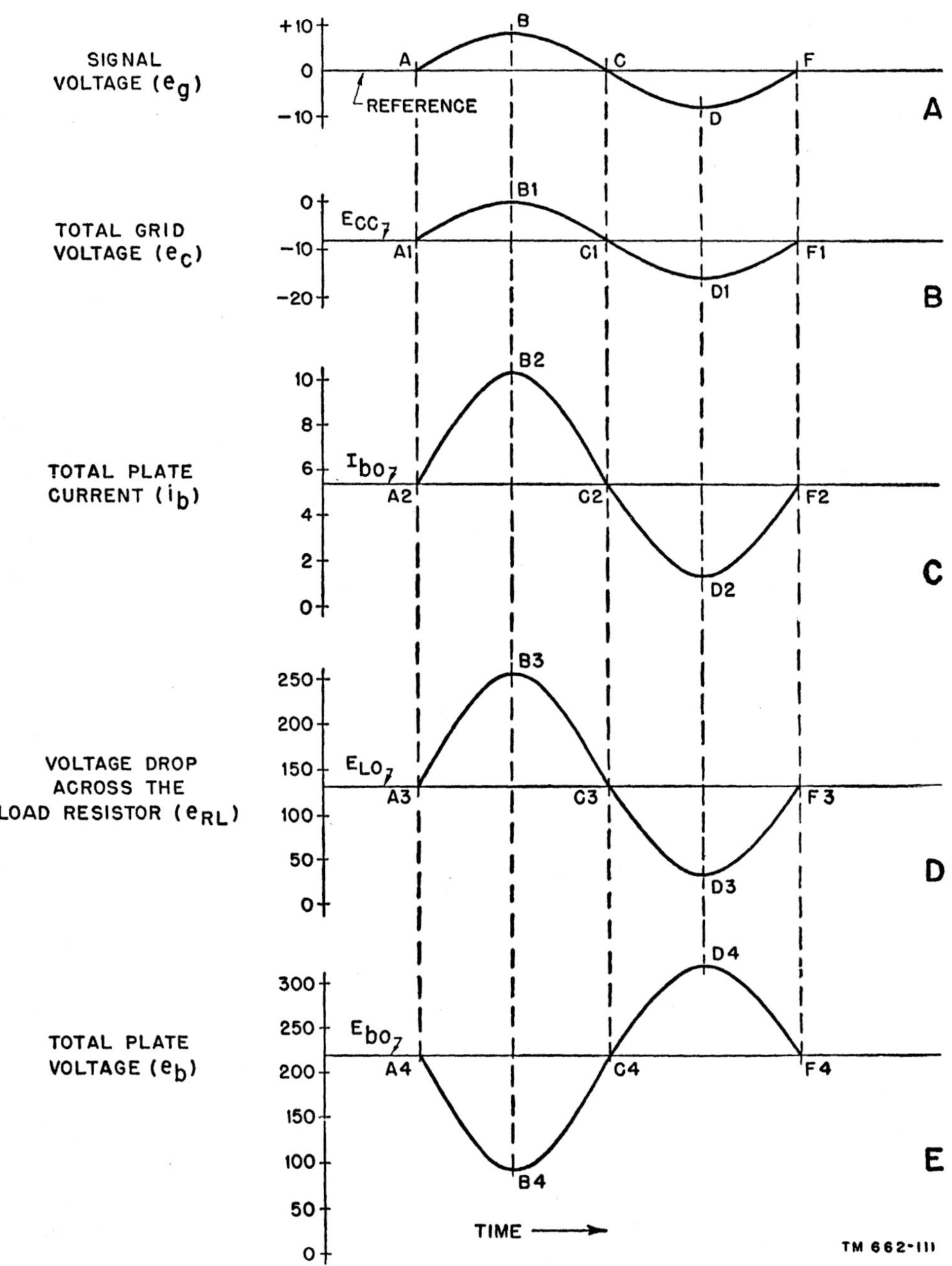

Figure 102. Waveforms in triode amplifier circuit.

age drop across this resistance is dependent upon the plate current of the tube or the grid signal. *Combination bias* is a combination of fixed bias and self-bias.

b. FIXED BIAS. A method of obtaining fixed bias is shown in A of figure 103. As previously stated, bias is the d-c voltage between grid and cathode. This voltage usually is negative, and it is used to establish the operating point. The bias voltage shown in the circuit is —5 volts and is developed by the bias battery. No grid current flows under quiescent conditions since the grid is negative with relation to the cathode. The grid is 5 volts negative with relation to the cathode, or the cathode is 5 volts positive with relation to the grid. Grid resistor R_g is part of a coupling network for the input grid signal. The voltage that appears from grid to ground

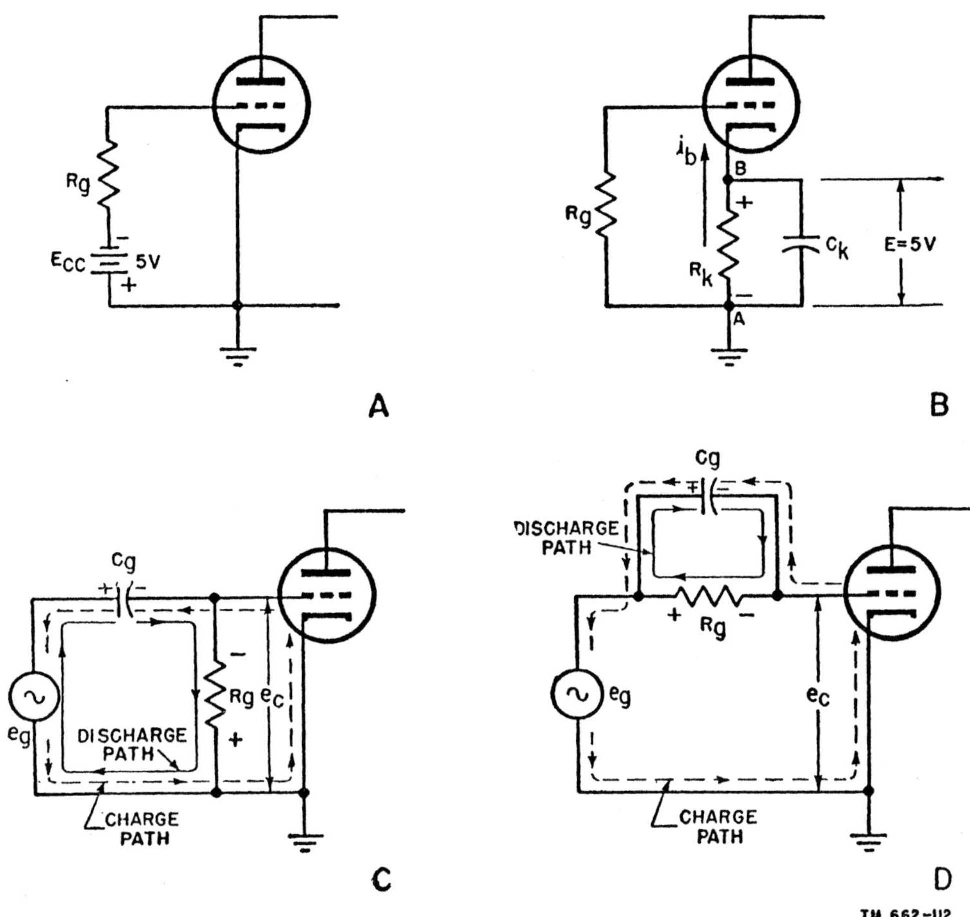

Figure 103. Various types of bias.

is equal to the bias voltage plus the instantaneous value of grid voltage. In this case the bias is independent of the input grid signal. This is always true in a fixed-bias system.

c. SELF-BIAS.

  (1) *Cathode bias.*

    (*a*) The most common method used to obtain self-bias is *cathode bias* (B of fig. 103). In this circuit the bias voltage is developed across cathode resistor R_k. Under quiescent conditions, plate current i_b flows continuously from cathode to plate and back to the cathode through resistor R_k. Since the plate current flows from points A to B, point A is negative in respect to point B. Assume that the voltage drop across R_k equals 5 volts. This makes the cathode 5 volts positive with relation

to the grid or the grid 5 volts negative with relation to the cathode.

    (*b*) As explained previously, resistor R_g is part of the coupling network for the input grid signal. If a sinusoidal grid signal is impressed across R_g, it causes the plate current to vary sinusoidally about an average d-c value. The varying plate current flows through cathode resistor R_k. Since the required bias is a fixed voltage, the a-c component of plate current through resistor R_k must be removed. This is accomplished by capacitor C_k. The value of this capacitor is large so that its capacitive reactance is small compared with the resistance of R_k at the frequency of the input grid signal. This low value of capacitive reactance effectively short-circuits or *bypasses* the a-c voltage component around R_k. The result is that the voltage drop across R_k does not vary and the bias voltage remains fixed at —5 volts.

(*c*) The value of C_k in a-f amplifiers is approximately 10 to 50 μf (microfarads). In r-f amplifiers, it is considerably smaller. Smaller capacitors are used since higher frequencies are involved. The value of R_k is usually from 250 to 3,000 ohms. R_k can be calculated by Ohm's law if the desired bias and plate-current values are known. For example, assume that an average d-c plate current of 10 ma flows and that a negative bias of 5 volts is required. Then by Ohm's law, $R_k = E_{cc}/I_{bo} =$ 5/.01 or 500 ohms.

(2) *Grid-leak bias.*

(*a*) Another self-bias method is grid-leak bias, illustrated in C. Grid capacitor C_g is large in value so that its capacitive reactance is small compared with the resistance of R_g at the input signal frequency. Resistor R_g is large in value. The voltage, e_c, appearing across R_g consists of the grid signal voltage, e_g, plus the bias voltage, E_{cc}.

(*b*) With no input signal applied to the circuit, the difference in potential between the grid and the cathode is 0. Therefore, the bias voltage is 0. The bias voltage can be obtained only when a grid signal is applied. This grid signal appears across R_g since the low capacitive reactance of C_g makes it a virtual short circuit for the grid signal. The sinusoidal grid signal alternately makes the grid positive and negative with relation to the cathode.

(*c*) During the positive half of the grid signal, the grid becomes positive with relation to the cathode. The grid draws current and charges C_g to the peak value of e_g so that the

plate of the capacitor connected to the grid becomes negative. The charge path is through the tube, shown by the dotted lines in the figure. During the negative half-cycle of e_g, the grid becomes negative with relation to the cathode. C_g discharges only slightly through R_g, since R_g has a large value. The discharge path is shown by a solid line. The top of R_g becomes negative with relation to the bottom or grounded end. During the next positive half-cycle of grid signal, C_g regains the charge that it lost and charges to the peak value of e_g once more. The discharge of C_g through R_g produces a pulsating d-c current. This current produces a pulsating d-c voltage. The average value of this voltage is the negative bias voltage.

(3) *Another form of grid-leak bias* (D of fig. 103). In this method of obtaining self-bias, R_g is connected in parallel with C_g. The operation of this circuit is similar to that shown in C, the only difference being in the discharge path of C_g, shown by the solid line. Here the discharge path of C_g is through R_g alone. The charge paths for C_g are identical in both cases as indicated by the dotted line. The voltage drop across R_g is the bias voltage produced by the discharging of C_g. The grid voltage, e_c, equals bias voltage E_{cc} plus grid voltage e_g.

(4) *Contact potential bias.* The circuit used for this method of obtaining self-bias is similar to that used for grid-leak bias. The grid signal, however, does not develop the bias voltage. In this circuit, R_g has a very large value —10 megohms. With no grid signal voltage applied to the circuit, a space charge forms near the cathode. Since the electrons forming the space charge are negative with relation to the grid, a difference in potential exists between the space charge and the grid. This difference in potential causes a

minute grid current flow through R_g. In turn, since R_g has a very high ohmic value, the current flow causes a small voltage drop to appear across this resistor. The voltage appearing across R_g is the bias voltage.

d. COMBINATION BIAS. Another type of bias is *combination bias* (fig. 104) which consists of both fixed bias and self-bias. In A of this figure, the fixed bias is —5 volts and the self-bias (cathode bias) is —4 volts. The total bias voltage is —9 volts. The grid is 9 volts negative with relation to the cathode and the cathode is 9 volts positive with relation to the grid. In B, fixed bias produces a voltage of —5 volts and self-bias (grid-leak bias) produces a voltage of —2 volts. The total bias voltage for this circuit is —7 volts.

e. Plate efficiency is defined as the ratio of a-c power output that is developed across the load to the d-c power supplied to the plate.

f. A class B amplifier is one that is biased at or near plate-current cut-off. Plate current flows during the positive half of the input grid signal and stops flowing during the negative half.

g. A class AB amplifier is one in which plate current flows for more than half the cycle but less than the entire cycle of input grid signal. In a class AB_1 amplifier, the positive peak of the grid signal does not exceed the fixed bias value so that grid current does not flow. In a class AB_2 amplifier the signal exceeds the bias, and grid current does flow.

h. A class C amplifier is one that is biased one and a half to four times cut-off. Plate cur-

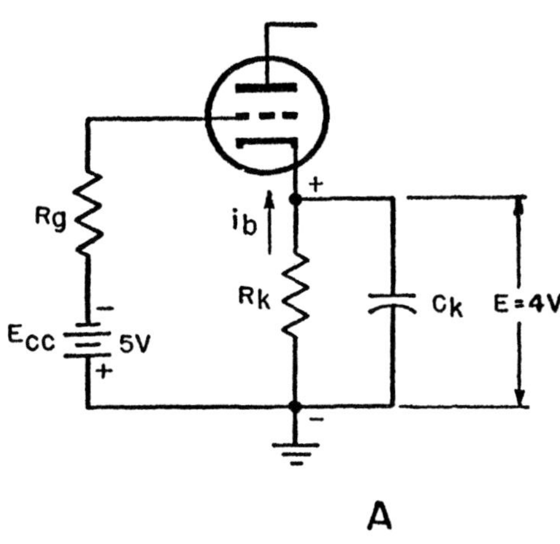
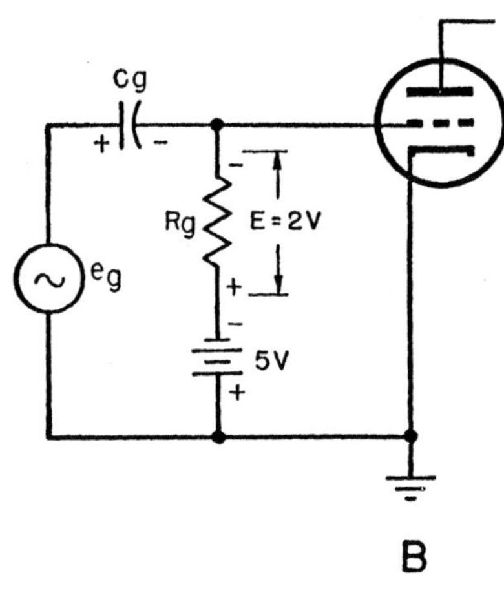

Figure 104. Circuits showing combination bias.

86. Summary

a. Gain is the ratio of output to input. Usually the greater the number of stages in an amplifier, the greater is its gain.

b. An a-f circuit amplifies frequencies in the audio range. An r-f circuit amplifies frequencies above the audio range.

c. The amount of distortion in an amplifier depends partially on the linearity of its dynamic characteristic. The more linear the characteristic, the less is the distortion.

d. A class A amplifier is one in which plate current flows all the time. The input grid signal operates along the most linear portion of the dynamic characteristic.

rent flows for appreciably less than a half-cycle.

i. The approximate plate efficiencies of the various classes of amplifiers are: class A, under 20 percent; class AB, 40 percent; class B, 40 to 60 percent; and class C, 60 to 80 percent.

j. A *push-pull* amplifier consists of two tubes arranged so that the plate current of one tube is 180° out of phase with the plate current of the other tube. The magnitudes of the currents are equal.

k. The input grid signals for a push-pull amplifier are obtained from a transfomer which has a center-tapped secondary or from a paraphase amplifier.

l. Compared to single-tube operation, less distortion is obtained in push-pull operation since its dynamic characteristic is more linear. Also, a greater grid signal swing is permissible.

m. In an electron-tube circuit, e_g, e_c, i_b, and e_{RL} are in phase with each other, but are 180° out of phase with e_b.

n. Fixed bias is obtained from a separate voltage source. Self-bias is obtained from a voltage drop across a resistor. This voltage drop is caused by the current of the tube. Combination bias is a combination of fixed bias and self-bias.

o. Grid-leak bias is obtained through the action of the input grid signal.

87. Review Questions

a. What is the primary function of an electron tube?

b. Does an i-f amplifier operate in the a-f or r-f range?

c. What type of distortion results in an amplifier when the positive peaks of the input grid signal cause the grid to draw current?

d. Where are the operating points of class A, AB, B, and C amplifiers?

e. Define plate efficiency.

f. Compare the plate efficiencies of class A, AB, B, and C amplifiers.

g. Compare the relative distortion of the different classes of amplifiers.

h. How can a push-pull amplifier circuit be identified?

i. What methods are used to obtain two grid signals 180° out of phase with each other for a push-pull amplifier?

j. Explain how a push-pull dynamic characteristic is obtained.

k. Compare the phase relationships of e_g, e_c, e_b, e_{RL}, and i_b in an electron-tube circuit.

l. What are the differences between fixed bias, self-bias, and combination bias?

m. Can grid-leak bias be obtained with no input grid signal? Explain.

n. Explain how the bias voltage for contact potential bias is obtained.

CHAPTER 7

AMPLIFIER GAIN AND COUPLING

88. Introduction

a. There are two principal types of amplifiers —*voltage amplifiers* and *power amplifiers*. A voltage amplifier produces a large voltage with a small current at its output. A power amplifier usually produces a large current with a small voltage at its output. The load resistor for a voltage amplifier is large—10,000 ohms or more —compared with the plate resistance of the tube. The larger the value of load resistance, the larger is the voltage drop across it, and the less is the current flow through it. The amplitude of the input grid signal is relatively small so that a large voltage and a small current are obtained at the output.

b. Power amplifiers often are referred to as *current* amplifiers. In a power amplifier, the value of load resistance usually is much smaller than in a voltage amplifier—under 10,000 ohms —so that its plate current is comparatively large. The relatively small value of load resistance produces a small voltage drop across it. The amplitude of the input grid signal to a power amplifier is made relatively large so that its output plate current is large.

c. Schematically, there is no way of distinguishing between voltage and power amplifiers except, perhaps, by the values of their loads. In practical applications, a voltage amplifier usually feeds its output to a power amplifier. For example, consider a small grid signal that is amplified by a voltage amplifier. The amplitude of output voltage is comparatively large. This output is fed or *coupled* to a power amplifier. Its large power output can be used to feed another power amplifier so that a greater power amplification is obtained, or it can be used to feed a load such as a transmitting antenna or a loudspeaker.

d. When the output of one stage is coupled to the input of another stage, a *coupling network* must be used for transferring this energy. The amplification, or gain, of a stage is dependent on several factors. These include the tube constants, μ, g_m, and r_p, the d-c operating potentials, E_{cc}, E_{bb}, and E_{c2}, the input grid signal, e_g, and the load, R_L. The characteristics of the coupling network also determine the amount of amplification that results.

89. Resistance-Capacitance Coupling

a. INTRODUCTION. Various coupling networks are used to transfer energy from one circuit to another. These arrangements consist of capacitors, resistors, inductors, and transformers. One of the most common types of coupling networks is the R-C (resistance-capacitance) network, usually known as resistance coupling.

b. R-C COUPLING NETWORK.

(1) Two triode amplifiers coupled by means of an R-C coupling circuit are shown in figure 105. The use of plate-load resistor R_L has been discussed in previous chapters. When a varying signal voltage, e_g, is applied to the grid of V1, it causes the plate current to vary through the tube and through R_L. The changing current through the load resistor produces a varying voltage drop across it. The amplified output of tube V1 is the varying voltage between its plate and ground. This output voltage is numerically equal to the fixed plate-supply voltage minus the varying voltage across R_L.

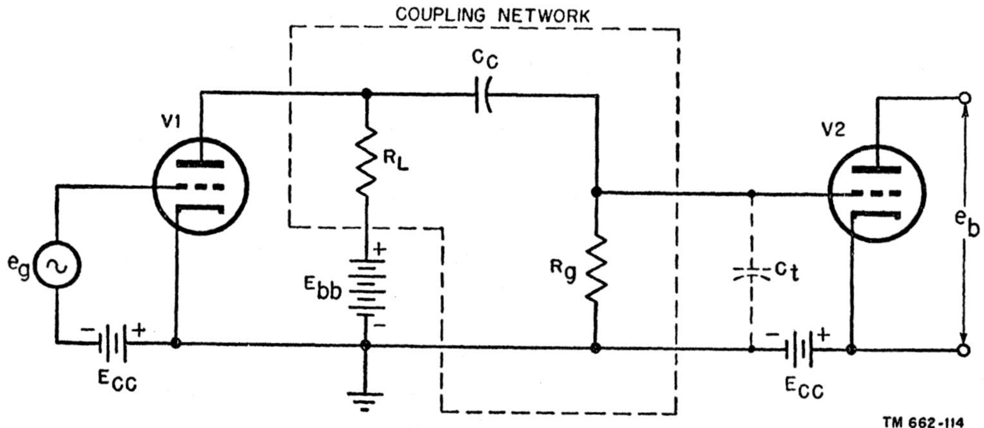

Figure 105. R-C coupling network between two triode amplifiers.

(2) It is desirable to make the resistance of R_L as large as possible. As the value of this resistor is increased, a larger signal voltage appears across it. As a result, the size of the output voltage from tube V1 is increased and the stage is said to have a larger amplification. There is a practical limit to the value of the plate-load resistor. If this resistor is made too large, it produces a large d-c voltage drop. This reduces the plate voltage on the tube and the resultant plate current so that the tube output is reduced. Typical values for the plate-load resistor range from about 25,000 to about 500,000 ohms.

(3) The output signal of tube V1 is coupled through coupling capacitor C_c to the grid of tube V2. This capacitor is necessary to prevent the high positive plate voltage of V1 from being applied to the grid of V2. If a direct connection were used instead of the capacitor, a positive bias would be applied to the grid of V2. As a result, this tube would not be biased properly, grid current would flow, and considerable distortion would be produced. Because this capacitor blocks or prevents the passage of dc, it sometimes is referred to as a *blocking capacitor*. This coupling or blocking capacitor also must pass the varying signal voltage readily. Consequently, the reactance

of C_c should be low. Therefore, the capacitance of the coupling capacitor is made fairly large. Typical values for the coupling capacitor range from about .001 to approximately .01 μf. Larger values usually are not used, because of their large physical size and their excessive stray capacitance to ground.

(4) The a-c signal coupled through C_c is applied to grid resistor R_g of tube V2. The signal voltage drop across R_g is applied in series with the bias voltage E_{cc} between the grid and cathode of V2. This resistor has several other uses besides acting as a load across which the input signal to V2 is developed. R_g also provides a *grid return* for the grid of V2; that is, it serves to connect the grid of the tube to its source of bias voltage—in this case, a battery. In this way, the grid is not allowed to *float*. The resistor also provides a discharge path for the coupling capacitor and prevents the improper accumulation of charge. Because this resistor provides a path through which electrons can leak off C_c, it often is referred to as a grid-leak resistor. R_g can be used also as a source of grid-leak bias or contact potential bias. Typical values for the grid resistor range from one-half megohm to several megohms.

(5) Coupling capacitor C_c and grid resistor R_g form a voltage divider. The output of this voltage divider is the voltage drop across resistor R_g which is the signal voltage applied to the grid of V2. Because of the infinite reactance of the capacitor to dc, all of the d-c voltage drop appears across it and no dc from the plate of V1 is applied to the grid of V2. The reactance of the capacitor to an a-c signal, however, is much smaller than the resistance of the resistor. Consequently, very little of the a-c signal is lost across the capacitor, and most of the signal appears across the resistor and is applied to the grid of tube V2. C_t represents the total shunt capacitance of the circuit. It is a stray capacitance made up of the interelectrode capacitances of the tubes and wiring capacitance.

c. R-C COUPLING USING PENTODES. R-C coupling is used also for pentode circuits. A typical R-C coupled amplifier of this type is shown in figure 106. The input signal, e_{in}, to V1 is amplified by V1. The output of V1 is coupled to V2 through the R-C coupling network, shown enclosed by dotted lines. The output of the amplifier, e_{out}, is the voltage appearing from plate to ground of V2. V1 and V2 use cathode bias which is obtained by the action of R_K and C_K. The letter B represents the plate-supply voltage source. The screen grids of both tubes obtain their d-c operating voltages through the series dropping resistors, R_{sg}. Capacitor C_{sg} *bypasses* any a-c signal voltage that appears in the screen-grid circuit, thereby preventing it from causing a fluctuation in screen voltage. The value of C_{sg} is selected so that it presents a minimum of reactance to the signal voltage.

90. Gain of R-C Coupled Amplifier

a. RESPONSE CHARACTERISTIC.

(1) An amplifier usually is rated by the amount of gain it has. However, the gain of an amplifier may be greater at one frequency than another. Some method must be used, therefore, to designate its gain over a band of frequencies. This can be done by plotting a voltage gain characteristic curve, sometimes called a *response characteristic* curve since it signifies the response of an amplifier over a wide range of frequencies (fig. 107). Here the voltage gain characteristic is that of a typical R-C coupled a-f amplifier. Frequency (X-axis) is plotted against relative voltage gain (Y-axis). The

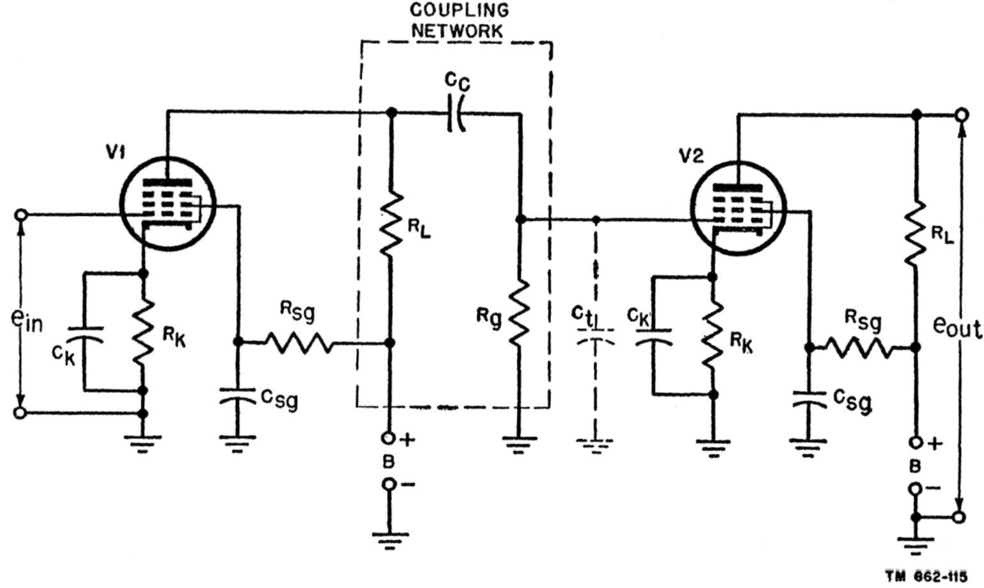

Figure 106. Typical R-C coupled amplifier using pentodes.

range of frequencies shown extends throughout and beyond the audio-frequency band, which is from about 16 to 16,000 cps. The dotted lines on the curve separate the relative frequency ranges into low frequencies, middle frequencies, and high frequencies. The voltage gain is maximum at the middle frequencies. The coupling capacitor C_c is designed to have a negligible reactance at the middle frequency of about 1,000 cps. Since C_c and R_g form a voltage divider and very little signal voltage is lost across C_c, most of the signal to V2 appears across R_g. Therefore, V2 produces a maximum voltage amplification. The reactance of C_c changes negligibly and V2 still provides a maximum voltage gain.

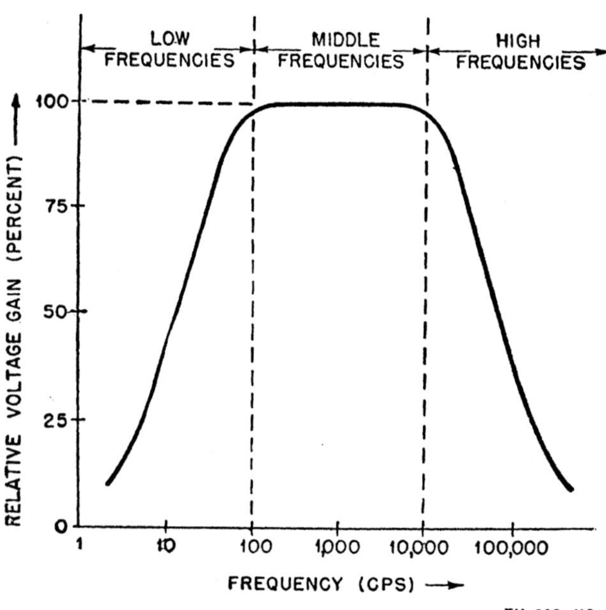

Figure 107. Voltage gain characteristic of typical R-C coupled a-f amplifier.

(2) The curve starts tapering off at about 100 cps to a minimum value at the lowest frequency. As the frequency decreases, the reactance of C_c increases. This causes more of the signal from V1 to appear across C_c and not across R_g. Consequently, the volt-

age gain of the amplifier decreases at low frequencies. At the higher frequencies, the reactance of the shunt capacitance, C_t, from grid to ground of V2 becomes very small. The smaller reactance causes the impedance from grid to ground of V2 to become smaller. As a result, a smaller grid signal voltage appears at the grid of the tube and a decrease in voltage gain occurs at the higher frequencies. Therefore, it is seen that the R-C coupling network helps to determine the voltage gain at various frequencies.

b. EQUIVALENT CIRCUIT OF R-C COUPLED AMPLIFIER.

(1) The effects at different frequency ranges can be explained by equivalent circuits. Figure 108 shows the equivalent circuit of an R-C coupled amplifier for the low-middle and high-frequency ranges. The equivalent circuit for the low-frequency range is shown in A. The output signal voltage for the amplifier is represented by an a-c generator designated as $-\mu e_g$. The minus sign preceding μe_g is used to indicate that the output voltage is of opposite polarity to the input. Resistor r_p represents the a-c plate resistance of the tube, R_L is the load resistance, and R_g is the grid resistor. Coupling capacitor C_c is included in the equivalent low-frequency circuit since it has an appreciable reactance at low frequencies. R_g and C_c form a voltage divider which delivers the output voltage, e_{out}. The drop across R_g is the output voltage of the equivalent circuit. The lower the frequency of e_g, the higher is the voltage drop across C_c and the lower is the voltage drop across R_g.

(2) In the middle-frequency range, shown in B, the reactance of C_c becomes neg-

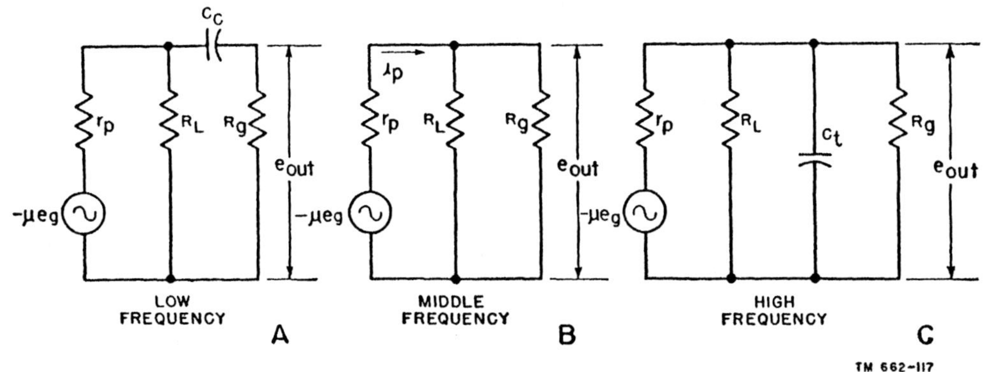

Figure 108. Equivalent circuit of R-C coupled amplifier.

ligible and is not included in the equivalent circuit. At the high-frequency range, in C, capacitor C_t (interelectrode and wiring capacitance) is included in the equivalent circuit. As the signal frequency increases, the reactance of C_t decreases, and more of the output voltage is short-circuited or bypassed to ground.

c. TRIODE AMPLIFIER GAIN.

(1) The mathematical relationship for gain of a triode amplifier at the middle frequencies can be determined from the equivalent circuit in B. R_g and R_L are in parallel with each other. Since the resistance of R_g usually is much greater than that of R_L, their equivalent resistance is approximately equal to R_L alone. The a-c plate current, i_p, flows from the a-c generator, $-\mu e_g$, through the two series resistors, r_p and R_L.

(a) Using Ohm's law,

$$i_p = \frac{\mu e_g}{r_p + R_L}.$$

(b) The minus sign preceding μe_g does not appear, since we are not concerned with phase relationships in the equation. The output voltage, e_{out}, appearing across R_L is—

$$e_{out} = i_p \times R_L.$$

(c) Substituting the value of i_p from equation (a) into equation (b), the output voltage is—

$$e_{out} = \frac{\mu\, e_g}{r_p + R_L} \times R_L = \frac{\mu e_g R_L}{r_p + R_L}.$$

(d) Voltage gain, A, is defined as the ratio of output voltage to input voltage—

$$A = \frac{e_{out}}{e_g}.$$

(e) Substituting the value of e_{out} from equation (c) into equation (d), voltage gain is—

$$A = \frac{\dfrac{\mu e_g R_L}{r_p + R_L}}{e_g} = \frac{R_L}{r_p + R_L}\mu .$$

The relationship of equation (e) is the voltage gain of an R-C coupled amplifier in the middle-frequency range.

(2) Consider the following numerical example: A 6J5 triode has a plate resistance of 7,700 ohms and a μ of 20. This tube is used as an R-C coupled amplifier, in which the value of R_L is 25,000 ohms. Its voltage gain is—

$$A = \frac{R_L}{r_p + R_L}\mu$$
$$= \frac{25,000}{7,700 + 25,000} \times 20$$
$$= \frac{500,000}{32,700}$$
$$A = 15.3 .$$

d. MATHEMATICAL RELATIONSHIP OF GAIN IN PENTODES.

(1) In pentodes or tetrodes, the voltage

gain at the middle-frequency range, A, is obtained in the following manner. It is known that—

(*a*)

$$\mu = g_m r_p.$$

(*b*) Substituting this for μ in equation in $c(1)(e)$ above, A is given as—

$$A = \frac{R_L}{r_p + R_L}\, g_m r_p = \frac{g_m r_p R_L}{r_p + R_L}.$$

(*c*) In tetrodes and pentodes, r_p is very much greater than R_L; therefore, the denominator of the fraction in the preceding equation, (*b*), becomes just r_p, or—

$$A = \frac{g_m r_p R_L}{r_p}.$$

(*d*) The values of r_p cancel in the preceding equation, (*c*), and the voltage gain at the middle-frequency range becomes simply—

$$A = g_m R_L.$$

(2) In a 6SK7 pentode, the transconductance equals 2,000 umhos. If a plate-load resistor of 25,000 ohms is used, then the voltage gain is

$$A = g_m R_L = 25,000 \times 2,000 \times 10^{-6} = 50.$$

91. R-C-L Coupling Network

a. GENERAL. If a coil is substituted for R_L in an R-C coupling network, an R-C-L (resistance-capacitance-inductance) coupling network is obtained. This coupling method is referred to also as *impedance coupling*. The R-C-L coupled amplifier using triodes in figure 109 is similar to the R-C coupling network in figure 105 except that a coil is substituted for the load resistor, R_L. This coil has a certain impedance value, Z_L, which is made up of the inductive reactance and the resistance of the coil. The input signal, e_{in}, to V1 is amplified by this stage and is coupled to V2 through the R-C-L coupling network shown inclosed by dotted lines. The output, e_{out}, of V2, appears from plate to ground of this stage.

b. CIRCUIT OPERATION.

(1) The operation of the R-C-L coupling network is similar to that of the R-C coupling network. Coil Z_L is so designed that its impedance is very high at the signal frequency. This results in a large output voltage being applied to the input of V2. Since the d-c resistance of the coil is very small, only a small d-c voltage drop occurs. This permits the use of a lower value of d-c plate-supply voltage.

(2) Z_L equals $\sqrt{R_L{}^2 + X_L{}^2}$, where R_L is the d-c resistance of the coil and X_L is the inductive reactance of the coil. The inductive reactance of the coil varies with the signal frequency. As the signal frequency increases, the inductive reactance increases, and as

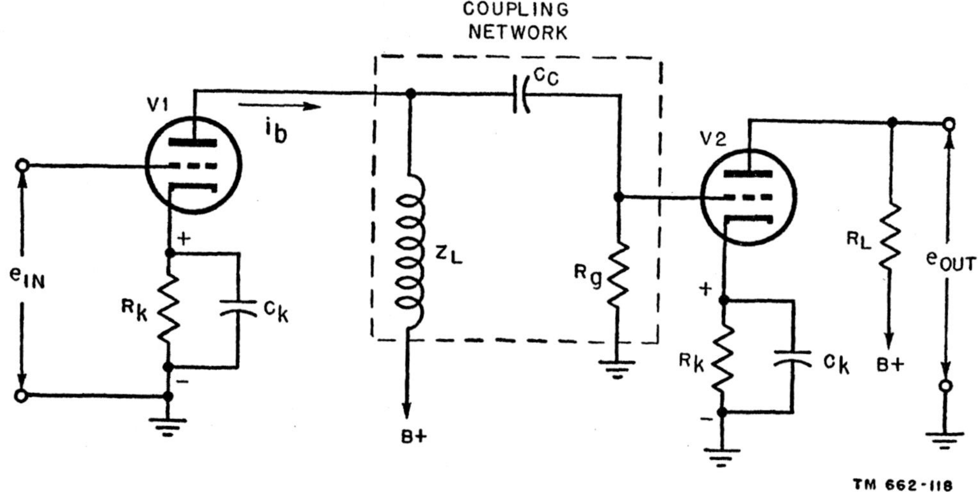

Figure 109. R-C-L coupled amplifier using triodes.

the signal frequency decreases, the inductive reactance decreases. Therefore, the impedance of the load rises with frequency.

(3) The gain of V1 depends on the voltage developed across Z_L. The greater this voltage the greater is the gain; the less the voltage the less the gain. At low frequencies, when the impedance of Z_L is small, the voltage developed across it is small and the gain is low. At high frequencies, when the inductive reactance of Z_L is large, the voltage developed across it is large and the gain is high.

(4) The same limiting factors to the gain at high frequencies exist in the interelectrode, wiring, and stray capacitances of the circuit. The distributed capacitance between the turns of the coil further limits the gain at high frequencies. The combination of this distributed capacitance and the inductance produces a parallel-resonant circuit. As a result, a peak in the gain may occur at the resonant frequency.

c. RESPONSE CURVE. Generally, it is possible to obtain a fairly uniform response characteristic from an R-C-L coupled amplifier over a limited frequency range only. The characteristic curve shows a higher gain than is obtained in an R-C coupled amplifier at the high frequencies and a lower gain at the low frequencies.

92. Transformer Coupling

a. CIRCUIT EXPLANATION.

(1) Another method of coupling is transformer coupling. In the typical transformer-coupled amplifier using triodes in figure 110, transformer T, which comprises the coupling network, is shown within dotted lines. The primary, P, of the coupling transformer is connected in the plate circuit of triode V1 and the secondary, S, is connected in the grid circuit of V2.

(2) The input grid voltage to V1 is amplified and varies the plate current, i_p, flowing through the primary of the transformer. The varying plate current causes an induced voltage, e_s, to appear in the secondary. Transformer T has a one-to-three step-up turns ratio. With negligible transformer losses and unity coupling, the secondary voltage is three times as great as the primary voltage. The secondary voltage is amplified by V2 and appears as e_{out} between the plate and ground. Figure 110 shows the phase relationships of the various voltages and their relative amplitudes.

b. EQUIVALENT CIRCUITS.

(1) The equivalent circuit of a transformer-coupled amplifier at the low-, middle-, and high-frequency ranges is shown in figure 111. In the low-frequency range (A of fig. 111) $—\mu e_g$ is

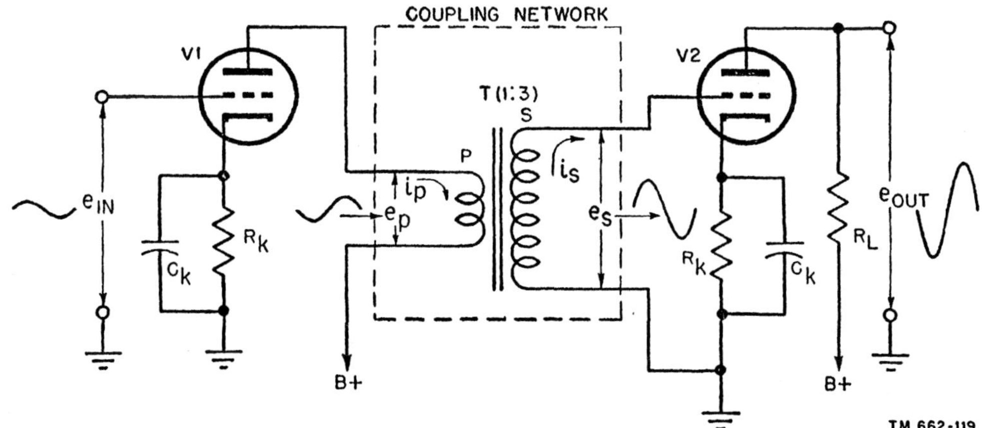

Figure 110. Transformer-coupled amplifier using triodes.

an a-c generator which represents the amplified output of V1, and r_p represents the a-c plate resistance of the tube. Coil L_p represents the inductance of the primary of the transformer. The d-c resistance of the primary winding, which is usually very small, can be considered as included in r_p. L_p and r_p form a voltage divider and the output voltage is taken across L_p. At very low signal frequencies, the reactance of L_p becomes small and most of the signal voltage appears across r_p, decreasing the output voltage. The magnitude of the output voltage is dependent also on the turns ratio of the transformer. The greater the turns ratio of the step-up transformer, the greater is the output voltage. In figure 110 the turns ratio is one to three and the secondary voltage, e_s, is three times as great as the primary voltage, e_p.

e_g, and the turns ratio of coupling transformer N.

(3) In C of figure 111 the equivalent circuit in the high-frequency range, C_t represents the distributed capacitance between the windings of the transformer in addition to the interelectrode, stray, and wiring capacitances. Coil L_e is the equivalent inductance of the transformer. The size of L_e depends on the leakage flux and the amount of mutual coupling between the windings. The equivalent circuit takes the form of a series-resonant circuit which resonates at some high frequency of signal voltage. At or near resonance, the magnitude of the voltage across C_t is extremely large. If the frequency is increased beyond resonance, the reactance of C_t decreases and the output voltage decreases.

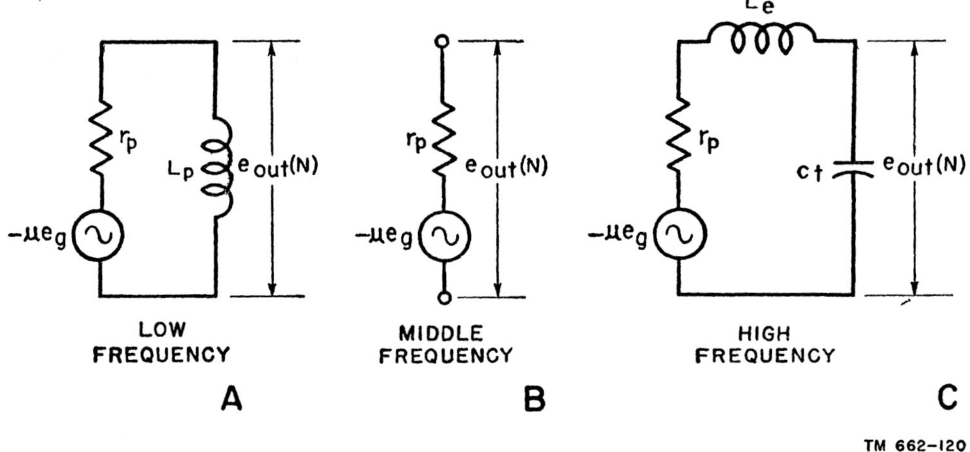

Figure 111. Equivalent circuits of transformer-coupled amplifier.

(2) In the equivalent circuit in the middle-frequency range (B of fig. 111) the reactance of L_p is large compared with r_p so that it can be considered an open circuit. The output voltage is taken across r_p and $-\mu e_g$ in series. In this circuit, the magnitude of the output voltage is dependent on the tube constants, the amplitude of signal voltage

c. RESPONSE CHARACTERISTIC.

(1) The response characteristic of a typical transformer-coupled a-f amplifier is shown in figure 112. In the middle-frequency range, the relative voltage gain is fairly constant. In the low-frequency range, the curve drops off as the frequency is decreased. This occurs since the reactance of L_p de-

creases with frequency. The result is a decrease in the output voltage. In the high-frequency range, the curve drops off as the frequency is increased because the reactance of C_t (C of fig. 111) decreases as the signal frequency increases. The result is a decrease in the output voltage.

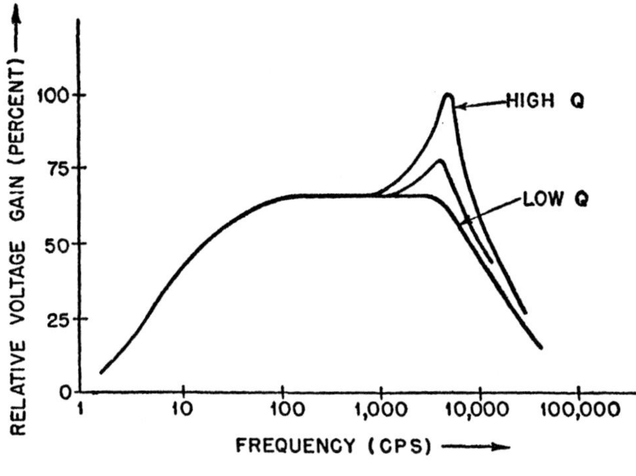

Figure 112. Voltage gain characteristic of typical trans- former-coupled a-f amplifier for different values of Q.

(2) Three curves are shown in the high-frequency range. They represent different values of Q (figure of merit) for the series-resonant circuit in C. Q is equal to the ratio of the reactance to the resistance. The actual Q is determined by the transformer used, the circuit constants, and the operation of the amplifier. A high Q usually is not desirable because of the peak in response which occurs.

(3) The frequency response of an R-C coupled amplifier usually is more uniform over a wider frequency range than the frequency response of a transformer-coupled amplifier. However, the magnitude of the relative voltage gain of a transformer-coupled amplifier usually is greater.

d. MATHEMATICAL RELATIONSHIP OF VOLTAGE GAIN. The relationship of voltage gain in the middle-frequency range, in A, is—

$$A = \mu N,$$

where μ is the amplification factor of the tube and N is the turns ratio of the transformer. As an example, consider a 6J5 triode ($\mu = 20$) that is used as the first stage in a transformer-coupled amplifier. If the number of turns in the primary of the coupling transformer is 100 and in the secondary is 200, the turns ratio is 200/100 or 2. The voltage gain is—

$$A = \mu N = 20 \times 20 = 40.$$

It is desirable to make N as large as possible to obtain a large voltage gain. However, increasing N has its limitations. If N is made too large, then the equivalent inductance, L_e, becomes too large and the frequency response falls off at the high frequencies.

e. CHARACTERISTICS.

(1) Transformer coupling has several distinct advantages as compared with R-C and R-C-L coupling. One advantage is that a greater gain is obtained in transformer coupling because a step-up transformer is used. Another advantage is that a lower value of d-c plate voltage can be used, since the d-c resistance of the primary winding is small. Also, the secondary winding of the transformer can be center-tapped and used to supply two grid voltages 180° out of phase to a push-pull amplifier. The impedance matching properties of a transformer, which was previously discussed, is a fourth advantage.

(2) The transformer-coupling system has several drawbacks. One disadvantage is the high cost of the transformer. Also, the stray fields introduced by the transformer can interfere with the operation of other stages, although this can be alleviated by shielding the transformer properly. In addition, a transformer usually is large and bulky and adds to the weight of the equipment.

93. Tuned-circuit Coupling

a. TUNED TRANSFORMER COUPLING.

(1) If capacitors are placed across the primary and secondary windings of

the transformer in a transformer-coupled network, a double-tuned transformer-coupling system is obtained (A of fig. 113). Coil L1 is the primary and coil L2 is the secondary of the transformer. C1 tunes L1 to resonance at the signal frequency. A large signal voltage is produced across the high impedance of the parallel-resonant circuit formed by $L1$ and $C1$. The large circulating tank current in the primary of the transformer creates a magnetic field which induces a voltage into the secondary winding, $L2$.

(3) Double-tuned transformer coupling is used where a single frequency or a narrow band of frequencies is to be amplified. The resonant conditions in the network results in a voltage gain characteristic that is very selective. I-f stages in a receiver and r-f output stages in a transmitter frequently use this type of coupling. In these stages, selectivity is of primary importance. In some practical applications, a manual adjustment changes the degree of coupling between the primary and secondary windings of the transformer. This adjustment per-

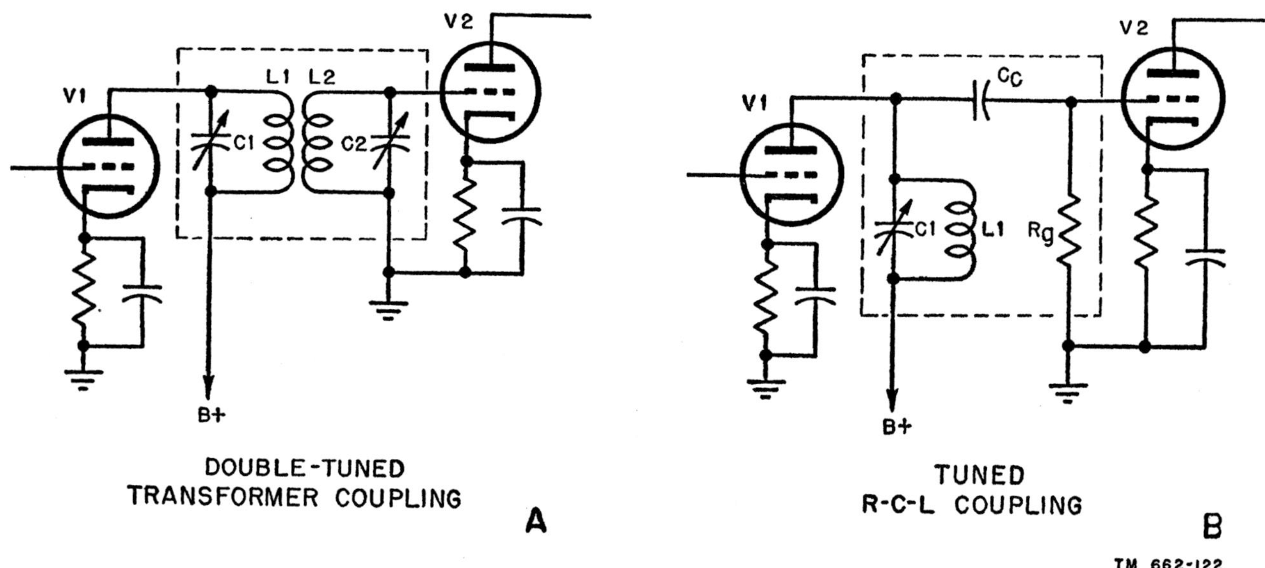

Figure 113. Tuned-circuit coupling system.

(2) The voltage applied to the secondary circuit, $L2C2$, by induction is considered to be in series with the components of this circuit. When the secondary circuit is tuned to resonance, a large current flows which is in phase with the induced voltage. Both this current and the induced voltage are 180° out of phase with the primary voltage as in any transformer. The large secondary current produces a large reactive voltage drop across $C2$, which is applied to the grid of V2. This reactive voltage lags the secondary current by 90°.

mits a maximum transfer of energy with specified amounts of selectivity. Occasionally, a double-tuned transformer may be overcoupled to produce a widespread characteristic.

(4) Sometimes *single-tuned* transformer coupling is used. That is, either the primary or the secondary contains a tunable capacitor. This type of coupling sometimes is used where a wider band of frequencies is to be amplified. Also, this method is used when it is desired to reduce the number of variable capacitors used.

b. Tuned R-C-L Coupling. If a variable capacitor is placed across the coil in an R-C-L coupling network, it becomes a tuned R-C-L coupling network (B of fig. 113), called also a tuned impedance coupling. C1 tunes L1 to resonance. At resonance, a large voltage appearing across the parallel-resonant circuit is transferred to the grid of V2 through coupling capacitor C_c. C_c has a small reactance at the resonant frequency and permits a maximum transfer of energy. All other circuit elements function similarly to previous circuits.

94. Direct Coupling

a. Introduction. Another type of coupling is direct coupling. A direct-coupled amplifier is one that can amplify dc and very-low-frequency voltages and currents. Its distinguishing feature is that the plate of one stage is coupled *directly* to the grid of the following stage without the use of a capacitor or a transformer. In figure 114, the plate of V1 is coupled directly to the grid of V2. In previous coupling networks, a capacitor or a transformer was used to block the d-c plate voltage of the first stage from appearing at the grid of the second stage. This was done to prevent the second stage from being improperly biased and from conducting too heavily.

b. Circuit Operation.

(1) For convenience, a voltage divider commonly is used to supply the necessary d-c operating voltages for the amplifier. The B-plus supply voltage is impressed across voltage-divider resistor R_d which is tapped at various points. Capacitor C_d is used to bypass any a-c variations of voltage that may appear across R_d. The input voltage, e_{in}, is amplified by V1 and V2 and appears as e_{out} at the plate of V2. The plate-load resistor of V1, R_{L1}, also acts as the grid resistor of V2, since the voltage developed across it appears at the grid of V2. R_{L2} is the plate-load resistor of V2.

(2) The cathode of V1 is connected to point A and the plate is connected to point B on the voltage divider. Point B is positive with relation to point A, making the plate positive with relation to its cathode. This permits V1 to conduct. The voltage developed from point A to ground is the bias voltage for V1.

(3) The plate of V2 must be positive with relation to its cathode for conduction to occur. In addition, the grid voltage of V2 must not be positive with relation to its cathode. The plate current of V1 which flows through R_{L1} produces a considerable d-c voltage drop across this resistor. As a result, the voltage at the plate of V1 and at the grid of V2 is less positive than is point

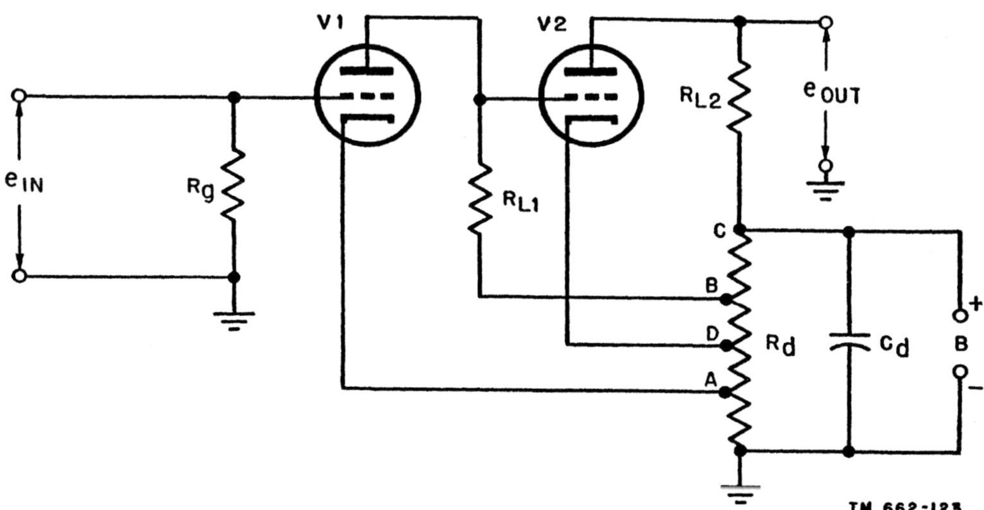

Figure 114. Typical direct-coupled amplifier using triodes.

B on the voltage divider. Tap D is so located on resistor R_d that the magnitude of the positive voltage on the grid of V2 is lower than the magnitude of the positive voltage on the cathode of V2. Therefore, the grid of V2 is actually less positive, or negative, with relation to the cathode of V2 and the tube operates normally. The voltage that appears from point C to point D is the plate voltage of V2.

c. USES. A direct-coupled amplifier is used to amplify d-c voltages and currents or very low frequencies. Since it does not contain any reactive elements, it has a fairly uniform response over a wide range. The absence of reactive elements also permits an instantaneous response to slowly varying grid signals. D-c amplifiers commonly are used when a large current is needed to operate such devices as d-c indicating instruments or relays.

95. Gain by Measurement

a. If the input and output voltages of a stage are measured, its voltage gain can be computed. For example, if the a-c input voltage to a stage is 5 volts and its a-c output voltage is 100 volts, the voltage gain is E_{out}/E_{in} minus 100/5 or 20. Similarly, if the input and output powers of a power amplifier are known, its power gain can be computed. Consider a power amplifier whose input voltage and current are 5 volts and 50 milliamperes (effective values), respectively. Its input power is E times I = 5 times .05, or .25 watt. If the signal voltage across the load is 20 volts and the current through it is 150 milliamperes, its output power is 20 times .150, or 3 watts. Its power gain is 3/.25, or 12.

b. A more common method of measuring power gain is by means of logarithmic ratios. In the common system of logarithms in which the base is 10, the logarithmic ratio of output power to input power is the unit of power gain, call a *bel.* Power gain, in bels, equals—

$$log\frac{P2}{P1}$$

where $P2$ is the output power and $P1$ is the input power. In practice, it has been found that the bel is too large a unit for the power ratios

ordinarily encountered. Therefore, one-tenth of a bel or the db (*decibel*) is the commonly used unit of gain used. The equation for gain in decibels is—

$$gain\ (decibels) = 10\ log\frac{P2}{P1}.$$

The decibel is 10 times the common logarithm of the power ratio. The two powers in the equation must be given in the same units (kilowatts, watts, milliwatts, microwatts, etc.). If $P2$ is larger in magnitude than $P1$, the logarithm of their ratio is positive. This is referred to as a positive gain. If $P1$ is larger in magnitude than $P2$, their logarithmic ratio becomes negative and this is referred to as a negative gain or a loss.

c. The following example illustrates how the preceding formula is used to calculate the db power gain of one amplifier stage. If a power amplifier with an input power of 1 watt delivers 2 watts to a loudspeaker, its *db* power again is computed as follows:

$$gain\ (db) = 10\ log\frac{P2}{P1} = 10\ log\frac{2}{1} = 10\ log\ 2.$$

From logarithmic tables, the log of 2 is found to be .301. Therefore,

$$gain\ (db) = 10 \times .301 = +3.01\ db.$$

The positive sign preceding the numerical answer indicates that the amplifier produces a power gain. Suppose the input power to the amplifier is 2 watts and the output power is 1 watt. The power ratio is then inverted to avoid the use of negative logarithms and a minus sign is used to indicate a power loss. In this case,

$$gain\ (db) = 10\ log\frac{P1}{P2} = 10 \times .301 = -3.01\ db.$$

d. When it is necessary to calculate the *overall* gain or loss of an amplifier system using two or more stages, the db gains or db losses of the individual stages are added algebraically. Consider the *db* gain of two amplifier stages that are coupled together. The first stage has an input power of 1 watt and an output power of 3 watts. The second stage has an input power of 3 watts and an output power of 6 watts. The *db* gain of the first stage is

$$gain\ (db) = 10\ log\frac{P2}{P1} = 10\ log\frac{3}{1} = 10 \times .477 = 4.77\ db.$$

The db gain of the second stage is

$$gain\ (db) = 10\ log\ \frac{6}{3} = 10\ log\ 2 = 10 \times .301 = 3.01\ db.$$

The over-all gain of the two amplifier stages is then 4.77 db plus 3.01 db or 7.78 *db*.

e. Although the decibel is founded on power ratios, voltage or current ratios can be used also, but only when the impedance is the same for both values of voltage or current. The gain equations then are—

$$gain\ (db) = 20\ log\ \frac{I2}{I1}\ and$$
$$gain\ (db) = 20\ log\ \frac{E2}{E1}.$$

f. The foregoing discussion was concerned with the application of the decibel to increases or decreases in power. Another significant use of the decibel is in power-level problems. In this case, the power that exists in a circuit is expressed as being so many db above or below a standard or reference level. However, the level in most common use in radio practice is a power of 1 milliwatt. Using this level as a standard, the formula can be rewritten as

$$power\ level\ (db) = 10\ log\ \frac{P2}{P_{level}}$$
$$= 10\ log\ \frac{P2}{.001}$$

where P2 is the output power in watts. For example, if a power amplifier delivers .01 watt to its output, its power output level is

$$db = 10\ log\ \frac{.01}{.001}$$
$$db = 10\ log\ 10$$
Since the $\quad log\ 10\ is\ 1$
$$db = 10.$$

The answer usually is written as 10 *db*/.001 watt or 10 *dbm* to indicate that the power amplifier delivers a power level of 10 *db* above the arbitrary reference level of .001 watt. The term dbm indicates a reference level of 1 milliwatt. It is important to note that the reference level used above is an arbitrary choice by the radio industry. The telephone engineer frequently uses a level of 6 milliwatts.

96. Distortion

a. TYPES OF DISTORTION. The output waveform of an ideal distortionless amplifier should be an exact replica of the input waveform. Therefore, distortion may be defined as a change in the waveform. However, all amplifiers introduce some distortion and it is the problem of the design engineer to keep this distortion to a minimum. The types of distortion found in vacuum-tube circuits classified according to their cause are *nonlinear, frequency,* and *phase*. Nonlinear (sometimes called amplitude) distortion was touched upon in the preceding chapter.

b. NONLINEAR DISTORTION. If the output of an amplifier contains frequencies that were not present in the input, nonlinear distortion has been introduced. Nonlinear distortion is caused in an amplifier when the input grid signal operates along the nonlinear portion of the dynamic characteristic. If a sinusoidal grid signal operates along the nonlinear portion of the characteristic, the output waveform will contain unwanted harmonics (new frequency components) generated by the nonlinear action of the tube. In A, figure 115, it is assumed that a waveform is fed to a two-stage amplifying device and an unwanted second harmonic is generated. The input waveform is the fundamental, and the distorted waveform caused by the addition of the undesired second harmonic is shown as the resultant. Distortion caused by the tube can be minimized by proper design to operate along the straight line portion of the transfer characteristic. Another source of nonlinear distortion results from the hysteresis effect in coils and transformers with iron cores.

c. FREQUENCY DISTORTION. Frequency distortion results when certain frequencies are amplified more than others; that is, frequency distortion exists in amplifiers when the voltage gain at one frequency is more or less than the gain at some other frequency. Frequency distortion generally is due to the inductive and capacitive elements in a circuit, because their impedance varies with frequency. In B of figure 115 a complex waveform composed of a fundamental and a third harmonic is passed through a two-stage amplifier. After passing through the stage which introduces frequency distortion only the fundamental has been amplified and the third harmonic component does not appear in the output. In some applications, particularly video amplifiers, it is important

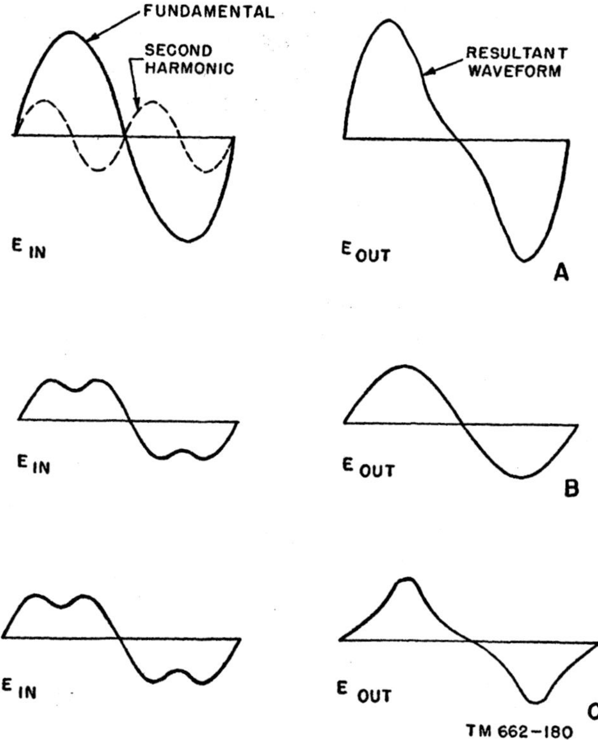

*Figure 115. Effects of distortion, A. Nonlinear,
B. Frequency, C Phase.*

that all frequencies be amplified in correct proportion. Compensation schemes are used to minimize frequency distortion in video amplifier applications.

d. PHASE DISTORTION. Phase distortion is a major problem when complex waveforms must be passed through an amplifier system. A good example would be a television signal which is amplified by a video amplifier. Such a signal may be resolved into sinusoidal components with finite amplitudes and phase displacements. If one frequency component of the complex input waveform takes a longer time to pass through the circuit than another, a time displacement or time delay has occurred, and the output will not be identical to the input. In C of figure 115 an input signal consisting of a fundamental and a third harmonic is passed through a two-stage amplifier. Although the amplitudes of both components are increased by identical ratios, the output waveshape is considerably different from the input because the phase of the third harmonic has been shifted in respect to the fundamental. Phase distortion arises because of the presence of reactance in

the tube circuit, and principally the coupling network between the stages of the amplifier. Special coupling circuits used in video amplifiers minimize phase distortion. In audio applications, phase distortion is not serious because the human ear cannot detect time-delay variations within the audio pass band.

e. OTHER TYPES OF DISTORTION. Other factors limiting amplifier performance include hum, microphonics, and noise. The frequency of hum in most amplifiers is 60 cps. It is introduced into amplifiers by stray fields, tube heaters, and inadequately filtered power supplies. Proper shielding, good ground connections, and short connecting leads reduce the possibility of hum. Microphonics result if the electrodes within a tube vibrate. The effects of tube vibrations can be reduced greatly by mounting a microphonic tube on springs or rubber cushions or by otherwise shielding the tube from mechanical vibrations.

f. RANDOM FREQUENCY SIGNALS. Noises in amplifiers are spurious random frequency signals which are amplified along with the desired signals. A common type of noise is thermal agitation, produced by random movements of electrons in a material. This produces minute pulses of current. The noises that are produced contain energy in the entire frequency band, and limit the lowest amplitude of signal voltage that can be amplified. Another source of noise is shot effect, produced by the irregular plate current which results from the irregular motion of electrons moving from the cathode to the plate of a tube.

97. Feedback in Amplifiers

a. In special amplifier applications, it is desirable to feed back a signal from the output stage to the input of the same stage or to a preceding stage. This *feedback* signal can take either of two forms, shown in A and B of figure 116, where the voltage amplitudes are plotted against time. The principle of feedback is shown in the block diagram of figure 117. The phase of the signal that is fed back (with reference to the input signal) determines the type of feedback that results. If the feedback aids the original input signal by increasing its amplitude, it is called *regenerative* or *positive* feedback. If it opposes the original input signal and decreases its amplitude, it is called *degenerative, negative,* or *inverse* feedback.

b. To produce regenerative feedback, the original and feedback signals must be in phase with each other, as in A of figure 116. Adding these two waveshapes produces the regenerative signal, which is larger in amplitude than the original signal. In B, the original and feedback signals are opposite in phase, and degenerative feedback is produced. The waveform is smaller in amplitude than the original signal. Circuitry involving regeneration and degeneration is discussed in another manual of this series.

c. In positive feedback, the voltage output of an amplifier is increased because the effective input voltage is increased. The greater amplification also usually increases the amount of distortion and noise in the amplifier. Sometimes the amount of regenerative feedback is so great that it produces sustained oscillations. In negative feedback, the voltage gain of an amplifier is decreased because the effective input voltage is decreased. In practical applications, this type of feedback is used to reduce the effects of distortion. Degenerative feedback also improves the frequency response and stability of amplifiers.

98. Summary

a. A voltage amplifier produces a large voltage with a small current at its output. A power amplifier usually produces a large current with a small voltage at its output.

b. When the output of one stage is coupled to the input of another stage, a coupling network is used for transferring the energy.

c. One of the most common types of coupling networks is the R-C coupling network. The plate of one amplifier stage is connected to the grid of the following amplifier stage through a coupling capacitor. Resistors are used as the plate load and in the grid circuit.

d. A voltage gain or response characteristic curve designates the amount of voltage gain an amplifier has over a wide range of frequencies.

e. The response characteristic of an R-C coupled amplifier drops off at the low frequencies because of the high reactance presented by the coupling capacitor.

f. The response characteristic of an R-C coupled amplifier drops off at the high frequencies because of the low reactance presented by the interelectrode, stray, and wiring capacitances.

g. The mathematical relationship for the voltage gain of an R-C coupled amplifier using triodes in the middle-frequency range is $A = \mu R_L / r_p + R_L$.

h. In an R-C coupled amplifier, it is desirable for large output voltages that the ohmic value of the load resistor be very large compared to the a-c plate resistance of the tube.

i. The mathematical relationship for the voltage gain of an R-C coupled amplifier using pentodes in the middle-frequency range is $g_m R_L$.

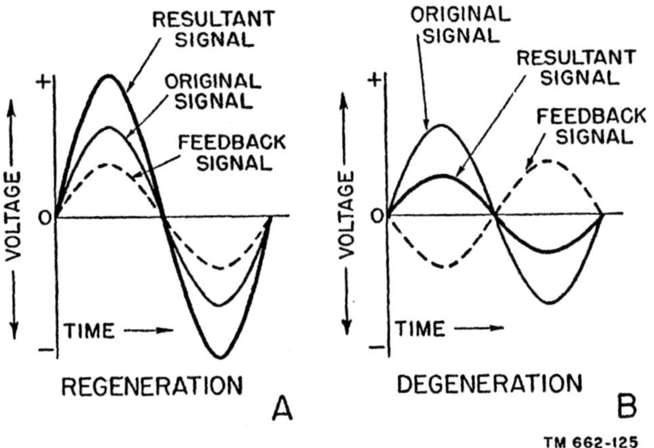

Figure 116. Phase relationships of signals producing regeneration and degeneration.

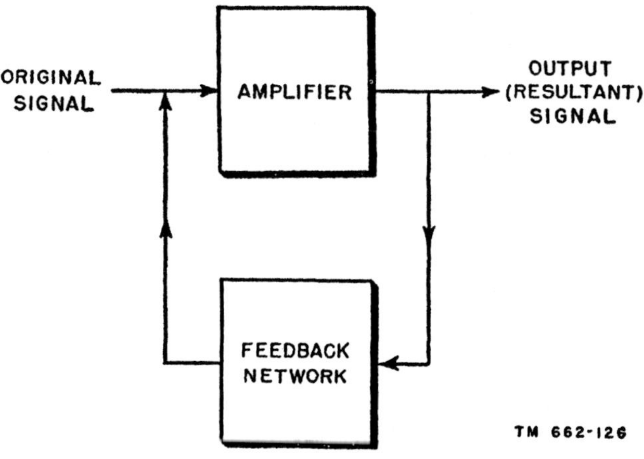

Figure 117. Block diagram of feedback amplifier.

j. If a coil is substituted for the load resistor in an R-C coupled amplifier, an R-C-L or impedance-coupled amplifier results.

k. In a transformer-coupled amplifier, the primary of the transformer is connected in the plate circuit of one tube and the secondary is connected in the grid circuit of the second tube.

l. The response curve of a transformer-coupled amplifier drops off at the low-frequency range because of the decreased reactance of the primary of the transformer. The response curve drops off at the high-frequency range because of the decreased reactances of the interelectrode, stray, and wiring capacitors. A peak in response may occur in the high-frequency range because of the formation of a parallel-resonant circuit.

m. The frequency-response characteristic of an R-C coupled amplifier is usually more uniform over a wider middle-frequency range than is a transformer-coupled amplifier.

n. In a transformer-coupled amplifier, the mathematical relationship of its voltage gain at the middle-frequency range is $A = \mu N$.

o. Transformer coupling can be used in both a-f and r-f amplifiers. In a-f amplifiers, an iron-core transformer is used; in r-f amplifiers, an air-core transformer is used.

p. In a double-tuned transformer-coupled amplifier, the secondary voltage is 90° out of phase with the primary voltage.

q. If a variable capacitor is placed across the coil in an R-C-L coupled network, it becomes a tuned R-C-L coupling network.

r. A direct-coupled amplifier is one that can be used to amplify d-c and/or very-low-frequency voltages and currents.

s. The response characteristic of a direct-coupled amplifier is fairly uniform over a wide range, since it does not contain any reactive elements.

t. A *decibel* is a unit used to express a logarithmic ratio of gain. The formula for power gain (in decibels) is $10 \log P2/P1$.

u. When calculating the over-all db gain or loss of several amplifier stages, the db gains or losses of the individual stages are added algebraically.

v. If the output waveshape of an amplifier is identical with its input waveshape, the amplifier is said to be distortionless. Types of distortion include amplitude distortion, frequency distortion, phase distortion, and harmonic distortion.

w. If the feedback signal in an amplifier aids the original input signal, the feedback is called *regenerative* or *positive*. If the feedback signal opposes the original input signal, it is called *degenerative*, *negative*, or *inverse*.

99. Review Questions

a. What are the fundamental differences between voltage and power amplifiers?

b. What is meant by a coupling network?

c. Name the factors that determine the amplification or gain of an amplifier.

d. Describe the circuity of an R-C coupled amplifier.

e. What is the function of a d-c blocking capacitor?

f. Relatively speaking, what is the capacitive reactance of a bypass capacitor?

g. What is a response characteristic curve?

h. What does $-\mu e_g$ represent in the equivalent circuit of an amplifier?

i. Explain why the voltage gain of an amplifier decreases at the low- and high-frequency ranges.

j. What is the mathematical relationship for the voltage gain of an R-C coupled amplifier?

k. What are the advantages obtained by substituting a coil for the load resistor in an R-C coupled amplifier?

l. What causes the response curve of a transformer-coupled amplifier to drop off at the low- and high-frequency ranges?

m. In a transformer-coupled amplifier, how does the value of Q of the transformer help determine its gain at the high-frequency range?

n. Which response characteristic is more uniform over a wider frequency range, an R-C-coupled or a transformer-coupled amplifier?

o. Give a few advantages and disadvantages of transformer coupling as compared to R-C and R-C-L coupling.

p. Explain how direct coupling is accomplished.

q. Why is the response characteristic of a d-c coupled amplifier fairly uniform over a wide range?

r. Define a decibel and give its mathematical relationship in terms of power.

s. What does the expression $+17$ *db/.001* watt indicate?

t. Name various types of distortion existing in amplifiers.

u. Explain the difference between regeneration and degeneration.

v. In a feedback amplifier, what phase relationship exists between the original signal and the feedback signal with regeneration and with degeneration?

Section I. POWER RECTIFICATION

100. General

a. In most of the circuits shown previously in this manual, the plate voltage was supplied by a battery. This d-c voltage was designated E_{bb}. Excepting in certain types of mobile equipment, batteries no longer are used for this purpose. They have been replaced by circuits which supply all the d-c voltages necessary to operate the equipment, including plate, screen, and bias voltages where required. A-c filament voltages also can be taken from the same circuit. A circuit supplying these voltages is called a *power supply*. The output of the power supply providing plate voltage is designated by the letter B, and the output is taken between B plus and B minus. On some schematics, only the B+ point is marked, with the B— point understood.

b. Alternating current can be sent over power lines more conveniently and efficiently than can direct current. Today, most power lines supply ac at frequencies from 25 to 240 cycles per second. Ac taken from the power line can be transformed to the dc necessary to operate electron-tube circuits. This transformation of a-c power to d-c power is called *power rectification*. The circuit which accomplishes the transformation is called a *power rectifier*.

101. Power Rectification

a. PROCESS. The term ac designates a current which flows first in one direction and then in the other. Current taken from a power line usually is sinusoidal in form. Its frequency depends on the characteristics of the a-c generator used. In power rectification, this sine wave of current must be changed to dc—that is, a current which flows steadily in only one direction. Consequently, a power-rectifier circuit has two functions. First, the current must be made to flow in one direction only; this function usually is designated by the term *rectification*. Second, this unidirectional current must be made to flow steadily; this function is called *filtering*.

b. RECTIFIER.

(1) A simple device for rectifying ac is a crystal. In figure 118 a crystal is inserted in an a-c line. Certain types of crystals, such as germanium, have the property of permitting current to flow through them in one direction but not in the other direction. The schematic symbol for the crystal in the figure indicates the direction of electron current flow by the arrowhead. When the a-c generator makes terminal A positive with relation to terminal B, current flows through R and the crystal in a counterclockwise direction as indicated. When the a-c generator reverses its polarity, current attempts to flow in the opposite (clockwise) direction. This current is blocked by the crystal. Consequently, current flows through R during the positive half-cycles only, and none flows during the negative half-cycle. The polarity of the voltage drop across R never changes.

(2) The current that flows through R is a fluctuating dc. The crystal, therefore, has rectified the ac supplied by the generator. The a-c voltage impressed across terminals A and B is compared with the resultant current through R in B of figure 118. The current consists of short surges occurring during

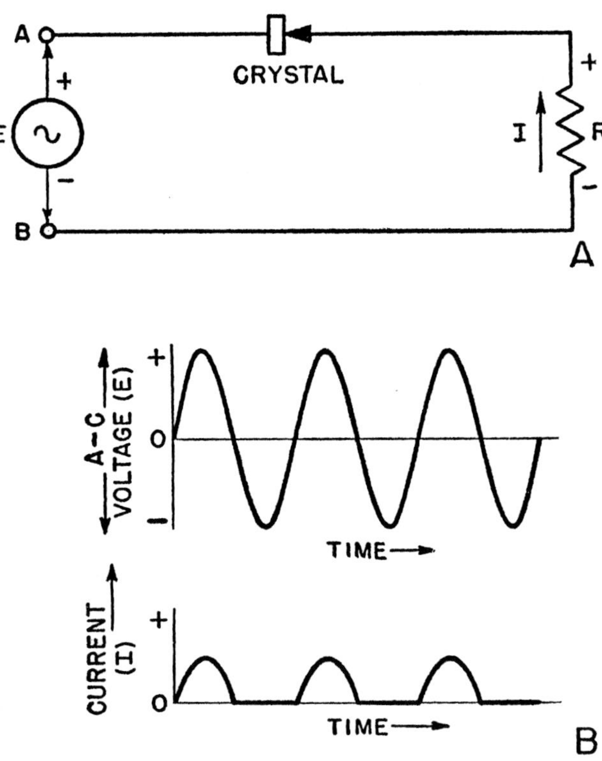

TM 662-127

Figure 118. Crystal rectifier circuit and operating waveforms.

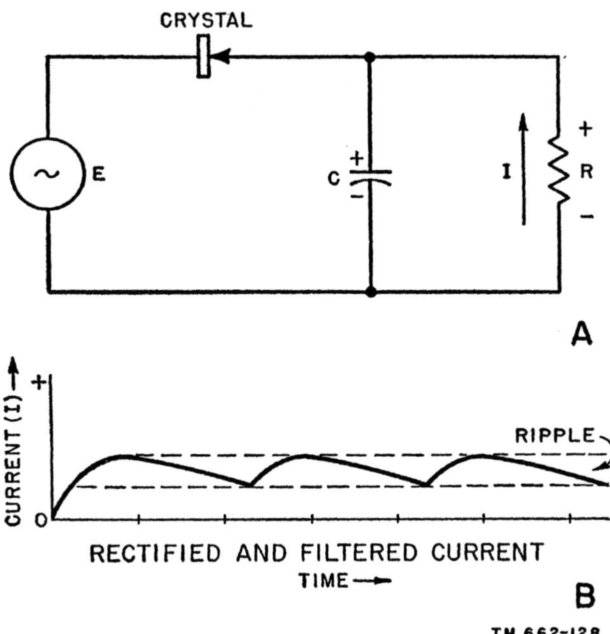

TM 662-128

Figure 119. Crystal rectifier and filter.

counterclockwise direction, maintaining the same polarity across R as when the a-c generator supplied current.

the positive half-cycles of the a-c input voltage. These surges are unidirectional.

c. FILTER.

(1) The d-c surges can be filtered or smoothed to provide a steady d-c current. A circuit accomplishing this filtering action is called a *filter.*

(2) In A of figure 119 a capacitor, C, is placed across R. When the crystal permits a surge of current in the counterclockwise direction, a division of current takes place; part of the current passes through R and part of the current charges the capacitor to the polarity shown in the figure. When the polarity of the generator reverses, current flow is blocked by the crystal. The capacitor, however, has accumulated a charge. It now discharges through R, the only available path. This flow of discharge current is also in the

(3) The total current flow through R when a filter is used is shown in B. Compared with the current surges across R without a filter, the filtered current is a relatively steady dc. The fluctuations that remain are referred to as *ripple*. With a good filter, the ripple amounts to less than 5 percent of the total current.

102. Half-wave Diode Rectifier

a. GENERAL.

(1) The crystal rectifier described above is a half-wave rectifying device. This means that the a-c voltage applied to the circuit causes current to flow only on every other half-cycle. It may be said that only half of the input wave has been rectified. This is because the crystal permits current flow in one direction only. Any device which permits current flow in one direction only can be used as a half-wave rectifier.

(2) An electron tube is a device which permits current flow in one direction

only. Current can pass from cathode to plate, but not from plate to cathode. Therefore an electron tube such as the diode can be substituted for the crystal in the power-rectifier circuit of figure 119. The new circuit is given in figure 120. This circuit is known as a half-wave diode rectifier.

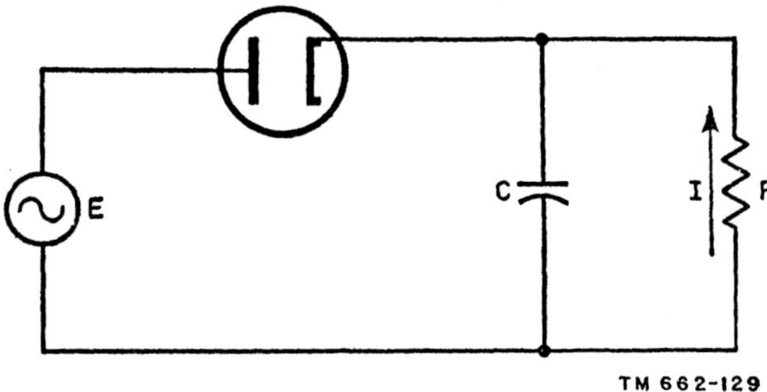

Figure 120. Half-wave diode rectifier circuit.

b. DIODE ACTION.

(1) The diode responds to the a-c voltage applied to the circuit in the same manner as the germanium crystal in the crystal rectifier. When the applied voltage makes the plate positive with relation to the cathode, current flows through the circuit in the direction indicated. Part of this current flows through R and a part charges C.

(2) When the applied voltage reverses polarity, making the plate negative with relation to the cathode, current cannot flow through the tube. The capacitor, however, now discharges through R, maintaining the current flow through the resistor in the same direction.

(3) The total current through the resistor is the same as the rectified and filtered current of the crystal rectifier (B of fig. 119). It is dc with a ripple component.

103. Full-wave Diode Rectifier

a. GENERAL.

(1) In the half-wave rectifier, the applied a-c voltage causes current flow only on every other half-cycle. This is because of the unilateral conductivity of the crystal and the diode. Only half the power available at the a-c source is utilized. By using two half-wave rectifiers, it is possible to have current flow produced by both half-cycles of a-c voltage. One half-wave rectifier allows current flow on the positive half-cycle, the other on the negative half-cycle. The only condition to be fulfilled is that these currents must flow through R in the same direction. This provides the required dc, which is filtered in the same manner as in the half-wave rectifier.

(2) Two half-wave rectifiers, combined in this manner, are known as a full-wave rectifier. When two diodes are used as the half-wave rectifying devices, the circuit is a full-wave diode rectifier (fig. 121).

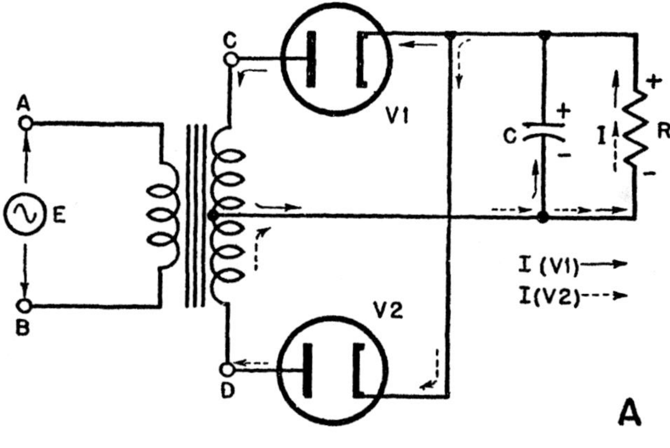

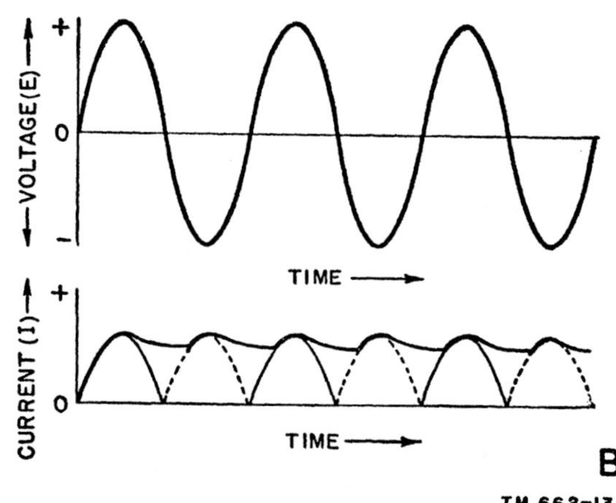

Figure 121. Full-wave diode rectifier circuit.

AGO 2244A

b. CIRCUIT.

(1) The a-c input to this power supply is applied through a step-up transformer. The ratio of the secondary voltage to the primary is chosen in this case so that the primary voltage is applied across each tube taken separately. If the a-c line impresses 117 volts across the primary, 234 volts appear across the secondary. The plates of the two diodes are connected to opposite ends of the secondary. The two cathodes are connected to each other and also are connected through R to the center tap of the transformer.

(2) Since R is in series with both cathodes, current flow through either tube must pass through R. Because of the unilateral conductivity of these tubes, current passes through R in only one direction. This fulfills the requirement for a d-c output.

(3) When the a-c input makes terminal A 117 volts negative with relation to B, the 180° phase shift in this transformer causes C on the secondary to be 234 volts positive with relation to D. Because of the center tap, the plate of V1 is 117 volts positive with relation to its cathode, and V1 conducts. At the same time, the plate of V2 is 117 volts negative with relation to its cathode, and V2 does not conduct.

(4) During the entire negative half-cycle of input, the circuit acts as a half-wave rectifier. Current flows through V1, the top half of the transformer secondary, and through the center tap. A portion of this surge of current charges C, and the rest flows through R and returns to the cathode of V1. This current is shown as solid arrows in the figure.

(5) When the a-c input reverses polarity, terminal C on the secondary is made 234 volts negative with relation to D. The plate of V1 is now negative with relation to its cathode and it stops conducting. The plate of V2, however, is now made positive with relation to

its cathode, and there is a surge of current through this tube. This surge takes the path shown by the dashed arrows.

(6) In the half-wave rectifier discussed previously, current surged through R only on the positive half-cycles of input. During the negative half-cycles, only the filter supplied current to R. In the full-wave rectifier, there are current surges on every half-cycle. It may be said that the full wave has been rectified. This means twice as many surges and twice the power available. Also, the filter does not have to provide current for an entire half-cycle between surges.

(7) In B of figure 121, the solid-line surges are supplied by V1 and the dashed-line surges are supplied by V2. If C is removed from the circuit, these surges constitute the current through R. The filter action results in current I flowing through R. Note that the current is higher and the ripple smaller than in the half-wave rectifier. The full-wave circuit is, therefore, more satisfactory than the half-wave rectifier for many applications.

104. Rectifier Tubes

a. HALF-WAVE.

(1) Several diodes are available for use in half-wave rectifier circuits. A directly heated tube which can fill moderate current requirements is the 81. Where an indirectly heated type is desirable, a 12Z3 can be used.

(2) For reasons of economy, the diode sometimes is incorporated in the same envelope with another set of elements. An example of such a multiunit tube is the rectifier-pentode 25A7–G. Beam-power diode types also are available. Examples are the 117N7–GT and the 117P7–GT.

b. FULL-WAVE.

(1) Full-wave diode rectifiers are much more common than the half-wave, and

many tubes are available for this purpose. Although two separate diodes can be used in a full-wave rectifier circuit, the usual practice is to place both diodes in a single envelope. Duo-diodes of both the directly and indirectly heated types are available for use in full-wave power supplies. The filaments or cathodes usually are in-ternally connected, thereby eliminating this connection from the external circuit.

(2) A directly heated duo-diode supplying moderate current requirements is the 5Y3–G. Directly heated types capable of supplying higher currents are the 5T4 and the 5U4.

Section II. SIGNAL RECTIFICATION

105. Radio Communication

a. TRANSMISSION BY WIRE.

(1) Until the advent of radio, long-distance communication was carried on by means of the telegraph and the telephone (fig. 122). In the basic telegraphic system a pair of wires extends from the transmitter to the receiver. The sender operates a key which controls a sounder at the receiver end. For simplicity of explanation, the conventional telegraphic sounder in the figure is replaced by an ordinary buzzer.

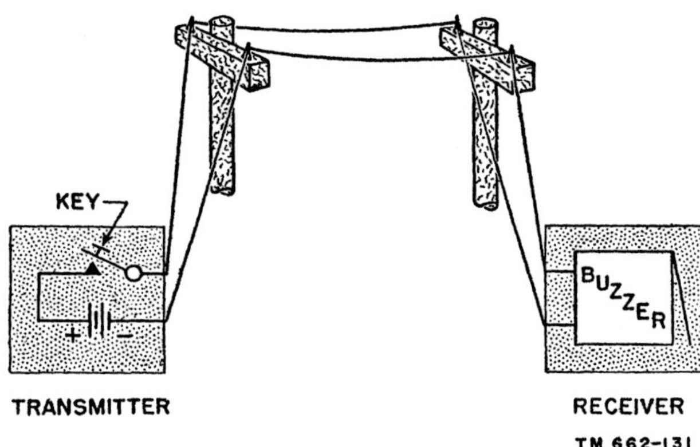

Figure 122. Simple telegraph circuit for communication.

(2) When the key is pressed down, the voltage of the battery is applied to the circuit. A current flows which operates the buzzer. So long as the key is held down, the buzzer emits an audible sound. If the key is held down for a short time, a sound of short duration is produced at the receiving end of the circuit; if the key is held down for a long time, a sound of long duration is produced. If these long and short sounds are produced in accordance with a system or a *code* known to both the sender and receiver, then *information* can be transmitted.

(3) The dashes and dots of the Morse code, for example, refer to the length of the sounds which represent a given letter of the alphabet. The letter A is represented by a dot and a dash, a short sound followed by a long sound. The letter N is a dash and a dot, a long sound followed by a short sound.

(4) The great disadvantage of wire transmission lies in the necessity of stringing telegraph lines to every single point with which communication is desired. Before the coming of radio it was necessary to place lines in cables which were laid across the bed of the Atlantic Ocean in order to connect New York to London. The difficulty and expense are obviously very great. Radio, on the other hand, requires no wires, and the transmitter can send information to a large number of receivers.

b. TRANSMISSION BY RADIO.

(1) Militarily, the vital function of radio is the maintenance of communication between mobile units. Tanks moving across open terrain cannot have telegraph wires running between them, yet they must maintain contact in order to coordinate their movements. Likewise, ships and planes depend on

radio for information about the positions of friend and enemy, and also general information such as weather reports.

(2) Pressing down a telegraph key causes a current to flow through the telegraph circuit. This current operates a sounding device at the receiving end. In radiotelegraphy pressing down the key also induces a current to flow through a sounding device in the receiver. In the telegraph circuit, the voltage is supplied by the battery, and the current is transmitted by means of wires. In the radio-telegraph circuit, the energy is supplied by a producer of radio waves, and the radio waves are transmitted through space by means of electromagnetic fields.

c. RADIATION.

(1) An alternating current flowing in a wire radiates energy into the space around it. This radiation is in the form of an electromagnetic field. If this wire forms the primary of a transformer, then the secondary of the transformer is cut by the magnetic lines of force which are part of the electromagnetic field. The magnetic lines of force expand and contract because of the alternating current in the primary. These magnetic lines, moving back and forth across the secondary, induce an alternating voltage in the secondary. Consequently, an alternating current flows in the secondary winding. It is this principle which is used in radio communication.

(2) If the primary of the transformer is replaced by a transmitting antenna and the secondary by a receiving antenna, the result is a radio circuit. An alternating current through the transmitting antenna induces an alternating current in the receiving antenna by means of the electromagnetic field. If this radio circuit is keyed in the same manner as the telegraph circuit just described, dots and dashes of current are induced in the receiving antenna.

(3) Figure 123 shows the current waveforms at the receiver when the letter C is transmitted. In A, the letter is sent by telegraph, in B, by radio. The telegraph current is simply dc supplied by the battery and turned on and off by the key. The radio current is alternating current (actually rf) supplied by the transmitter and turned on and off by the key. When the dash is sent by radio, 6 cycles of ac flow in the receiver antenna. When the dot is sent, 2 cycles of ac flow.

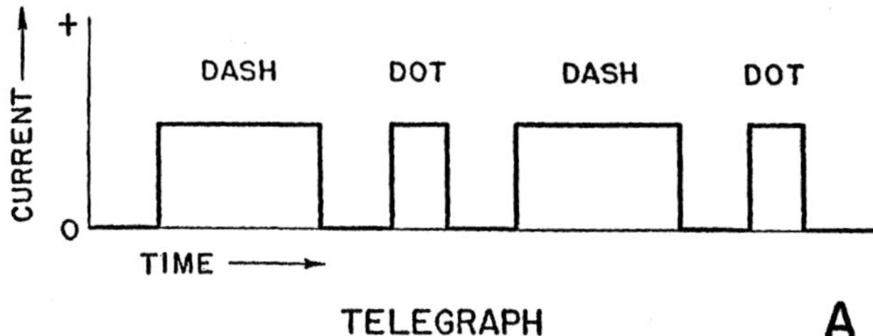

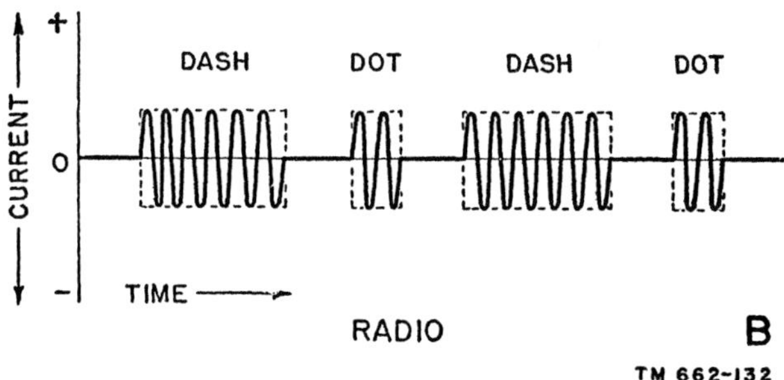

Figure 123. Current waveforms for code letter C.

d. MODULATION.

(1) The human voice or other sounds can be transmitted by wire. This is the simple telephone system. The telegraph key is replaced by a device which changes voice vibrations to current variations. The changing current operates a mechanical system in the receiver which reproduces the original voice vibrations. If a single sound is emitted at the transmitting end, the current through the wire varies at the frequency of the sound.

(2) In the telephone system, only a *variation* in current produces a sound at the receiver. When no sound is being transmitted, the current is simply

some d-c value. Consequently, when the single sound is transmitted, the current through the receiver has the form shown in A of figure 124. Since a sound varies at some audio-frequency rate, the current varies at the same audio-frequency rate.

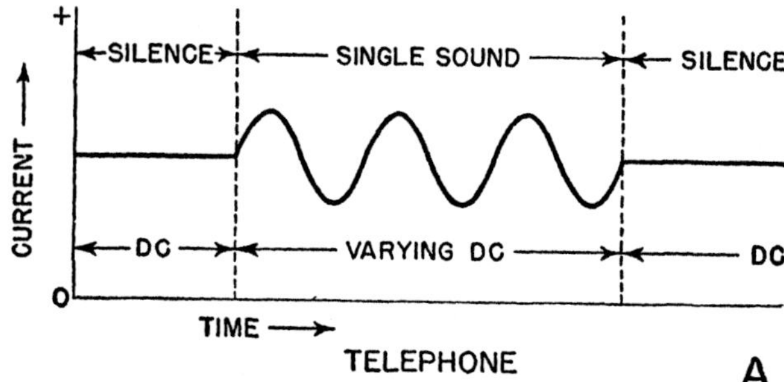

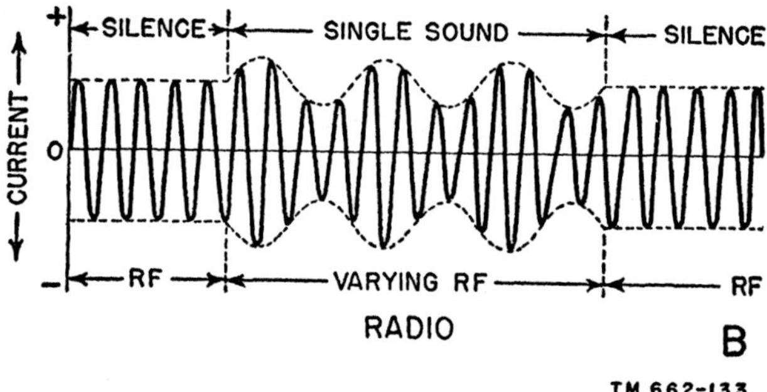

Figure 124. Transmission of single sound.

(3) It can be said that the direct current flowing continually through the system is made to *carry* the sound signal from the transmitter to the receiver. The carrier does not produce any sound at the receiver unless it is varied by a sound at the transmitter. In the telegraph system, the direct current itself carries information in the form of dots and dashes. This is no longer true in the telephone system.

(4) The sound causing the variation in carrier current flow is known as the *modulating* signal. A carrier on which modulation has been imposed is a modulated carrier, which carries information in the form of audio-frequency variations, usually produced by voice.

(5) A voice-modulated radiotelephone system functions in the same manner as the telephone system just described.

The alternating current in the transmitting antenna is modulated by the sound at the audio-frequency rate. The alternating transmitter current is the carrier. In radio transmission the frequency of this transmitter current is called the radio frequency, or rf. The frequency of the modulating signal is the audio frequency, or af. The radiotelephone carrier is referred to as the r-f carrier.

(6) A current induced by a modulated r-f carrier in a receiving antenna is shown in B of figure 124. This r-f carrier is said to be amplitude-modulated because the sound variation results in changes in the amplitude of the rf. If the r-f amplitude is constant, the receiver is silent. It can be seen that the amplitude variations of the carrier occur at the audio-frequency rate of the modulating signal.

e. DEMODULATION.

(1) Unlike the modulated telephone carrier, the modulated r-f carrier cannot be used directly to reproduce the sound. The telephone carrier is always some d-c value, whether it is modulated or unmodulated. The r-f carrier, on the other hand, is always some a-c value. Consequently, the r-f carrier must be changed to dc before it can be used. The process of changing ac to dc is the process of rectification.

(2) When ac is changed to dc for the purpose of obtaining d-c power, the process is called power rectification. When a modulated r-f carrier is changed to a d-c signal, the process is called signal rectification. Since the signal is imposed on the carrier by a process called modulation, the removal of this signal is called *demodulation,* or more commonly, detection. Circuits which accomplish signal rectification are called signal rectifiers or detectors.

106. Detection

a. In power rectification, after the a-c signal is rectified, it is filtered to obtain a fairly constant value of dc. In signal rectification, the modulated r-f carrier also is rectified and filtered, to obtain a dc varying at the audio-frequency rate of the original sound. This dc varying at an a-f rate then is used to reproduce the original sound.

b. In paragraph 102, a half-wave diode rectifier circuit was used for power rectification. The tube, a device having the characteristic of unilateral conductivity, performed the function of rectification. The capacitor across a load resistor performed the function of filtering. This circuit can be used in the same manner to accomplish signal rectification. When the circuit is used for signal rectification, it is called a half-wave diode detector.

107. Diode Detector

a. HALF-WAVE DIODE DETECTOR.

(1) The purpose of this circuit (fig. 125) is to convert the modulated r-f carrier to a direct current varying at the a-f rate of the original modulating sound.

(2) The amount of current flow induced in the receiver antenna by the r-f signal is small. Consequently, several stages of r-f amplification often precede the detector. The output of the last r-f amplifier usually is trans-

former-coupled to the detector circuit. The diode rectifies the rf by allowing current to flow on every other half-cycle only. This results in a unidirectional current flow through R_L.

(3) If an unmodulated r-f carrier is applied to the detector, the output is similar to the output of the power rectifier with an a-c input. The rf is rectified by the diode and filtered by the capacitor. The result is a relatively smooth dc with a ripple varying at the r-f rate.

(4) If a modulated r-f carrier is applied to the detector, the rf is rectified and filtered as before. However, the resultant d-c output has two ripples, shown in B. One ripple is caused by the rf, the other by the af. The r-f carrier frequency is 128 kc, and the single modulating sound is at an audio frequency of 4,000 cps. For all practical purposes, this waveform can be considered as dc varying at the a-f rate of the original modulating sound.

(5) This waveform can be used to operate headphones, as was the practice in the early days of radio. Modern receivers, however, apply this waveform through a stage of a-f amplification to a loudspeaker. This enables the sound to be heard by more than one listener.

b. TUBES USED.

(1) In modern receivers, the diode used in the detector stage sometimes is incorporated with another stage as a multiunit tube. Diode-triode combinations such as the 1H5–GT and 1LH4 are used in battery-operated receivers. Diode-pentodes 1N6–G, 1S5, and 6SF7 are used also.

(2) The diode is found also in other combinations. In the duo-diode-triode type, of which there are about twenty examples in the tube manual, the two diode plates frequently are connected to serve as a half-wave diode detector. Other multiunit types which incorporate diode detectors include the duo-diode-pentode combination, examples

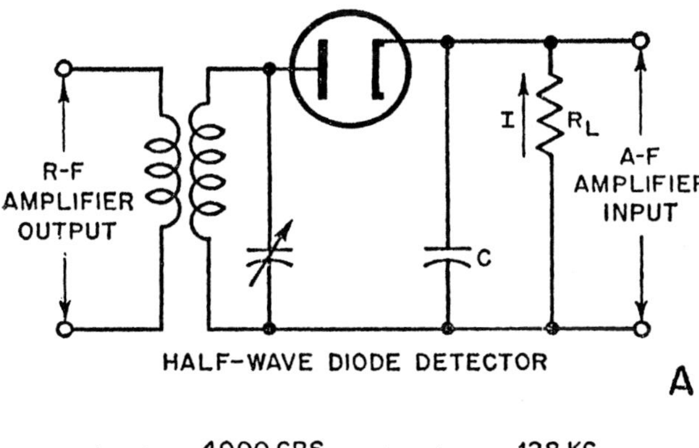

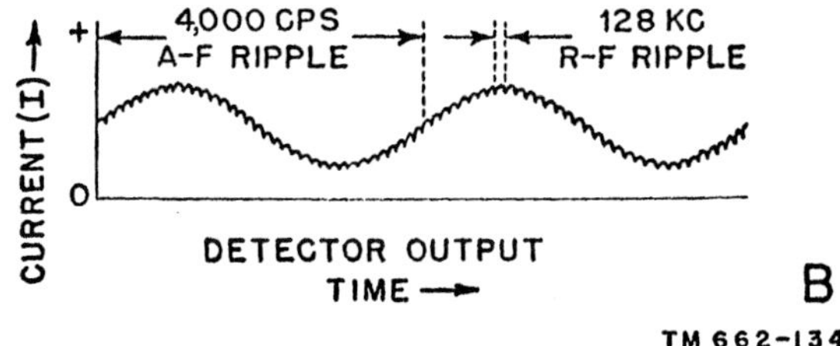

Figure 125. Half-wave diode detector circuit.

of which are the 6B7, 7R7, and 12C8. The duo-diode type 6H6 frequently is used also in the detector stage.

c. FULL-WAVE DIODE DETECTOR. The full-wave diode detector is related to the half-wave in the same manner that the full-wave power rectifier is related to the half-wave. The advantages of the full-wave over the half-wave detector, however, are not great enough to justify the enlarged circuit. This is particularly true since power output is only a minor consideration in a detector circuit. Consequently, the full-wave detector is used rarely.

108. Other Types of Detectors

a. GENERAL.

(1) All electron tubes have the characteristic of unilateral conductivity. Consequently, any one of them can serve as a rectifying device. Combined with a filter, a triode, for example, can serve as a detector in the same manner as the diode detector.

(2) A diode, having no grid, cannot amplify a signal. As a matter of fact, the output taken from a diode stage is less than its input. This disadvantage is overcome by using a tube with a grid, such as a triode or pentode. The signal is not only detected, but also, is amplifed in one stage. In early receivers, this arrangement was necessary because of the lack of r-f amplification. As much gain as possible had to be achieved.

b. GRID-LEAK DETECTOR.

(1) The operation of the grid-leak detector (fig. 126) is similar to that of the half-wave diode detector. The signal voltage applied to the grid of the triode tube is alternately positive and negative. Grid current flows during the half-cycles in which the grid is positive with relation to the cathode. During the negative half-cycles, no grid current flows. As a result, a unidirectional pulsating direct current flows through R_g. Capacitor C_g serves as a filter to smooth the r-f pulses. A d-c voltage is produced across R_g

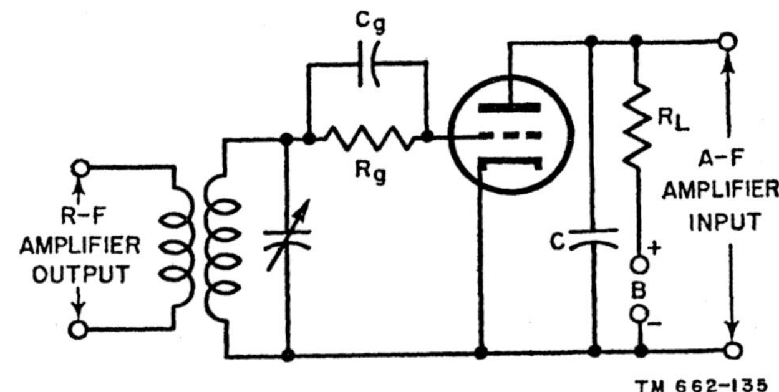

Figure 126. Circuit of grid-leak detector.

which varies at an audio rate just as in the case of the diode detector.

(2) The a-f voltage across the grid resistor now can be used as a signal voltage for the triode amplifier. As a result, an amplified a-f signal appears in the plate circuit of the grid-leak detector. Capacitor C is an additional r-f filter.

(3) The grid-leak detector operates as a *square-law* device. A square-law detector is one whose characteristic curve is entirely nonlinear. The a-f output voltage varies as the square of the r-f input voltage. In the diode detector, by comparison, the output varies directly with the input.

(4) The development of higher-gain r-f amplifiers led to the replacement of the grid-leak detector by the half-wave diode detector. The diode detector distorts the a-f signal much less than the grid-leak detector.

c. PLATE DETECTOR (fig. 127).

(1) This circuit usually is known as a plate detector because detection occurs in the plate circuit. The operation of the circuit is somewhat similar to that of a class B amplifier. Although cathode bias, provided by R_k, cannot bias a tube to plate-current cut-off, operation at the lower end of the dynamic characteristic is possible.

(2) Normal plate current flows during the positive half-cycles of the input signal voltage. However, most of the negative half-cycles are cut off. As a result, a unidirectional plate current flows. The average value of this cur-

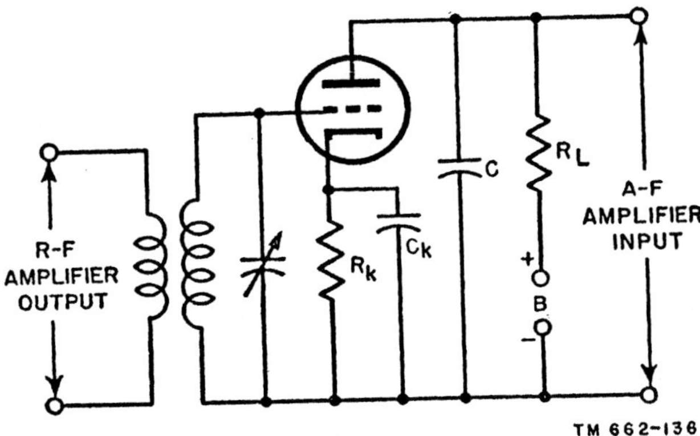

Figure 127. Diagram of plate detector.

rent varies in accordance with the a-f variations. Capacitor C acts as an r-f filter.

d. INFINITE-IMPEDANCE DETECTOR.

(1) If the plate of the plate detector is connected directly to B plus, and the output is taken across the cathode network, the result is an infinite-impedance detector. Although there is no amplification of the signal in this circuit, it has the advantages of good reproduction of the a-f signal for large inputs.

(2) The modulated r-f signal varies the d-c plate current through the tube. This current returning through the cathode network is filtered by C_k, and the current passing through R_k is dc varying at the a-f rate with a negligible r-f ripple.

e. TUBES USED. Any typical triode or pentode can be used in the circuits discussed in this paragraph. When a pentode is used, its suppressor is connected to the cathode, and its screen grid is connected to B plus through a suitable dropping resistor. A capacitor usually is connected from screen grid to ground in order to bypass rf and keep the screen grid at zero potential with relation to the signal voltage.

109. Detector Characteristics

a. GENERAL.

(1) Detectors can be rated according to four general characteristics: sensitivity, linearity, selectivity, and signal-handling ability.

(2) If some output is chosen as a reference, the signal input necessary to obtain this output gives the *sensitivity* of a detector. A detector stage providing some measure of amplification has greater sensitivity than one providing no amplification.

(3) The amount of distortion in the a-f signal as compared with the original sound is a measure of the *linearity* of a detector. A nonlinear device results in the greatest distortion of the a-f signal.

(4) In the detector circuits shown, the input consisted of a tuned circuit adjusted to the frequency of the r-f carrier. The sharpness with which this circuit can be tuned determines the *selectivity* of the detector. Selectivity in a detector depends on the current drawn from the input circuit. The lower this current the greater is the selectivity.

(5) *Signal-handling ability* is an indication of the amount of signal amplitude that can be handled without overloading the circuit or producing improper operation. If the detector can handle a large signal and operate properly, it has good signal-handling ability.

b. COMPARISONS. In the tabulation below, a comparison is made between the qualities of the various detectors discussed. For weak input signals, the grid-leak detector is obviously the best detector to use because of its high sensitivity, and despite its poor linearity. With efficient r-f amplifiers, the most popular detector is the half-wave diode circuit. This detector combines good linearity, high signal-handling ability, and simplicity, resulting in its almost universal use in modern receivers. The use of the infinite-impedance detector is not common.

Detector	Sensitivity	Linearity	Selectivity	Signal-handling ability
Diode_________	low	good	poor	high
Grid-leak______	high	poor	poor	limited
Plate_________	medium	fair	good	medium
Infinite-impedance.	low	good	good	high

110. Summary

a. The transformation of a-c power to d-c power is called power rectification.

b. A circuit accomplishing power rectification is called a power rectifier, or a power supply.

c. The output of the power supply used for the plate voltage of an electron tube is designated by the letter B. The positive terminal is B plus, the negative terminal B minus.

d. Power rectification is accomplished by rectifying and filtering the a-c input. The output is dc.

e. A crystal, such as germanium, can be used to rectify ac because of its characteristic of unidirectional conductivity.

f. A suitably chosen capacitor acts as a filter when placed across the rectifier output load. With a good filter, ripple amounts to less than 5 percent of the total rectified current.

g. An electron tube has the characteristic of unilateral conductivity and can be used as a rectifying device.

h. The half-wave diode rectifier utilizes only one-half of the a-c input wave.

i. Two half-wave rectifiers can be combined to utilize both halves of the a-c input wave. Such a circuit is known as a full-wave rectifier.

j. Information can be transmitted over a telegraph system by means of a code consisting of dots and dashes. The dots and dashes are the result of opening and closing the telegraph circuit.

k. Dots and dashes can be sent in the form of radio waves by keying a radio transmitter.

l. Current is induced to flow in a receiving antenna by the electromagnetic field radiated by the transmitting antenna.

m. Radio can transmit sound by modulating the r-f current producing the radiated electromagnetic field. The r-f current is known as the carrier.

n. The modulated r-f carrier must be demodulated in order to reproduce the original sound at the receiver.

o. Demodulation is signal rectification. More frequently, it is called detection.

p. Detection consists of rectifying the modulated r-f carrier and filtering the rf.

q. The half-wave diode detector consists of a diode rectifier and capacitive filter. This detector is used almost universally in modern receivers.

r. The grid-leak detector is a square-law device whose sensitivity is good.

111. Review Questions

a. How is d-c power obtained from an a-c line?

b. What are the two steps in power rectification?

c. Draw the circuit of a typical full-wave power rectifier.

d. What are the advantages of full-wave as compared to half-wave power rectification?

e. What is modulation?

f. Compare signal to power rectification.

g. Draw and explain the operation of a half-wave diode detector.

h. Why is a triode used in the plate detector circuit?

i. Compare the grid-leak detector to the diode detector.

j. Compare the plate detector to the diode detector.

CHAPTER 9

OSCILLATORS

112. Introduction

A-c generators supplying current at line frequencies are devices which produce oscillations mechanically. The frequency at which the current oscillates depends, among other factors, on the speed of the generator. Special a-c generators have been designed to operate at frequencies as high as several hundred kilocycles. However, this frequency is not high enough for most types of communication. Because of the high frequencies required, a nonmechanical system is needed to produce the oscillations which form the carrier. Electron-tube circuits have been designed to fulfill this function. These circuits are called oscillators. There are several basic circuit arrangements which produce oscillations.

113. Conditions for Oscillation
(fig. 128)

a. L-C CIRCUIT.

    (1) If C, a charged capacitor, is placed across L, an inductor, the capacitor discharges through the inductor, as shown in A. The capacitor acts as a source of emf (electromotive force) and forces current through the circuit. The flow of current through L creates a magnetic field about the coil. This magnetic field opposes the rise in current through the circuit. As C discharges, the current reaches a maximum and then declines. Between the time the current is 0 and the time the current is maximum, the magnetic field builds up. In other words, energy is stored in the magnetic field.

    (2) When the capacitor is discharged, the current attempts to fall to 0. This attempted change in current flow is opposed by the inductance. The magnetic field now acts as a source of emf and continues to force current through the circuit in the same direction, as in B. This current serves to charge the capacitor to a voltage whose polarity is opposite to the polarity it had when it was first placed in the circuit. As the magnetic field collapses, the current slowly falls to 0.

    (3) The capacitor, in C, now discharges in the opposite direction, reversing the original path of current flow. This current flow again is opposed by the inductor, which builds up a magnetic field opposite in direction to the original magnetic field. The current reaches a maximum and then declines.

    (4) When the capacitor has discharged, the current attempts to fall to 0. This change is opposed by the inductor, which uses the energy stored in the magnetic field to prevent the current from dropping to 0 at the instant the capacitor is discharged. Therefore, the current is sustained, and it serves to charge the capacitor to a voltage whose polarity is the same polarity that it was originally, as shown in D.

    (5) The entire process now is repeated for as long as energy remains in the circuit. The current in the L-C circuit, in E, has the form of a sine wave. It has changed from 0 to a maximum in one direction, through 0 to a maximum in the opposite direction, and back to 0. Several cycles of output are shown in F.

    (6) The output described above is possible only in an L-C circuit which loses no energy. If any resistance is present

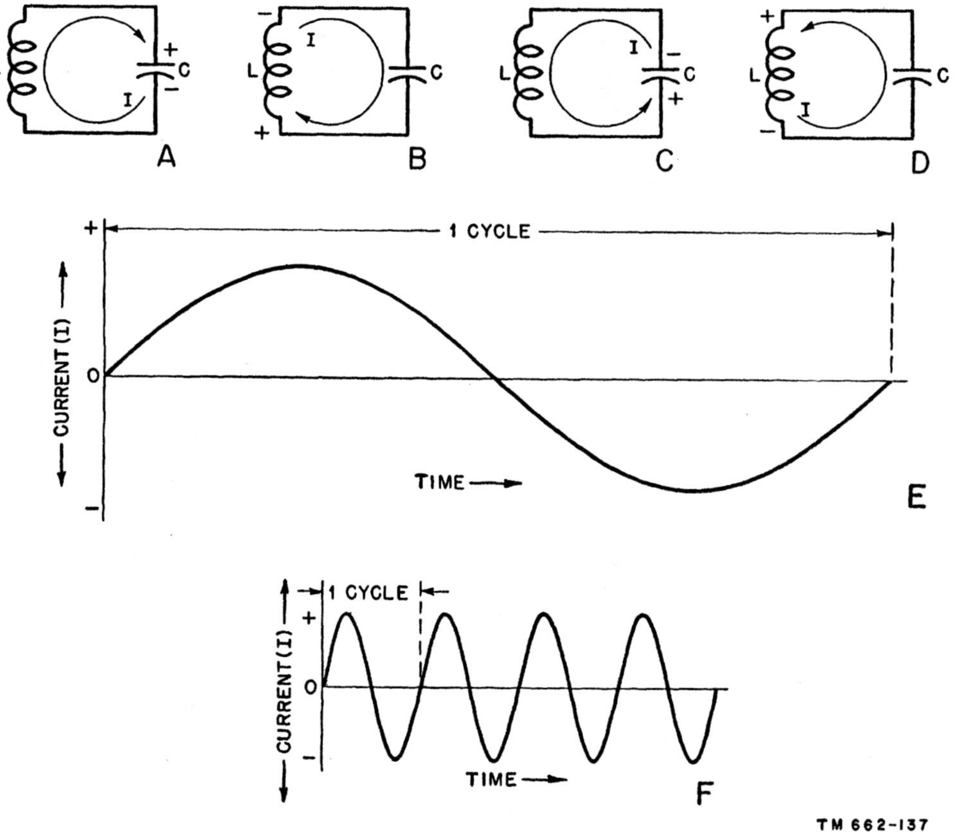

Figure 128. Oscillatory current in an L-C circuit.

in the circuit, energy is lost in the form of heat. Because no actual L-C circuit is free of resistance, the circuit can be represented as in A, figure 129. The resistance of R is the d-c resistance of the wire forming the coil, plus the leakage resistance of the capacitor, plus the resistance of the connecting leads.

(7) In the ideal L-C circuit, energy is transferred back and forth between the inductor and the capacitor without loss. In the actual R-L-C circuit, each transfer of energy involves a loss. The current which accomplishes the transfer of energy must pass through R, and each time the current flows, a part of the energy is dissipated. Consequently, less and less energy is transferred on each succeeding alternation of current flow. This means that the oscillations die out as the energy is dissipated by the resistance. The cur-

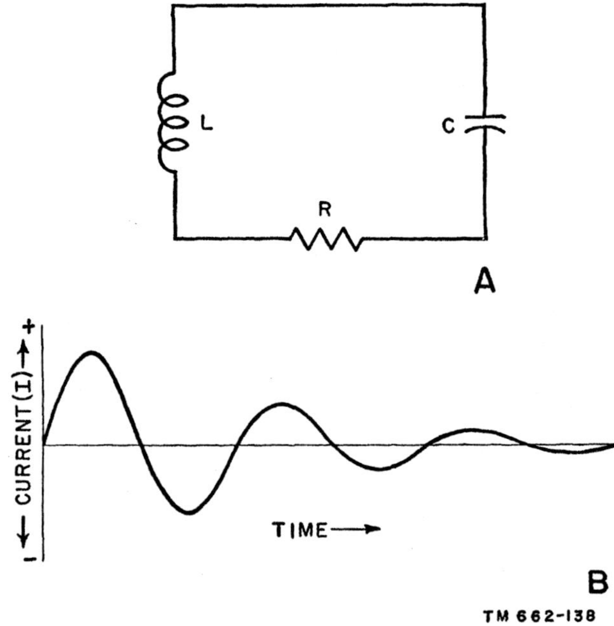

Figure 129. Damped oscillatory current in R-L-C circuit.

rent in the R-L-C circuit is shown in B. Note how the amplitude of each succeeding alternation is decreased. This is known as a *damped* oscillatory current.

b. ENERGY.

(1) If a fully charged capacitor is substituted for C on each cycle, the oscillations can be sustained. In other words, if the circuit can be supplied with sufficient energy to make up for the resistive loss, the current continues to oscillate so long as the extra energy is supplied.

(2) This extra energy, however, must be injected into the circuit so that it *aids* the flow of current. The current flows in one direction for a time, then in the other direction. Consequently, the source of energy must cause current to flow in the right direction and at the right time. The source, therefore, must be an emf producing a current of the same frequency and time relations as the oscillatory current. This current must also be large enough to replenish the energy loss. Energy of this nature can be obtained from a similar R-L-C circuit. How this is done is described below.

c. AMPLIFIER.

(1) A of figure 130, shows a transformer-coupled triode amplifier. R_k and C_k provide the tube with the proper bias for class A operation. A signal applied to the input circuit, L1–C1, is coupled to the grid circuit, L2–C2, by the transformer action of L1 and L2. A signal at the grid of the triode causes a variation in plate current. The current flowing in the plate circuit flows through L3–C3, from which the signal is coupled to the input circuit of the next stage, L4–C4. The resistance associated with each of these L-C circuits is not shown, but it is assumed to exist.

(2) The output of an amplifier is larger than its input. If an oscillatory current is made to flow in the grid circuit by means of a signal applied at the input, an oscillatory current of larger amplitude flows in the plate circuit. If only 1 cycle of input is applied, the signal at the grid is the result of a damped oscillatory current in the L2–C2 circuit. This causes a damped oscillatory current of larger amplitude in the plate circuit.

d. FEEDBACK.

(1) Both the grid and the plate circuits consist of L-C circuits, as described previously. By a process known as *feedback*, it is possible to take energy from the plate circuit and couple it back to the grid circuit.

(2) In considering the energy relations in an L-C circuit, it was noted that oscillations die out if energy is not supplied to replenish the losses. It also was noted that, if the signal input to the amplifier is limited to 1 cycle, a damped oscillation occurs in the grid circuit which is amplified in the plate

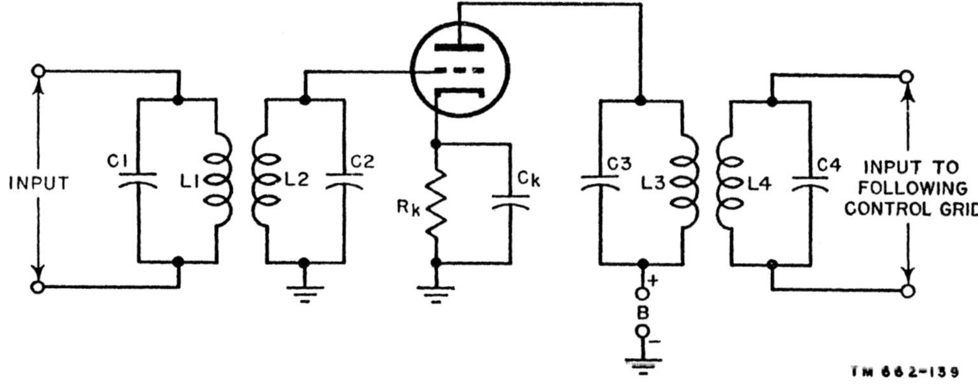

Figure 130. Transformer-coupled triode amplifier.

circuit. This damped oscillation can be sustained by coupling a portion of the amplified plate signal back to the grid circuit. The signal energy coupled in this manner must induce a current at the right time and in the right direction in the grid circuit. In order to insure that this feedback occurs at the right time, the resonant frequency of L3–C3 is made approximately the same as L2–C2. This means that 1 cycle of oscillation in the plate circuit takes place in the same period of time as 1 cycle of oscillation in the grid circuit.

(3) The feedback must occur at the right time. If the current in the grid circuit is flowing in one direction and the feedback induces a current in the opposite direction, the oscillatory current is damped more quickly than if no feedback were present. This type of feedback is *degenerative* (A of fig. 131). Note that the oscillatory current is more highly damped than the naturally damped oscillation shown in B of figure 129.

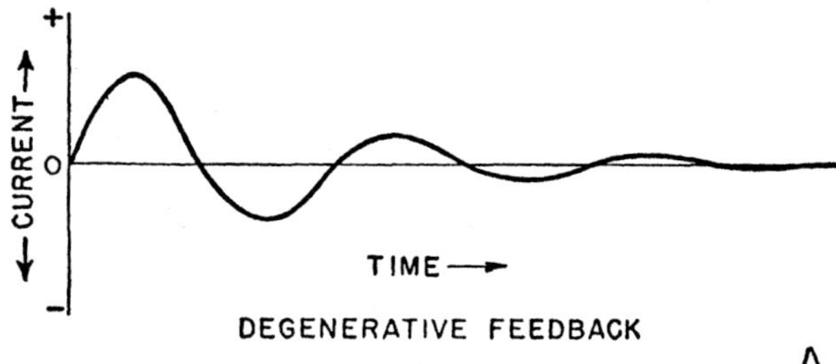

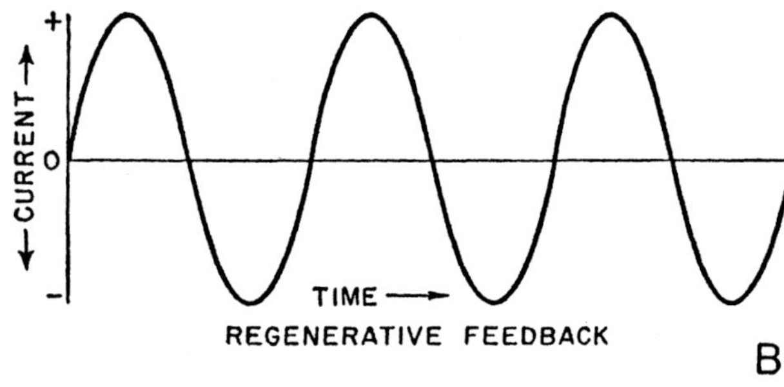

TM 662-140

Figure 131. Waveforms showing effect of feedback.

(4) If the feedback induces a current in the same direction as the oscillatory current in the grid circuit, the feedback is called *regenerative*. Its effect is to sustain the grid-circuit oscillations. If the feedback is just sufficient to replace the losses, each oscillation is of the same size (B of fig. 131).

(5) If the regenerative feedback is too small, the oscillations in the grid circuit die out, although the oscillations are not damped as highly as when there is no regenerative feedback. If the feedback is too large, the oscillations in the grid circuit build up until the plate current is swung alternately from 0 to saturation. This results in a highly distorted oscillation. The regenerative feedback, therefore, must have the proper amplitude.

e. OUTPUT.

(1) An oscillator must produce a useful output. The output can be small, provided it is sufficient to drive the grid circuit of the following stage. The following stage, or stages, can be used to build up the oscillations to the desired amplitude. This means that the oscillations in the plate circuit must be large enough to provide both the proper amount of feedback and a useful output.

(2) A certain amount of energy is needed to drive the grid of the triode in figure 130. This energy must be replaced by energy taken from the plate circuit in an oscillator. Since the output also is taken from the plate circuit, the energy available in the plate circuit must be greater than the energy necessary in the grid circuit. In other words, the total output must be greater than the input. This means that the tube must be capable of amplifying. Consequently, any tube with an amplification factor greater than one can be used in an oscillator.

f. CONCLUSION. The following must exist in order that an electron-tube circuit can be made to oscillate:

(1) An L-C circuit containing the proper amounts of inductance and capacitance to oscillate at the desired frequency.

(2) A tube capable of amplifying a signal applied to its control grid. The amplification factor must be greater than one.

(3) A means of coupling part of the output energy to the input circuit (feedback). This energy must produce a current in the input circuit which is in phase with the input signal (regenerative), and of sufficient amplitude to replace any losses in the oscillating L-C circuit.

114. Types of Oscillators

a. TUNED-GRID OSCILLATOR (fig. 132).

(1) The L-C tuned circuit producing oscillations is placed from grid to ground. The tuning of the L-C circuit through a range of resonant frequencies is possible by means of variable capacitor C1. The circuit sometimes is called a tickler oscillator, and inductor L2 often is called a tickler coil. It is known also as an Armstrong oscillator, after the name of its inventor.

(2) Oscillations in this circuit begin spontaneously. No external source is required to trigger it. Assume that the cathode is sufficiently heated for normal operation, but that switch SW is open, keeping B plus off the plate. No

current can flow. If the switch now is closed, B plus is applied to the plate and there is a surge of current through the tube. This surge is sufficient to cause the grid circuit, L1–C1, to begin to oscillate. These oscillations are amplified by the triode and appear across tickler coil L2. The tickler coil is coupled to L1 in such a manner that there is regenerative feedback from L2, plate circuit, to L1, grid circuit. This feedback is sufficient to sustain oscillations in the L1–C1 circuit. C_g and R_g provide proper grid bias by means of grid-leak action.

(3) Various methods of coupling the output to the next stage are available. These include capacitive-, transformer-, and impedance-coupling networks. The choice of coupling depends on the specific application of the oscillator. The output, in any case, usually is taken from the oscillating L-C circuit. This has the effect of loading down the circuit, and is equivalent to increasing R in the R-L-C circuit. For this reason, the load can be considered to increase the losses, increasing the amount of regenerative feedback required. Too great a load damps the oscillations in the L-C circuit, causing the oscillations to die out and the output to drop to 0.

(4) This oscillator can be used to produce oscillations in both the a-f and r-f ranges. Too high a frequency of oscillation, however, introduces undesirable effects as a result of the high interelectrode capacitance of the triode, reducing the output available. For this reason, low-μ triodes such as the 6C5 and 6J5 commonly are used.

b. TUNED-PLATE OSCILLATOR (fig. 133). This oscillator can be considered essentially the same as the tuned-grid oscillator discussed above. The only difference between the two oscillators is that here the L-C tuned circuit is placed on the plate side of the tube. In this case tickler coil L2 is in the grid circuit. The amplified grid signal provides feedback at the plate circuit which is inductively coupled to the grid circuit.

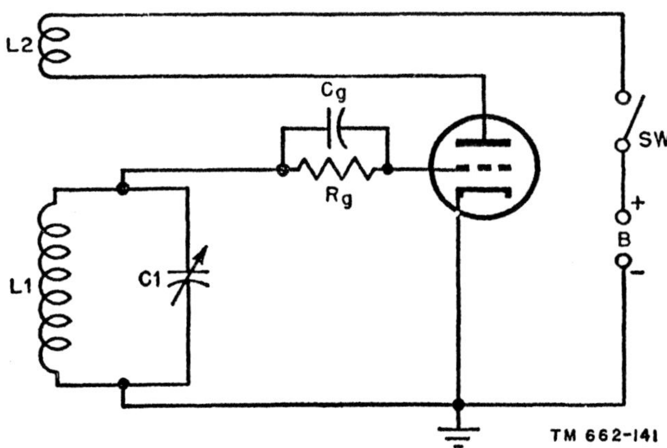

Figure 132. Tuned-grid oscillator.

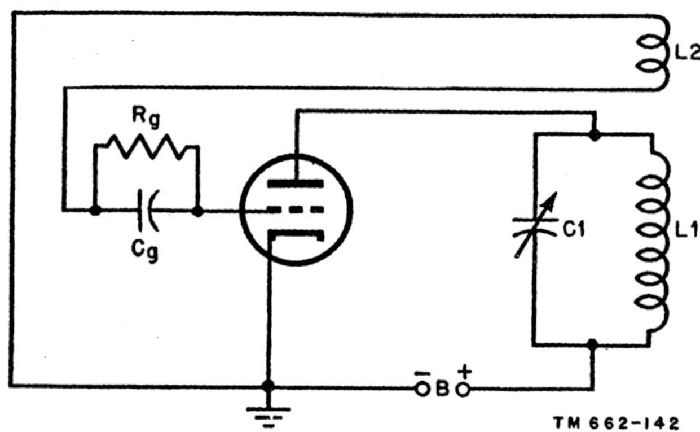

Figure 133. Tuned-plate oscillator.

c. TUNED-PLATE TUNED-GRID OSCILLATOR (fig. 134).

(1) It was noted in the discussion of feedback that one tuned circuit can be made to supply the extra energy necessary to sustain oscillations in another tuned circuit. The tuned-plate tuned-grid oscillator illustrates this principle.

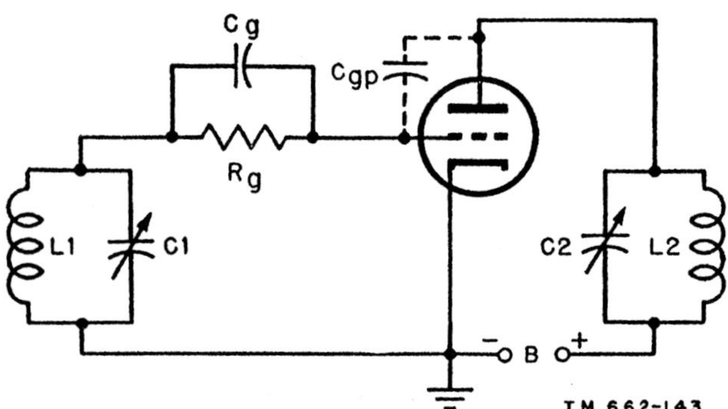

Figure 134. Tuned-plate tuned-grid oscillator.

(2) The grid circuit L1–C1 is tuned to the resonant frequency desired. When the first surge of current starts this circuit oscillating, the oscillations appear at the grid and are amplified in the plate circuit. The plate circuit consists of L2–C2.

(3) The feedback path in the tuned-plate tuned-grid oscillator occurs through the plate-to-grid capacitance of the triode. Energy is coupled from the plate circuit to the grid circuit. If L2–C2 is tuned to the same frequency as L1–C1, the phase of the feedback is not proper to sustain oscillations. For

this reason, the plate circuit is made inductive at the frequency of oscillation of the grid circuit to make the feedback regenerative. This is done by tuning the plate circuit to a slightly higher frequency.

d. SPLIT-TANK OSCILLATORS (fig. 135). An L-C circuit sometimes is called a tank circuit, after the term storage tank, because energy can be stored in the form of charge on the capacitor, or in the form of a magnetic field around the inductor. A split tank is an L-C circuit in which either the capacitance or the inductance is composed of two or more capacitors or inductors. There are two basic types of split-tank oscillators—the Hartley (split-inductance) and the Colpitts (split-capacitance). As in the oscillators previously discussed, low-μ triodes such as the 6C5 and the 6J5 can be used for both the Hartley and the Colpitts oscillators.

(1) *Hartley oscillator.* In the Hartley oscillator, one tank circuit is made to serve both as grid and plate circuit. The grid is coupled to one end of the tank and the plate to the other end. The cathode is connected to a point on the inductor. This divides the inductor between the grid and the plate circuits in the form of an inductive voltage divider, as shown in A. The voltage across L1 is between the grid and cathode, thereby applying a signal to the grid. The amplified voltage at the plate appears across L2. This provides the necessary feedback.

(2) *Colpitts oscillator.* This circuit also uses a split tank. The capacitance of the tank circuit, provided by two capacitors, C1 and C2, is divided between the grid and plate circuits, as in B. C1 and C2 form a capacitive voltage divider, in which C1 provides the grid signal and C2 provides feedback from the plate circuit. By adjusting C1 and C2, it is possible to control the frequency and amount of feedback.

e. PLATE-VOLTAGE SUPPLY.

(1) There are two methods for applying plate voltage to the oscillator tube. The d-c plate-voltage supply can be

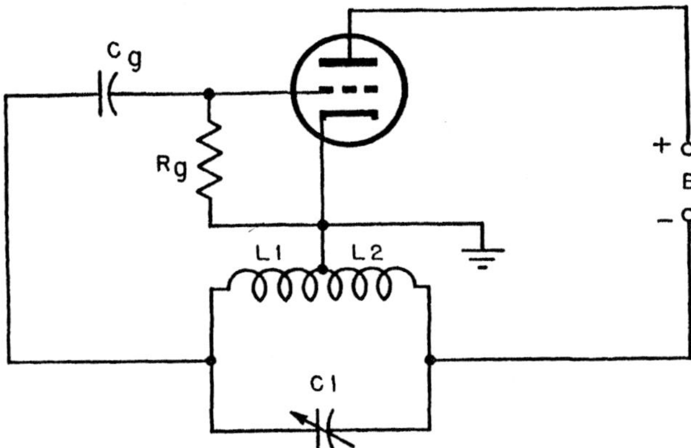

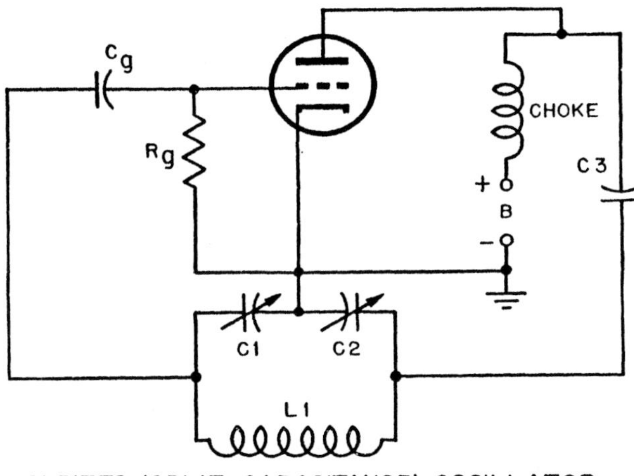

Figure 135. Split-tank oscillators.

placed in series with the oscillating plate circuit, in which case the circuit is referred to as series-fed. The d-c plate supply also can be placed in parallel with the oscillating plate circuit, and the circuit then is referred to as shunt-fed. In either case, there must be a d-c return path from plate to cathode for the plate current. The circuits in figures 132, 133, and 134 are all examples of series-fed oscillators. This is the simplest circuit arrangement.

(2) The Hartley oscillator, in A of figure 135, is also series-fed. The d-c plate current must pass through inductor L2 before it can return to the cathode.

The disadvantage in this arrangement is that the plate supply is placed at a high a-c potential with relation to the cathode. Also, the supply has a large distributed capacitance to ground, and this capacitance is shunted across the tank inductor L2.

(3) The disadvantage of the series-fed circuit can be overcome by keeping the d-c plate supply and the oscillating plate current separate. This is accomplished in the shunt-fed Hartley oscillator (fig. 136). The plate-current oscillations are coupled to the split-inductance tank by means of capacitor C2. The capacitor prevents the d-c plate current from returning to the cathode through the tank. The plate current, therefore, can return only through the choke in series with the supply. This choke prevents any oscillations from appearing in the supply, because its reactance is very large.

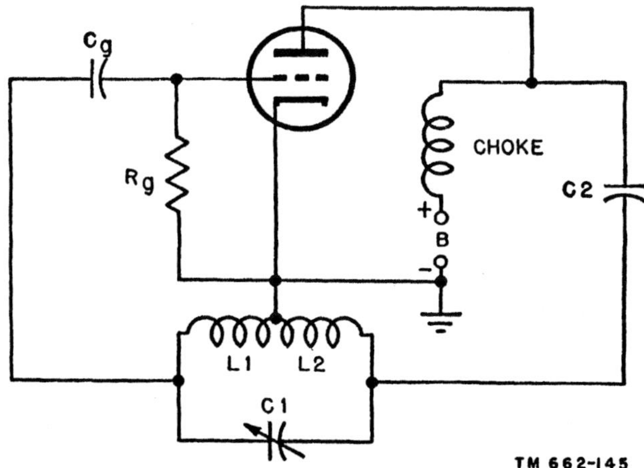

Figure 136. Shunt-fed Hartley oscillator.

f. ELECTRON-COUPLED OSCILLATORS.

(1) *Buffer amplifier.* If the output of an oscillator is coupled directly to a power amplifier, undesirable loading effects occur. There can be distortion of the output waveform or even a stopping of oscillation. In addition, the frequency of the oscillation will not be stable. A buffer amplifier, therefore, is used to couple the oscillations to the power amplifier. An ordinary triode

voltage amplifier can serve as a buffer, since it draws little power from the oscillator.

(2) *Electron-coupled oscillator.*

(a) By using a multielectrode tube, the oscillator and buffer stages can be replaced by one circuit which performs both functions. Such a circuit is called an electron-coupled oscillator. Figure 137 is a typical circuit arrangement, using a 6F6 pentode.

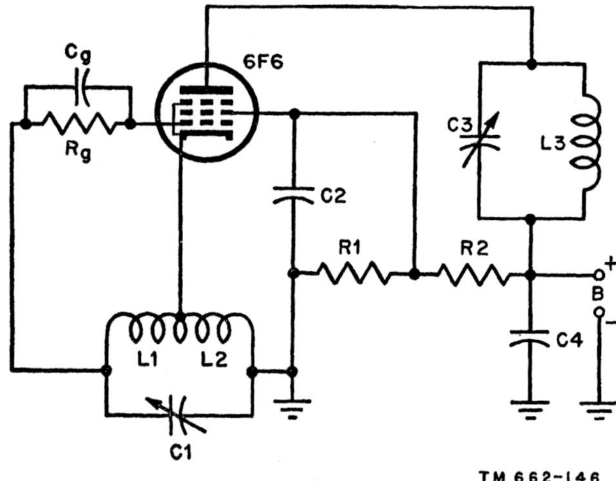

Figure 137. Electron-coupled oscillator.

(b) In this circuit the cathode, the control grid, and the screen grid perform the function of the triode in a Hartley oscillator. The cathode of the 6F6 taps the split-inductance tank consisting of L1, L2, and C1. The control grid is coupled to one end of the tank, and the screen grid takes the place of the triode plate. The screen voltage is taken from the voltage divider consisting of R1 and R2 across the B supply. This part of the circuit can be compared to the Hartley oscillator in figure 136.

(c) The signal appearing at the grid causes the current through the tube to oscillate. In the ordinary Hartley oscillator, this current is collected at the plate, where one portion of it is used for feedback and the rest for output. In the electron-coupled oscillator, however, the screen grid collects only that portion of the current needed for feedback. The output portion of the current passes through the screen grid to the pentode plate, where it is collected and passed through the output tank circuit consisting of C3 and L3. Capacitors C2 and C4 serve to bypass oscillations around the power supply.

(d) The only connection between the oscillator and the output circuit is the electron stream itself. This serves to isolate the oscillator from the load. The electron-coupled oscillator, therefore, has all the advantages of a separate oscillator and buffer.

(e) Typical pentodes utilized in the electron-coupled oscillator are the 6AG7, 6F6, 6K7, and 6SK7. The cathode, the control grid, and the screen grid can be connected to form any of the basic triode amplifiers. The plate circuit then is coupled to the following stage.

g. CRYSTAL OSCILLATORS.

(1) Certain types of crystals, such as quartz, Rochelle salts, and tourmaline, have the ability to generate small voltages when a mechanical force is applied. Conversely, when a voltage is applied to such crystals, the physical shape of the crystals is changed, and mechanical vibrations are produced. This property, referred to as the piezoelectric effect, permits the crystal to be substituted for a tuned tank circuit of an oscillator. The tuned-grid tuned-plate oscillator, for example, can have its grid tank replaced by a crystal held between plates (fig. 138).

(2) The circuit action is the same for both the tank and the crystal. The crystal, however, has two great advantages. A crystal can be cut with great precision to resonate within a very small percentage of a given frequency. It also possesses a far higher Q than any

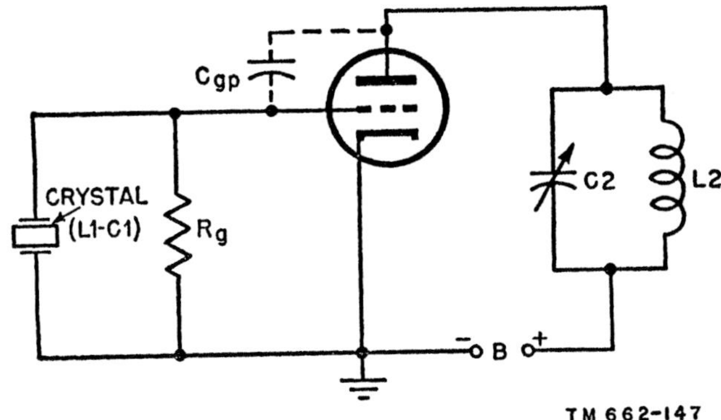

Figure 138. Crystal oscillator.

L-C network. This means that the circuit can have both frequency precision and frequency stability when using a crystal.

h. OTHER TYPES. Many other types of oscillators are used. These include the tri-tet, the dynatron, and the transitron. The last two named can be grouped under the heading of negative-resistance oscillators. In ultra-high-frequency work, such oscillators as the klystron, and the magnetron generally are used. A discussion of these oscillators is beyond the scope of this manual.

115. Heterodyne Principle

a. If the outputs of two oscillators which have different frequencies are fed to a common nonlinear circuit, the two sets of oscillations combine to produce several frequencies of oscillation. Four important frequencies appear at the output of the common circuit. They are the two original frequencies, the sum of the two, and the difference of the two. This is called heterodyning, and it is used in the superheterodyne receiver.

b. The circuit common to both oscillators can be a mixer or frequency-converter stage, described in paragraph 73. One set of oscillations is a received signal; the other set is produced by a local oscillator. The output selected is usually the difference frequency, also called the intermediate frequency. The difference frequency sometimes is called the beat frequency. In some applications, the sum of the two frequencies is selected instead of the difference frequency.

c. In the superheterodyne receiver, one set of oscillations is supplied by the r-f carrier. These oscillations originate in an oscillator located in the transmitter. The second set of oscillations is supplied by an oscillator located in the receiver, known as the local oscillator.

d. If a mixer stage, such as the pentagrid 6L7, is used, the local oscillator consists of an entirely separate circuit with its own tube, previously described. The pentagrid mixer has two separate control grids which are isolated from one another. The carrier signal is impressed on one control grid and the output of the separate local oscillator is applied to another grid. Both grid voltages affect the plate current, which contains the four frequencies mentioned above. The plate circuit is tuned to the desired frequency, which is the intermediate, and the other frequencies are rejected.

e. If a pentagrid converter, such as the 6A8 or the 6SA7, is used, the local oscillator tube is replaced by equivalent elements in the converter tube itself, and the necessity for an extra tube is eliminated.

116. Summary

a. Mechanical systems for generating oscillations generally are not high enough in frequency for use in communications and other types of electronic equipment.

b. Electron-tube circuits can perform the function of producing oscillations.

c. In order to produce oscillations, an electron-tube circuit must have these characteristics:

 (1) A tuned circuit having the proper amounts of inductance and capacitance to oscillate at the desired frequency.

 (2) A tube capable of amplifying a signal at its control grid.

 (3) A means of providing the tuned circuit with sufficient regenerative energy to sustain oscillations.

d. The tuned-grid oscillator obtains regenerative feedback by coupling the plate circuit to the tuned-grid circuit.

e. The tuned-plate oscillator has its tuned circuit on the plate side. Regenerative feedback is obtained by coupling a part of the oscillation to the grid. The oscillation is amplified and returned to the plate circuit.

f. In the tuned-plate tuned-grid oscillator, regenerative feedback occurs through the grid-to-plate capacitance of the tube.

g. There are two basic types of split-tank oscillators—the Hartley and the Colpitts.

h. The Hartley oscillator has a split-inductance tank divided between the grid and plate circuits.

i. The Colpitts oscillator has a split-capacitance tank divided between the grid and plate circuits.

j. An oscillator is shunt-fed when its d-c plate supply is in parallel with the oscillating plate circuit.

k. An oscillator is series-fed when its d-c plate supply is in series with the oscillating plate circuit.

l. A buffer amplifier is used to isolate the oscillator from undesirable loading effects.

m. The electron-coupled oscillator replaces the separate oscillator and buffer tubes with a single multielectrode tube, such as a pentode. The circuit performs the functions of both the oscillator and the buffer.

n. Certain types of crystals, such as quartz or tourmaline, can be used as tuned circuits in oscillators. Crystals are used to give frequency precision and stability.

o. If two oscillations of differing frequency are mixed, oscillations occur at four important frequencies. They are the two original frequencies plus their sum and their difference.

p. The heterodyne principle is used in the superheterodyne receiver. One frequency is supplied by the r-f carrier, the other by a local oscillator in the receiver.

q. If the local oscillator is a separate stage, then a mixer tube is used to beat the two frequencies.

r. In the pentagrid converter, the functions of the local oscillator and mixer are combined in one tube.

117. Review Questions

a. Why does a damped oscillation occur in an R-L-C circuit?

b. Under what conditions can oscillations be sustained in an R-L-C circuit?

c. What is the function of the amplifier in an electron-tube oscillator?

d. Explain regenerative and degenerative feedback.

e. What are the proper conditions for producing regenerative feedback in the tuned-plate tuned-grid oscillator?

f. How is energy coupled from the plate circuit to the grid circuit in the tuned-grid tuned-plate oscillator?

g. Name the basic types of split-tank oscillators.

h. Draw schematic diagrams of the Hartley oscillator to illustrate shunt feed and series feed.

i. What is a buffer amplifier?

j. What is an advantage of the electron-coupled oscillator?

k. What is the function of a crystal as used in an oscillator?

l. What four important frequencies appear at the mixer output if two sets of oscillations of differing frequency are mixed?

m. How is the heterodyne principle used in receivers?

n. What is a local oscillator? How is it used?

o. What is the function of the pentagrid converter?

CHAPTER 10

TRANSMITTING TUBES

118. Difference Between Transmitting and Receiving Tubes

a. GENERAL. Electron tubes used in the high-power stages of transmitters differ in design from tubes used in receivers because of the different function which they must perform. The basic principles of operation of these tubes, however, are not different from those that govern the tubes used in receivers. Transmitter tubes, like receiving tubes, are classified as diode, triode, tetrode, pentode, and beam power types. The active elements in a transmitting diode are the cathode, or filament, and the plate. In the triode, tetrode, or pentode there are one, two, or three grids between the plate and the cathode or filament. All are active elements. The beam power type can have one, two, or three grids plus beam forming plates at cathode potential. The pentode becomes a beam power tube when its screen and suppressor grids are physically alined, and the tetrode provides beam power action when its screen and control grids are alined.

b. POWER CONSIDERATIONS. Although transmitter power tubes must handle considerably greater amounts of power than receiver tubes, some receiving tubes are suitable for use in transmitter circuits not requiring high power. Transmitter power tubes, on the other hand, find application in all types of radio transmitters, audio-frequency circuits requiring high power output, and in other varieties of high power electronic equipments. Some of these applications will be discussed here.

c. HEAT DISSIPATION.

    (1) Conversion of energy from one form to another never is completely efficient, and there is always a loss of energy in the process. The energy lost usually is dissipated in the form of heat. In electron tubes, these losses occur mainly at the plate of the tube. If the grid of a tube draws current, as is common in transmitters, heat also is produced at the grid. In the largest tubes, great amounts of heat are generated, and it is necessary to provide elaborate means of heat removal to prevent damage to the tube.

    (2) Heat removal can be accomplished in several ways, depending on the amount of heat generated and the physical size of the tube. If plate dissipation is less than 200 watts, natural air currents usually will carry off the heat. Because black bodies radiate heat most efficiently, the plates of low-power transmitting tubes frequently are blackened to assist the heat-dissipation process. This method will be discussed more extensively. Where the plate must dissipate substantial amounts of heat energy, up to about 4,000 watts, air is forced through special cooling fins attached to the tube. The cooling air is provided by blowers or fans. For still higher power, it becomes almost impossible to circulate the volume of air necessary, and water cooling also is used. Water can carry off more heat than air for an equivalent volume because it has a higher heat-absorbing capacity. In extremely high-power transmitting tubes where a great deal of power is dissipated at the grids, the grids are constructed with internal tubing in which water is circulated.

    (3) Three typical transmitting tubes are shown in figure 139. In the 4–250A tetrode, designed for operation at plate

dissipations up to 250 watts, small holes can be seen in the metal jacket over the base. A small electric fan is mounted to blow air up from under the tube and through these holes. In the 893A–R triode, with a maximum plate dissipation of 20 kilowatts, a large number of copper cooling fins surround the tube. A heavy-duty blower forces air through these fins to provide the necessary removal of heat. The 5831 is a beam power triode which is cooled by the circulation of a considerable quantity of water through its plate and grid structures. The outside of this tube is machined from a heavy copper fan. Copper is used because it radiates heat efficient-

ly. It is capable of dissipating heat energy in excess of 125 kilowatts. The water-cooled grid dissipates up to 2 kilowatts, which is considerably more than many smaller tubes dissipate at the plate.

119. Construction of Transmitter Tubes

To provide high power output from a tube, very high voltages and currents must be used. Therefore, the tube must be built to withstand the application of these voltages and currents. To prevent arcing from one tube element to another when the tube is operating, the electrodes are separated by a substantial distance. In addition, special insulation is used at the points where the electrodes are mounted. These

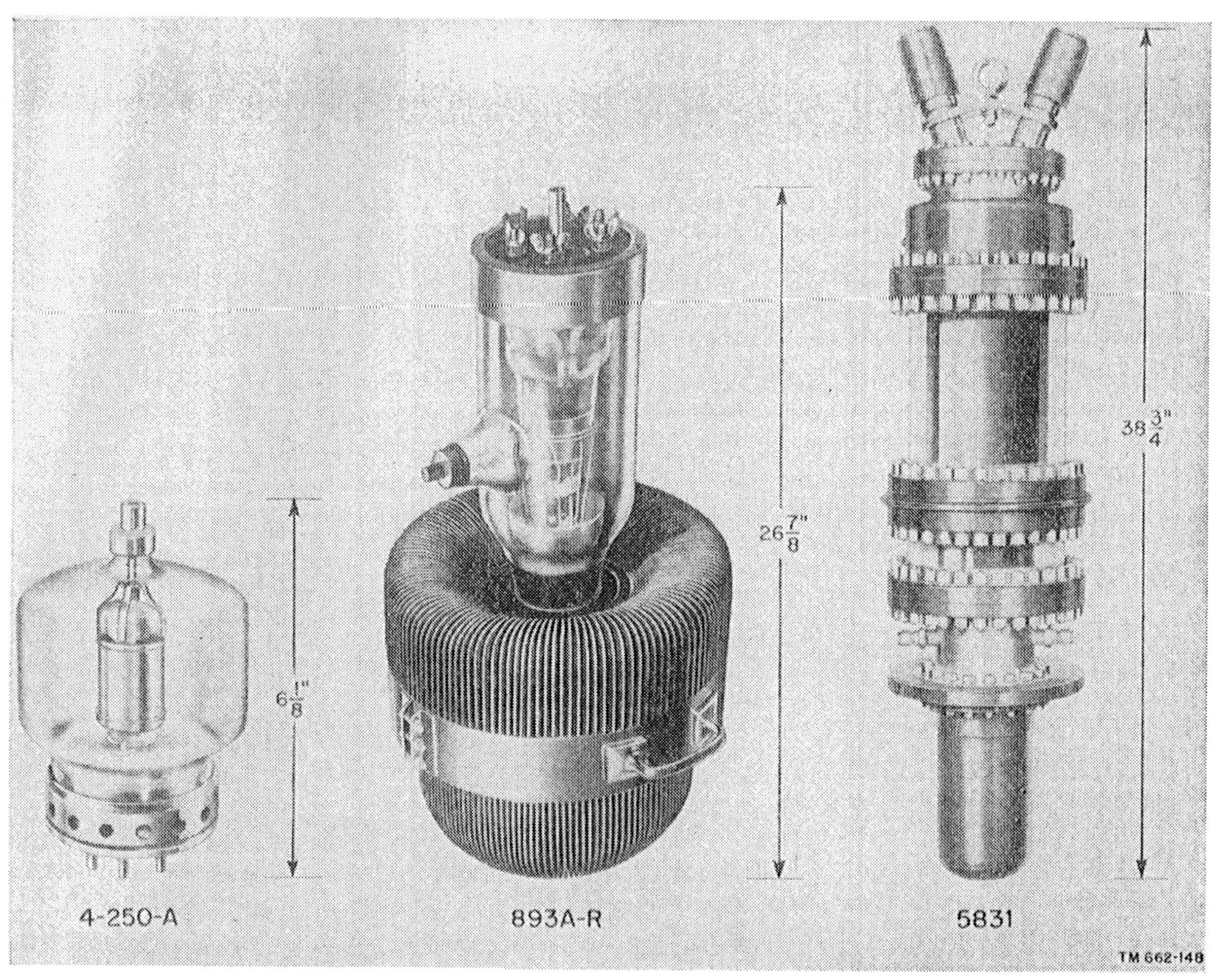

Figure 139. Three typical transmitting tubes.

precautions are vital because arcing can destroy the tube.

a. CATHODES.

(1) *Coated cathodes.*

(*a*) The flow of electrons from the cathode to the plate of a power tube is greater than in a receiving tube because of the high power output. This means that the cathode or filament emits a more plentiful supply of electrons. Considerable heating power must be applied, and the construction of the cathodes or filaments is larger and more complex. The majority of receiving tubes use indirectly heated or directly heated oxide-coated cathodes or filaments. This is not true of transmitting tubes since the oxide-coated cathode cannot withstand high voltages. It is used therefore only in relatively low-power tubes rated up to about 50 watts of plate dissipation.

(*b*) Many small transmitting tubes are larger versions of receiving tubes with slightly changed connections and bulb design. These use standard receiving-tube heater cathodes. In some cases, the elements are scaled up slightly in size, but modern practice is to design these tubes specifically for transmitting purposes. Recent advances in the design of oxide-coated cathodes of the indirectly heated type have permitted their use in specialized tubes with plate dissipation of several hundred watts.

(2) *Thoriated-tungsten cathodes.*

(*a*) It is necessary for the filament of a transmitting tube to be capable of very high emission; at the same time it must be able to withstand the high voltages that are present. In addition, transmitting tubes must have as much gas removed from the envelope as possible. Consequently, in early transmitting tubes, filaments made of pure tungsten were used, because tungsten contains very little gas, has a high melting point, and is extremely strong. Since tungsten does not have a high electron emission for a given amount of filament heating power, modern tubes use thoriated-tungsten filaments.

(*b*) The discovery that a small quantity of thorium oxide added to tungsten results in a thoriated-tungsten filament capable of greatly increased emission makes it possible to design filaments requiring much less power than pure tungsten. During the manufacturing process, the filament is heated momentarily to a higher than normal operating temperature. The oxide is converted into pure metal and rises to the surface as a layer one atom deep. This completes the process, and a layer of metallic thorium now coats the surface of the tungsten.

(*c*) In very large tubes, the thorium oxide is not changed completely into thorium by the simple application of heat alone. Consequently, a gas which is a compound of the element carbon is introduced during the heating process. The carbon penetrates the surface of the tungsten, forming a hard layer of tungsten carbide. The metallic thorium then diffuses to the surface of this strong tungsten carbide layer. Filaments made in this way are called *carburized* filaments. They overcome some of the disadvantages of the delicate layer of thorium on a simple thoriated filament, which can be damaged by a little absorbed gas in the tube. In very large transmitting tubes, such carburized filaments are woven together, with suitable insulation, much as many small fibers are twisted into a rope. The result is a *multistrand* thoriated-tungsten filament that is used in the highest power tubes.

(*d*) Thoriated-tungsten filaments must be operated within 5 percent of their specified voltage rating. If the filament voltage on the tube is allowed to rise too high, the layer of thorium will boil off, deactivating the tube. It can be reactivated by a repetition of the original activation process, until the supply of thorium oxide is exhausted. To reactivate a thoriated-tungsten filament, normal filament voltage is applied for several hours. The filament voltage then is raised 30 percent above normal for 10 minutes, after which it is reduced to normal for 20 minutes.

b. PLATES.

(1) Most of the heat losses in a transmitting tube occur at the plate. This heat must be removed in one of three ways —by natural radiation, or by forced-air- or water-cooling. Plates which are cooled by natural radiation of heat from their surfaces work at temperatures between 400° and 1,000° C. Tubes using these plates are required to dissipate from 4 to 10 watts per square centimeter of cooling surface. If forced-air cooling is used, the surfaces operate at temperatures from 150° to 200° C. The outside of the tube and the associated cooling fins are usually of large area, as can be seen from the photograph of the 893A–R in figure 139. They dissipate one-half to one watt per square centimeter of surface. The air must be supplied at a rate of 50 to 150 cubic feet per minute for each kilowatt of heat generated at the plate.

(2) Very high-power water-cooled tubes operate at surface temperatures between 30° and 150° C. Because water is an effective cooling agent, the dissipation per square centimeter is high, averaging 30 to 110 watts, and ¼ to ½ gallon of water must be supplied each minute per kilowatt of plate dissipation. These figures apply only if the water flow is rapid to prevent boiling or bubbling at the surface. If the surfaces are unclean, the ratings are reduced still farther. Coatings such as scale, rust, or dirt will reduce the cooling effect to a small fraction of the

rated amount. These coatings are good heat insulators and prevent the efficient transfer of heat to the cooling medium. Therefore, water-cooled tubes must be kept clean, and the water must be free of impurities. Distilled water is preferable. The use of water cooling permits the tube to be made somewhat smaller in size for a given power rating.

(3) The temperature developed at the plate in any tube depends on the physical properties of the metal of which it is made. Some metals are able to radiate heat better than others. The ability of a metal to radiate heat is called its *thermal emissivity.* This is a constant which describes this ability for any given material with an untreated or naturally formed surface. The constant varies with the degree of surface roughness. In the chart below are the thermal emissivity constants for some typical materials used in power tubes. They are measured at the temperatures at which these materials generally are used.

Material	Temperature	Thermal emissivity
Graphite	700°C.	.9
Copper	30°C.	.07
Molybdenum	1,000°C.	.13
Molybdenum with surface roughened by sand-blasting with quartz.	1,000°C.	.5
Nickel	300°C.	.09
Tantalum	1,100°C.	.18
Tungsten	2,300°C.	.30

(4) The choice of plate materials is determined partly by their ability to emit heat by radiation. In addition, different metals possess varying mechanical properties at high temperatures. Tungsten, for example, is far stronger than nickel at temperatures above 300° C., but it is far more difficult to fabricate into the shapes required. It is also much more expensive. Plates need to be strong enough to withstand mechanical shocks produced by vibra-

tion and handling. They must also withstand the severe mechanical stresses set up when they are raised to red heat.

(5) When materials approach their melting point, the molecules are able to leave the surface and go into the surrounding atmosphere. An analogy would be steam evaporating from boiling water. The pressure that these molecules exert in the surrounding atmosphere is called vapor pressure. Materials which vaporize easily have high vapor pressures. In vacuum tubes, it is desirable to use materials which have low vapor pressures at the operating temperature to prevent the vacuum from being contaminated with vapor molecules. When contamination with vapor molecules occurs, the tube is said to be *gassy*, it is erratic in behavior, and must be replaced.

(6) As shown in the chart, some typical materials used for power-tube plates are tungsten, molybdenum, graphite, nickel, tantalum, and copper. Nickel is cheap and easy to fabricate into a large variety of shapes; also it has low thermal emissivity and a relatively low melting point. It has the disadvantages of not being very strong, and of having a high vapor-pressure constant at high temperature. Therefore, its use is limited largely to low-power transmitting and receiving tubes.

(7) For medium-power, radiation-cooled tubes, graphite occasionally is used as the plate material. It is a very good radiator of heat, and therefore can dissipate considerable heat from a small area. It is not easy to fabricate and also it is rather fragile. It cannot be used where a great deal of mechanical vibration is present.

(8) Molybdenum is usable at temperatures slightly higher than graphite. When it is sandblasted with quartz, it possesses a satisfactorily high thermal emissivity without the mechanical disadvantages of graphite. It is also much easier to shape. Its melting point, which is higher than that of nickel but not as high as that of tantalum or tungsten, restricts its application to medium-power tubes which are not subject to severe overload.

(9) Tantalum is one of the most useful plate materials. It is easy to fabricate, possesses good thermal emissivity, especially if the surface is roughened, and is mechanically strong at high temperatures. In addition, it has the great virtue of being able to absorb undesirable residual gases when raised to a temperature of about $1,400°$ C. This property makes it possible to maintain a very high vacuum in tubes with a tantalum plate. It is one of the most commonly used of modern plate materials. The principal application is in radiation and forced-air-cooled tubes with plate dissipations up to several thousand watts.

(10) Tungsten has the highest melting point of any metal. Unfortunately, it is one of the most difficult to shape. Its good thermal emissivity permits its limited application in tubes operated at very high voltages where stability of the plate material at high temperatures is important. Tungsten plates sometimes are coated with a thin layer of zirconium, a metal that absorbs gases well at red heat.

(11) The plate material in very high-power tubes using water cooling usually is copper. It is the best conductor of heat among the common metals, and therefore, provides the best transfer of heat to the cooling medium. Its excellent machineability and relative cheapness make it desirable when large structures are required. Because it is usually operated below the boiling temperature of water, its high vapor pressure is not serious. It has a tendency to release absorbed gases if its temperature rises beyond $300°$ C. This is highly undesirable because these gases may cause severe arcing between plate and cathode at the high voltages commonly used.

c. GRIDS.

(1) In normal transmitting applications, power tubes frequently are operated so that the voltage applied to the control grid is positive during a part of the operating cycle. As a consequence, the control grid draws current, and heat is generated at the grid as well as at the plate. In multigrid tubes, the screen grids operate at relatively high temperatures; therefore, materials with a high melting point, such as molybdenum, tantalum, and tungsten, usually are used. If the grid voltage is appreciably positive, it intercepts many electrons destined for the plate. The heating of the grid is roughly proportional to the grid voltage for a given time interval. As noted previously, grids dissipating large amounts of heat are constructed with internal tubing through which water is forced.

(2) If the grid voltage is high enough, the electrons will strike the grid with sufficient energy to produce secondary electron emission, much like that which occurs at the plate of a tetrode (par. 70). Where oxide-coated or thoriated-tungsten cathodes are used, some of the cathode coating may evaporate and be transferred to the grid, increasing its ability to emit secondary electrons. This is undesirable in a power tube, as it may cause the grid voltage to go highly positive, which probably would result in its destruction. This secondary emission behavior is reduced by treating the surface of the grid with a chemically nonactive material, such as a thin plating of gold.

d. ENVELOPES AND BASES.

(1) Very low power tubes frequently are constructed according to standard receiving-tube techniques. They use similar glass-bulb or *envelope* construction, though slightly enlarged to handle the increased power. Their bases are made of a plastic, usually bakelite or micarta. Higher power

means that the glass must be able to withstand a greater amount of heat. Because the tube usually is large, it must have greater mechanical strength at the higher temperatures. Up to powers of several hundred watts, the tubes are usually of glass construction with the exception of the base, which is made of metal-clad ceramic or the plastics mentioned above. The glass must be a much better insulator than that used in receiving tubes; because power tubes require a high vacuum, it must be entirely freed of absorbed gases during the manufacturing process. Glass is designated as either hard or soft depending on its softening point. Hard glass with a high softening point and good electrical insulation properties is used most frequently. Soft glass bulbs are used only in receiving or very low-power transmitting tubes.

(2) The base of a transmitting tube is required to withstand larger amounts of heat and voltage than receiving tubes. For this purpose, ceramic materials such as steatite are used. These may be metal clad, as are some plastic materials that are used. Tubes operated at very high frequencies use no base at all, but instead, the leads are brought out through glass-to-metal seals. Such a tube, the 829B, is shown in figure 140. Very high-power tubes, either water- or forced-air-cooled, are made of a complex composite structure of glass and metal. They seldom have bases as such, but instead, have connections made to large-diameter rods or rings that are actually a part of the various electrodes and extend right through vacuum-tight seals.

120. Transmitting-tube Applications

a. GENERAL.

(1) The purpose of a transmitter is to transform power as supplied by a-c lines, batteries, or generators into radio-frequency power which is made to carry intelligence. The intelligence

Figure 140. Transmitting tube used at high frequencies.

can be in the form of speech, Morse code, pictures, controlled pulses for remote-control devices, or any other special form. Power tubes are used in these devices for the production and amplification of voltages and currents at radio and audio frequencies.

(2) Transmitters for speech and code are by far the most frequently encountered. In figure 141 a block diagram of a representative radiotelephone and radiotelegraph transmitter is shown. Its two selective functions are the transmission of speech and code.

(3) Operating power is supplied by vacuum tubes or motor-generator devices. These furnish the d-c operating voltages required by the tubes used in the transmitter proper. The basic radio frequency used in the transmitter is generated by the oscillator. The buffer isolates the oscillator from the following stages, and so improves oscillator stability. Another buffer amplifier isolates the frequency-multiplier section, which provides the desired operating frequency. The modulator superimposes the intelligence—that is, the speech—upon the carrier. The power amplifier increases the strength of the signal before it is impressed upon the antenna. In a radiotelegraph transmitter, a key which turns the transmitter on or off when pressed by the operator is used to transmit the information.

(4) The radio frequencies used for communication cover a wide range of the over-all electromagnetic spectrum (fig. 142). They extend from about 30 kc to 30,000 mc. Transmitting tubes are of different kinds in order to cover this enormous range. For the low frequencies (30 to 300 kc) used principally by long-wave fixed point-to-point stations, transoceanic, government, and direction-finding stations, large amounts of power are required, and large triode amplifiers, such as the 862A producing several thousand watts, are common. In the medium-frequency range (300 to 3,000 kc) where broadcasting, ship and harbor, government, amateur, and police stations are found, tubes such as the 893A–R triode are used. In the high-frequency range (3,000 to 30,000 kc) are found coastal ship, international broadcasting, government, aviation, and amateur stations using tubes like the 4–1000 triode. The very highest-power tubes cannot be used in this range, but many tubes are used with ratings up to several thousand watts.

(5) In the very-high-frequency range (30 to 300 mc), different tubes with

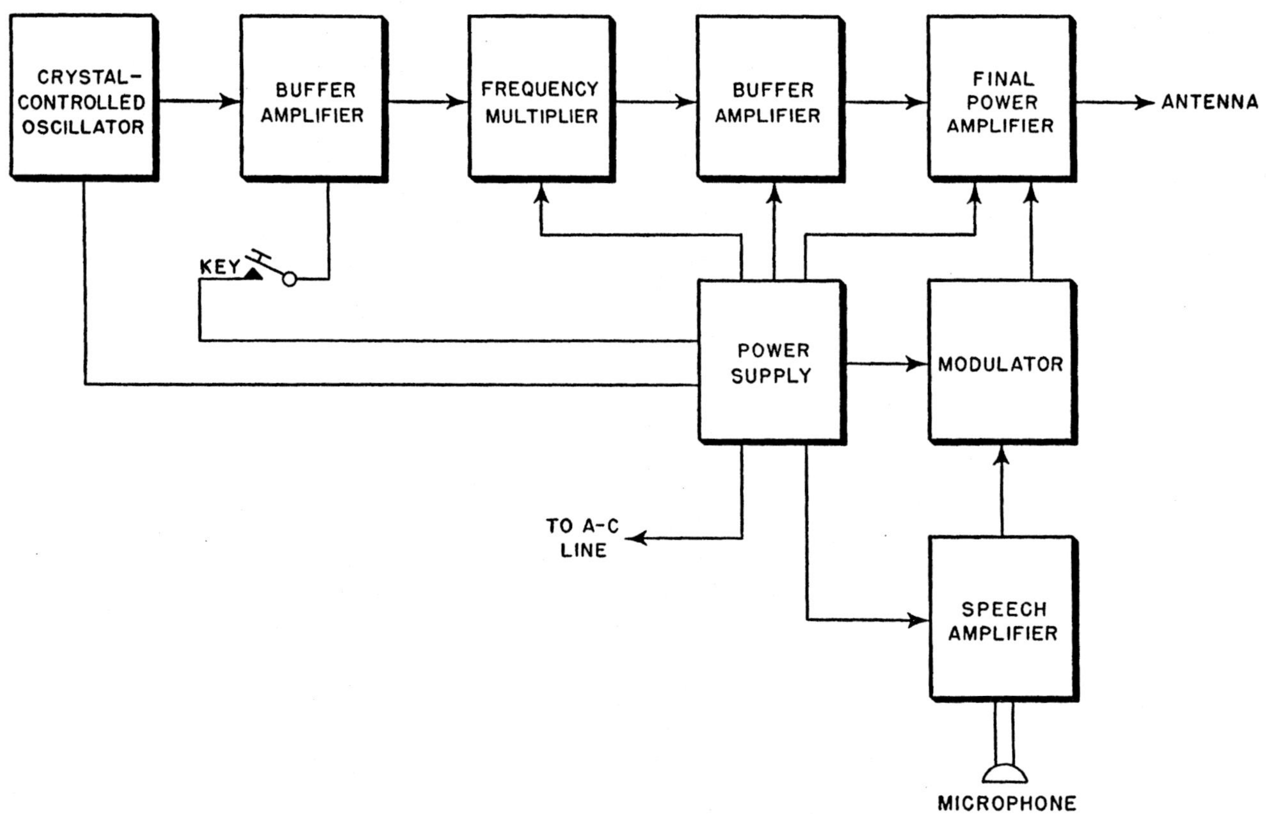

Figure 141. Block diagram of typical transmitter.

smaller structures, like the triode 4X500A, are necessary. Such services as the relay broadcast, television, mobile, and government operate in this region. In the ultra-high-frequency range (300 to 3,000 mc) radically different tube designs, such as the 5588 power triode, are necessary. This is because low lead inductance and low interelectrode capacitance are necessary. Air navigation, facsimile, television, citizen's radio, meteorological, amateur, and other services operate in this range. The super-high frequencies starting at 3,000 mc are used largely for radar and television relay service. Normal tube types no longer operate in this region, and specially designed devices such as the 2K26 klystron are required. This tube and the super-high-frequency magnetron are described in chapter 11.

b. OSCILLATORS.

(1) *General.* The oscillator used in transmitters converts a d-c energy into r-f energy. Receivers and transmitters use similar oscillator circuits: the Hartley, the Armstrong, the Colpitts, the electron-coupled, tuned-plate tuned-grid, and crystal-controlled oscillators. For operation above the medium frequency range, the oscillator in the transmitter generates the r-f carrier at a fairly low power level so that the tubes which are used do not differ substantially from those used in receivers. In fact, receiving tubes serve adequately in many transmitter-oscillator applications. In transmitter practice, the stability of the oscillator with relation to frequency is of paramount importance. This means that extremely little or no change in the car-

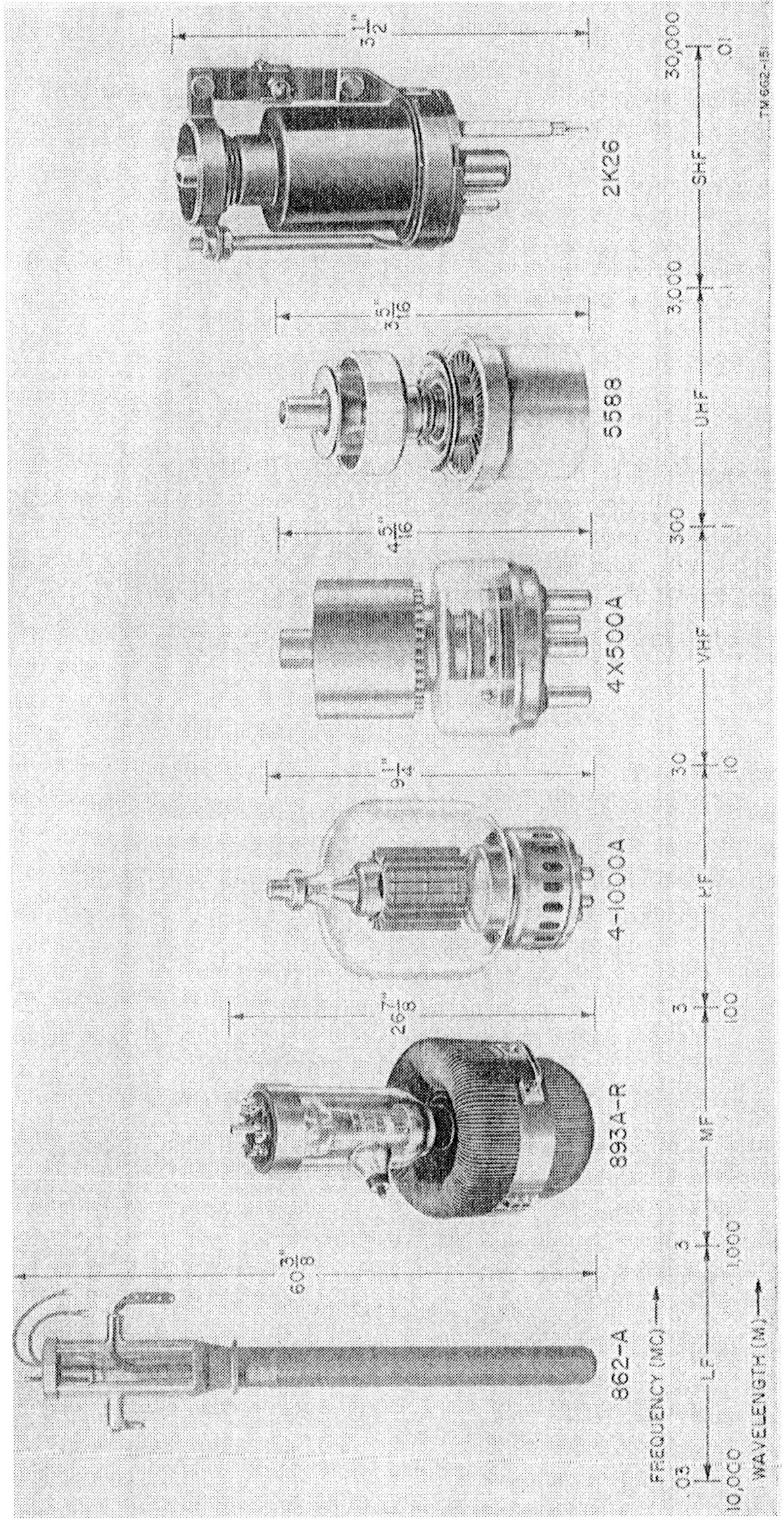

Figure 142. Typical transmitting tubes used in r-f spectrum.

rier frequency is tolerated under operating conditions.

(2) *Crystal oscillators.* To insure the highest degree of frequency stability, the crystal-controlled oscillator is used most frequently. The quartz crystal is specially housed so that a constant temperature is maintained, also, it is mechanically isolated to protect it from any shock or vibration. These precautions are necessary in exacting transmitter applications such as broadcasting and large fixed radio-telegraph stations. In the v-h-f (very high frequency) band, .002-percent shift from the assigned frequency usually is considered stable operation. At an operating frequency of 100 mc, this means that a 2-kc shift is tolerated. In some cases, it should be possible to change the transmitter frequency. One method is the use of a number of crystals, each differing in frequency from the others. These are switch-controlled, thereby permitting rapid changes in frequency.

c. CLASS C AMPLIFIERS.

(1) For efficient amplification of the radio-frequency carrier, class C power amplifiers are used most frequently. The class C amplifier can operate successfully from low to very high frequencies. It is possible in normal service to obtain plate-circuit efficiency from dc to rf of 65 to 75 percent. In a class C amplifier, the grid is biased negatively far below cut-off, and plate current flows for much less than one-half of the r-f cycle—that is, for less than 180° (fig. 143). Because the grid voltage goes positive with relation to the cathode, grid current also flows.

(2) When triodes are used as radio-frequency amplifiers, they must be neutralized to prevent energy from feeding back from the plate to the grid circuit and causing undesirable self-oscillation. This feedback is caused by grid-to-plate capacitance. It can be of a frequency different from that of the desired operating frequency, and this can result in unnecessary interference

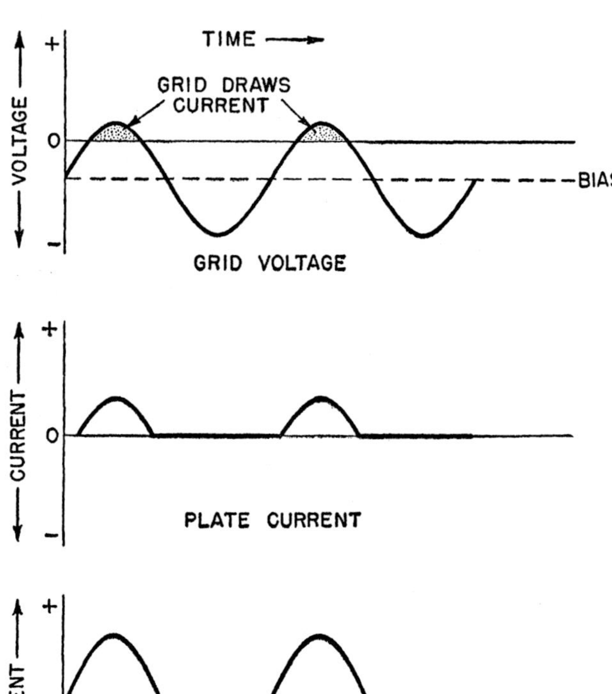

Figure 143. Voltage and current relationships in class C amplifier.

to other radio services. Neutralization of triodes usually is accomplished by feeding a small amount of the output from the plate circuit into the grid circuit *out of phase* with the feedback caused by the grid-plate capacitance. The neutralized energy cancels out the feedback in the tube. Where tetrodes or pentodes are used as r-f amplifiers, neutralization normally is not required unless the tube is operating at a frequency so high that the shielding effect of the screen grid is insufficient. In this case, tetrodes and pentodes also must be neutralized.

(3) The neutralizing circuit chosen for triode, tetrode, or pentode amplifiers depends largely on the design of the rest of the circuit and on the range of operating frequencies to be covered. The principle, however, remains the same: namely, the feeding back of an

amount of energy equal and opposite to that which tends to cause oscillation. The phase and amplitude of this neutralizing energy must be adjusted when the transmitter is tuned.

d. BUFFERS.

(1) One of the characteristics of class C amplifiers is the consumption of power in the input circuit caused by the grid voltage going positive with relation to the cathode and drawing current during part of the operating cycle. This power, called *excitation power,* must be supplied by the preceding stage in the transmitter. For the power amplifier to operate efficiently, a minimum amount of excitation is required. This is determined by the type of tube and the d-c voltage applied to it. The stages before the power amplifier must be able to supply this excitation without overloading. Because the frequency of an oscillator depends to some extent on the load impressed upon it, it is undesirable from the standpoint of frequency stabilization to attempt to supply excitation power directly from the oscillator. Another consideration is the modulation impressed on a carrier wave in the power amplifier. In this case, a varying amount of excitation is demanded by the power amplifier as the modulation changes. This changing load also can seriously affect the frequency stabilization of the oscillator if the oscillator is used to drive such a modulated amplifier directly.

(2) Therefore, a *buffer* amplifier is introduced between the oscillator and the power amplifier to isolate the two stages from each other. The buffer amplifier usually is operated class A, so that it will not affect the oscillator. In this condition, no power is drawn in its grid circuit. For class A service, the efficiency is low, and tubes of fairly high ratings must be used in buffer circuits for high-power final amplifiers. In broadcasting service, many buffer stages are used to build up the low-level output from the oscillator to a

value sufficient to provide excitation to the power amplifier. In general, the buffer must supply from 5 to 20 percent as much power as the final amplifier will produce.

(3) The buffer amplifier must supply this excitation and have considerable reserve power so that its output does not vary with changing load. This is termed *good regulation.* Since the efficiency of the class A buffer is low, its plate dissipation can be as much as one-half that of the tube used as the final power amplifier. Excitation requirements increase as the frequency of the operation is increased. This is because losses in the input circuit are greater at higher frequencies.

(4) Because the power level at which the oscillator operates usually is low, the buffer for the oscillator tube generally is of a similar power rating operated class A (fig. 144). In this circuit, a receiving tube triode is used as a low-power, crystal-controlled oscillator. Its output is coupled through a resistance-capacitance network to the buffer-amplifier grid, which is self-biased for class A operation by the resistor and capacitor in its cathode circuit. The buffer amplifier is a small receiving type pentode. The plate of the buffer is connected to the B plus of the power supply through a parallel-resonant circuit tuned to the crystal frequency. The output of the buffer amplifier is coupled to the following stage in the transmitter by the transformer action of the link coil coupled to the resonant output circuit.

(5) In the practical transmitter, buffer amplifiers are used between the oscillator and the frequency multiplier, and also between the frequency multiplier and the power amplifier (fig. 141).

(6) In figure 145 a high-power buffer supplies excitation to a final amplifier and isolates it from the preceding stages. Generally a power tetrode is used here with a link-coupled tuned input circuit. The grid bias is provided by a

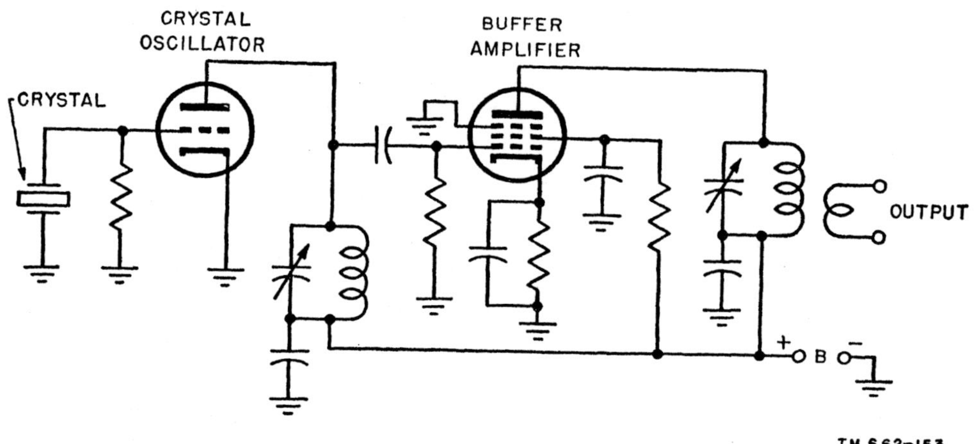

Figure 144. Crystal oscillator and buffer amplifier.

fixed d-c voltage supply. Voltage for the screen grid is derived from the plate voltage supply by a drop across the series resistor connected between B plus and the screen. The capacitors at the lower end of each tuned circuit and from the screen to ground are radio-frequency bypass capacitors. They keep r-f energy from the power supply and the screen grid.

e. FREQUENCY MULTIPLIERS.

(1) Crystals used in transmitters to generate the carrier are available only in the range of frequencies between about 100 kc and 30 mc. High-stability crystals are available only in the lower part of this range, between 100 and 10,000 kc, approximately. Therefore, some means must be used to raise the fre-

quency of the crystal if it is desired that the final carrier be of a higher frequency with high stability. For this purpose, a special type of amplifier, called a frequency multiplier, is used, which always operates class C. One of the characteristics of this type of operation is that it is rich in harmonics. It can use ordinary low-power transmitting tubes, but almost all transmitting tubes can be made to work in a frequency-multiplying circuit. Such a circuit using two small triodes is shown in figure 146. In this circuit, the electrical connections of each stage are nearly the same as those of the class C amplifier. The difference lies in the frequency to which the parallel-resonant circuits

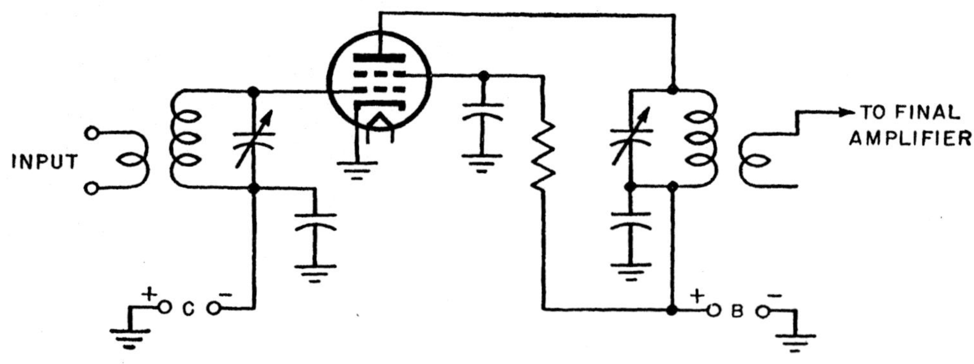

Figure 145. High-power buffer amplifier

are tuned. The output of the plate circuit in each stage is tuned to two times the frequency of the input circuit. If desired, greater multiplication may occur in each stage. When it is tuned to twice the frequency of the input circuit, the stage is called a doubler; three times the frequency, a tripler; four times, a quadrupler; etc.

The voltage and current relations in the grid-plate and tank circuit are illustrated in figure 147, which also shows the steep pulses of plate current and resultant current in the tank output circuit tuned to twice the input frequency (a doubler).

(4) Since the input and output circuits of frequency multipliers are tuned to fre-

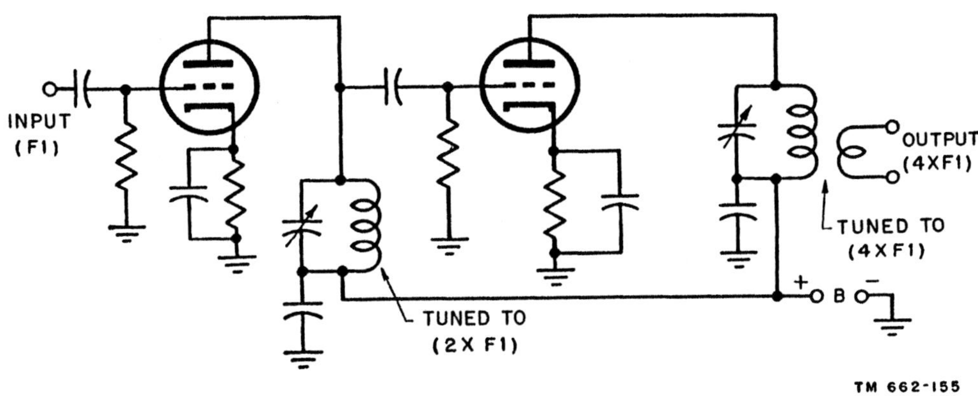

Figure 146. Frequency multiplier system with output frequency four times input frequency.

(2) In general, a multiplier of more than four seldom is used. The input circuit drives the tube under class C conditions, which produce pulses of current in the plate circuit considerably shorter than a half-cycle of input. These pulses shock excite the tuned circuit to oscillation at its resonant frequency. Its resonant frequency can be any whole number times the frequency of the input wave. When the pulses are exactly one-half, one-third, or one-quarter of the frequency to which the output circuit is tuned, they reinforce its current swing every second, third, or fourth cycle.

(3) Consequently, they will help to sustain the oscillation in the tuned, or *tank*, circuit. To do this most efficiently, the pulses must be made short and sharp by biasing the tube even farther beyond cut-off than is common for a normal class C power amplifier, and by using a larger amount of excitation.

quencies which differ by a large amount, there is little danger of undesirable feedback and oscillation taking place through the grid-plate capacitance. Therefore, no neutralization is necessary. The operating condition with high negative grid bias and a large grid-voltage swing to obtain steep pulses of plate current means that the frequency multiplier requires more excitation than the same tube operated as an ordinary class C amplifier. The efficiency is lower than that of the corresponding amplifier. A doubler runs at less than 50 percent plate efficiency. The tripler and quadrupler operate at even lower efficiency. It is not practical to multiply more than four times because the efficiency drops off too much. As the multiplication increases, the grid bias must be made greater and the grid-voltage swing correspondingly larger. Therefore, a tripler requires more excitation than a doubler, and so on. In a fre-

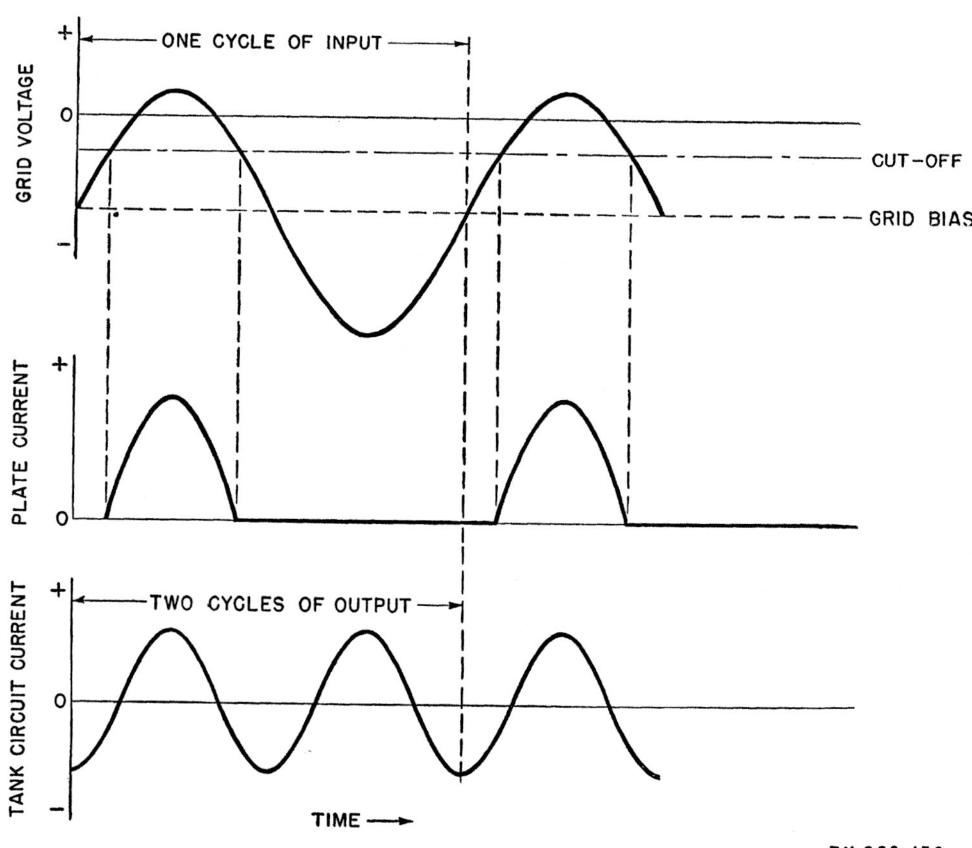

Figure 147. Voltage and current relationships in class C doubler.

quency multiplying circuit, the Q of the tank must be higher than the Q of an ordinary amplifier so that the multiplier action will be more efficient. Then there will not be too much loss of amplitude of oscillation between pulses of plate current.

f. POWER AMPLIFIERS.

(1) The generation of power at radio frequencies requires many varied types of tubes and circuits, some of which have been described earlier in this chapter. At this time, their applications to the generation and amplification of power will be considered. The losses anticipated and the frequency and mechanical considerations generally determine, within one or two choices, the type of tube to be used in any given case. Power tubes of both the air- and water-cooled types are available in a wide range of plate dissipation, up to several hundred kilowatts for water-cooled types.

(2) In more recent equipment, the beam power design is embodied in the tubes used. For transmitting applications the principle of beamed electron flow is useful when applied to almost any tube. Beam triodes such as the 5831 recently have come into use as modifications of the high-power water-cooled triode. Beam tetrodes are among the most frequently used of transmitting types because of their low excitation power requirements, ability to operate without neutralization in many cases, and high power efficiency. The beam principle, which also includes carefully alined control and screen grids, is applied to many tetrodes where beam-forming plates are not used, as in the 4–250A.

(3) Generally, power amplifiers are oper-

ated under class C conditions. For special applications, however, class B amplification sometimes is more suitable. In a class C amplifier, if sufficient excitation is supplied, the output waveform is relatively independent of the input waveform. A class B amplifier, on the other hand, reproduces the input waveshape and amplitude with a fair degree of faithfulness. Moreover, in r-f applications, when tuned circuits are used in the output, class B amplifiers can be used with only one tube. This is in contrast to audio applications where two tubes are necessary in push-pull connection to prevent excessive distortion. The use of one tube in r-f applications is permissible because the tuned circuit in the output supplies the missing half-cycle caused when the plate current stops as a result of the grid being driven beyond cut-off. Amplifiers operated in this service are useful when it is desired to make the output signal dependent upon the input. They also produce lower distortion of the waveform than class C amplifiers do and correspondingly lower unwanted harmonic output. Class B amplifiers require very stable sources of d-c voltage.

(4) A power amplifier that uses one tube is called a single-ended circuit. Two tubes, with the grids connected to the opposite sides of the input tank circuit and the plates to opposite sides of the output tank, form a push-pull circuit.

(5) The circuit usually is the same whether operated class B or C. The difference is in bias and excitation voltages and in values of components. The single-ended circuit generates more harmonic output which sometimes is a disadvantage. It is, however, the simplest circuit to use over a wide range of frequencies because switches can be inserted easily to change the tank circuits. Twice as many switches would be required in a push-pull circuit to accomplish the same purpose.

(6) Several types of tank circuits provide a wide range of impedance matching (fig. 148). These circuits are necessary where a variety of different antennas must be used. In A, L1C1 forms the input tank circuit. C2 is an r-f bypass capacitor which effectively grounds the bottom of the tank to rf. Grid bias is taken from a fixed bias supply. Capacitors C3 and C4 are filament r-f bypass capacitors, and T1 is the filament transformer. The plate tank circuit is composed of C6, a variable capacitor with a single rotor and two stators, and L2. C6 is called a split-stator capacitor, and is used to obtain the out-of-phase voltage which is fed back through C7 for neutralization. The output tank circuit is coupled to the antenna by a link coil around L2. This circuit is used with high-power triodes.

(7) In B, a tetrode is connected in a single-ended circuit without neutralization. The functions of the parts are much the same as those in A. Because no neutralization is necessary, a simpler type of output tank can be used. C7 is merely an r-f bypass capacitor which permits grounding the rotor of C6 to rf. This is occasionally desirable even though a higher voltage capacitor will be necessary.

(8) In C, a small pentode is utilized in a small transmitter for aircraft or liferaft service. In this case the type of antenna can vary considerably. The tube is capacitively coupled to the preceding stage through C1R1. R1, in addition, acts as a grid-leak resistor and provides grid bias. Cathode bias also is provided by R2 as a protective measure in case the excitation fails, which would result in a loss of grid bias if R1 alone were used. C2 is an r-f bypass capacitor for R2. C3 is a screen bypass capacitor used in conjunction with R3 to obtain the proper dc and r-f ground at the screen grid. The r-f choke and C4 isolate the output circuit from the power supply. This arrangement is known as shunt or paral-

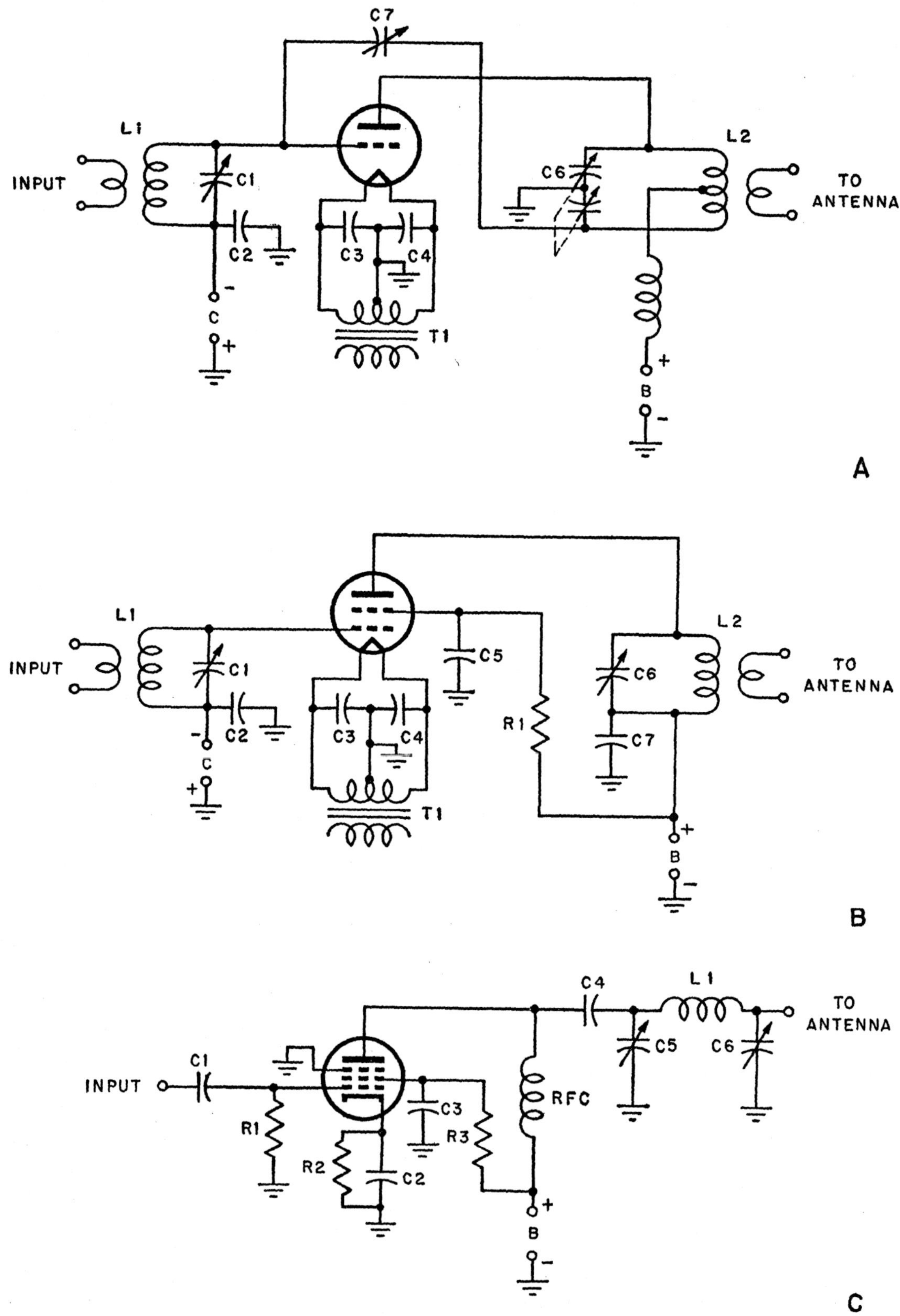

Figure 148. Single-ended power amplifier circuits.

lel feed and is used when it is undesirable to have dc in the tank circuit. The tank in this figure is a special impedance-matching circuit known as a pi network. It can be used with almost any length of antenna. This network (L1, C5, and C6) is a common feature of much portable equipment.

(9) Since a push-pull circuit uses two identical tubes, advantage is sometimes taken of this characteristic, and the two tubes are combined in one glass envelope. This is especially useful at very high frequencies. In this band, the length of connecting leads becomes important, and this is one method of keeping them short. A dual beam power tetrode such as the 829–B (fig. 140), can be air-cooled and is useful for up to about 100 watts output. It has no base and the leads are brought out through seals in the glass. It can be used at full power input up to 200 mc. Push-pull circuits have the advantage of lower harmonic output, ease of neutralization, and greater power output for a given amount of excitation. Such a circuit requires only as much excitation as one of its tubes does when run by itself in single-ended connection. This is because both halves of the grid-voltage cycle are used alternately by each tube. The output is recombined in the plate circuit. A and B, figure 149, shows two typical push-pull circuits for a power triode.

(10) In A, the input tank circuit provides equal and opposite voltages to the grids of the two triodes. The r-f choke provides a means of inserting grid bias from a fixed bias supply into the grid circuit. The plate tank circuit is made with a split-stator capacitor, C2, and recombines the output from the two triodes. Since opposite sides of a tank circuit are 180° out of phase, neutralizing voltages may be obtained simply by feeding back some of the energy at these points to the opposite grids. Capacitors C3 and C4 do this.

(11) The 829–B push-pull circuit in B has a grid tank, L1 and C1, which provides drive for the tube. Grid bias is furnished from a fixed supply. In the tube, the screens of both sections are connected, and one capacitor, C2, serves as a bypass for them. R2 provides screen voltage by dropping the plate voltage because of the current that passes through it. The output tank circuit is the same as that in A. The r-f choke permits feeding plate voltage to both halves of the tube.

g. SPEECH AMPLIFIERS.

(1) In a transmitter producing radio-telephone signals, it is necessary to take the extremely weak output of a microphone and build it up to the point where it can be used to modulate the carrier wave. This requires considerable amplification, depending on the type of microphone used. In general, the methods of coupling between the amplifier tubes are similar to the methods described in discussions of various types of amplifiers in earlier chapters. Special problems arise, however, in connection with the speech amplification stages of transmitters that do not occur in other cases.

(2) To begin with, the speech-amplifier tubes used in transmitters can be ordinary receiving types, although these are unsatisfactory in some applications. The average receiving-type tube does not have sufficient internal mechanical bracing to prevent minute vibration of the grid and other parts which cause noise on the signal. Therefore, tubes with high mechanical stability are manufactured especially for such applications. These are the *non-microphonic* tubes in the 1600 and 5500 series. They are electrically equivalent to many commonly used receiving tubes, but have the added internal bracing to prevent vibration of the tube parts.

(3) Speech stages operating at low levels in a transmitter usually are biased class A, since class A operation results

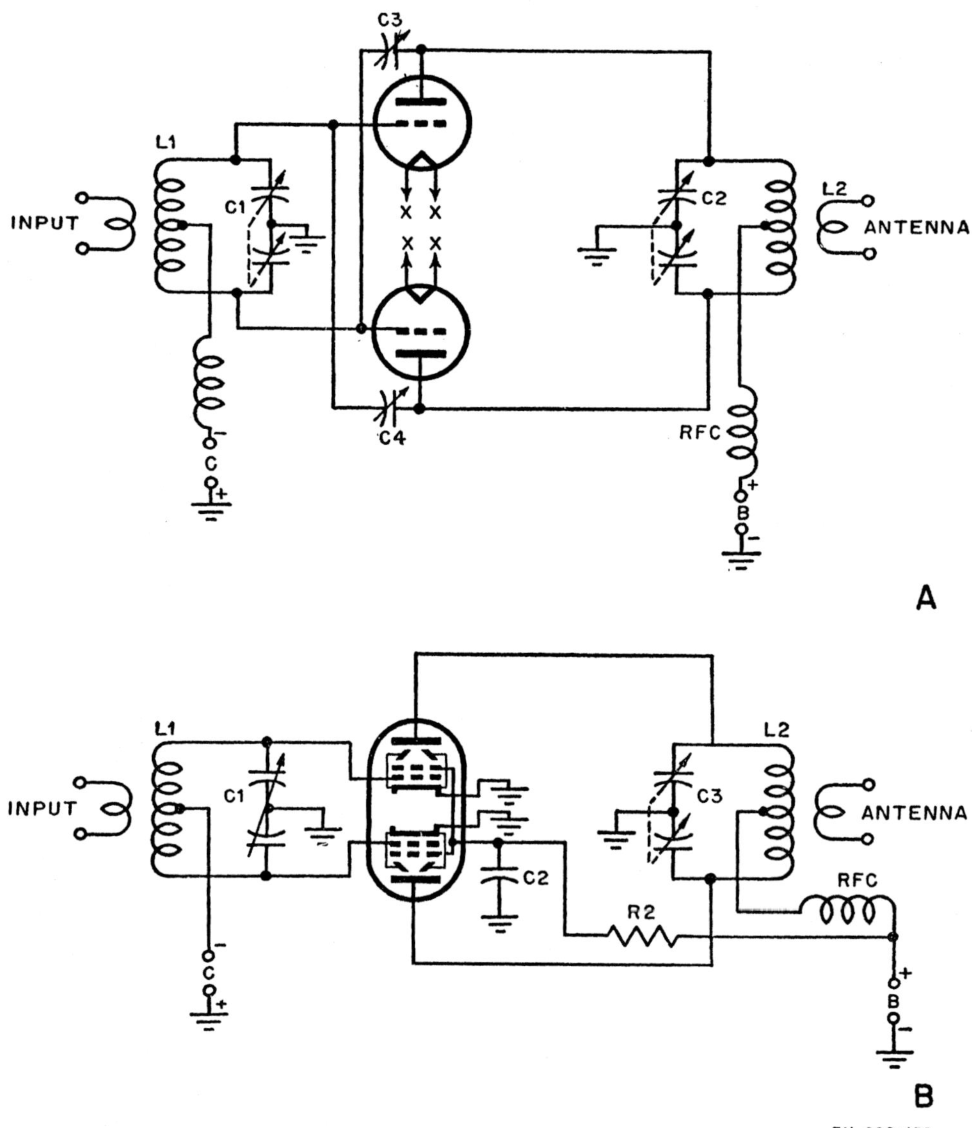

Figure 149. Push-pull power amplifier circuits.

in minimum distortion. Where a larger amount of power is necessary, operation with class AB conditions or even class B is used. In good transmitter design, some form of inverse feedback is used to improve the quality of the speech stages and lower the distortion. This means that extra amounts of gain will be necessary to compensate for the loss through feedback.

(4) In general, the first tube in a speech amplifier circuit is used as a microphone amplifier. It takes the form of a voltage amplifier, frequently a high-gain pentode. In this stage, high gain, low circuit noise, and low hum pick-up are desirable. Depending on the type of microphone used, it will be either transformer- or resistance-coupled. If a signal of reasonable amplitude is produced by this stage with low noise and hum content, it will tend to override any noise or hum that may be produced in subsequent amplifier stages.

(5) In figure 150, typical microphone

AGO 2244A

stages are shown for use with the three principle types of microphones encountered. These are the carbon, the crystal, and the magnetic microphone stages. The carbon microphone has the highest output of these three types, averaging around .1 volt for normal speech. The crystal is next, with approximately .03 volt, and the magnetic is lowest, with less than .01 volt. Greater fidelity is obtained with the lower-level microphones. There are other types of microphones in use, most of which use some type of variable resistance, variable magnetic field, or crystal effects.

(6) The carbon microphone amplifier, in A, is a simple transformer-coupled triode of low amplification factor. The transformer is necessary to allow the application of a small d-c voltage to the microphone. A carbon microphone uses loosely packed carbon grains mechanically coupled to a diaphragm which moves in response to sound. The resistance of the carbon varies with sound intensity. Therefore, if a current is allowed to flow through the carbon, a varying voltage is produced.

(7) The crystal microphone, in B, is used with resistance coupling to a high-gain pentode. The resistance-capacitance network shown in the input of the crystal microphone amplifier is used to prevent r-f current from being detected between the grid and the cathode and causing unwanted feedback or oscillation between the speech stages and the rest of the transmitter. The crystal microphone develops audio voltage as a consequence of the properties of certain mineral crystals that produce a current when a mechanical stress is applied to their surfaces. A diaphragm is coupled to the crystal to provide a transfer of force from the sound to the crystal.

(8) The magnetic microphone is coupled to the tube through a transformer which matches its low impedance to the high-impedance grid circuit, as in

C. It operates because of the voltage induced in a pick-up coil attached to a diaphragm, which moves in a magnetic field.

(9) The microphone amplifier usually is followed by one or more stages of resistance- or transformer-coupled class A amplifiers, generally using triodes. In broadcast transmitters, all stages, including the microphone amplifier, are push-pull to minimize distortion and hum. The output of these stages then is amplified by push-pull power amplifiers which operate in class AB_1. These AB_1 power amplifiers also can use ordinary receiving-tube power amplifiers, since they seldom operate at levels above 25 watts. Here again, consideration of ruggedness, long life, or nonmicrophonism can determine the use of special types which are basically the same electrically as receiving tubes. Figure 151 shows a typical speech amplifier with crystal microphone amplifier, intermediate triode amplifier, and push-pull AB_1 power amplifier circuits. The separate stages are all the same as those described in the text.

(10) In frequency-modulation transmitters, very little power is required for the incorporation of speech in the carrier wave, and for this application low-power class A voltage amplification is sufficient. The tubes and circuits are identical with those described in the foregoing material, with the exception that no class AB_1 power stage is necessary.

h. MODULATORS.

(1) For the generation of large amounts of audio power required in some methods of modulation, regular transmitting-type power-amplifier tubes such as those we have been discussing can be used. Most high-power audio amplifiers are operated class B, because this is the most efficient method of amplification that still preserves the input waveshape. In general, transformer coupling is used both in the

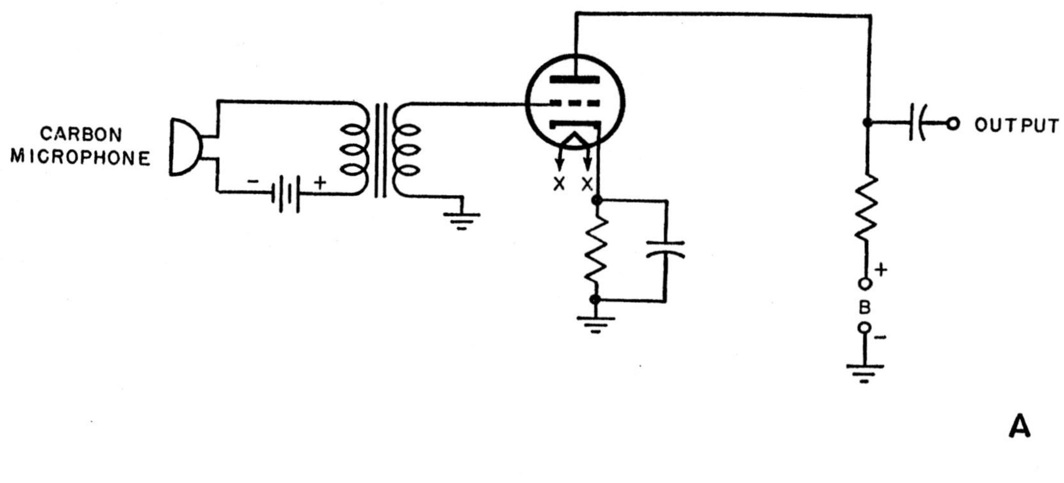

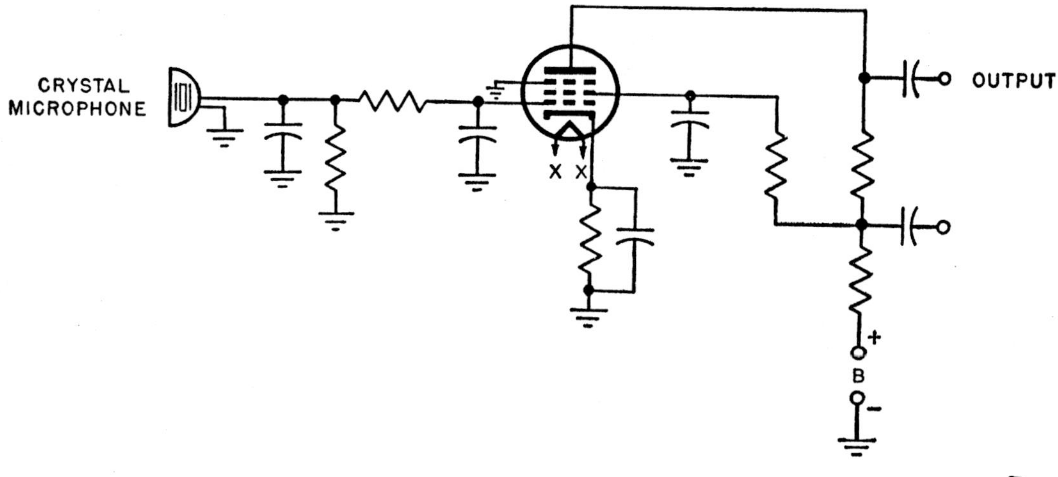

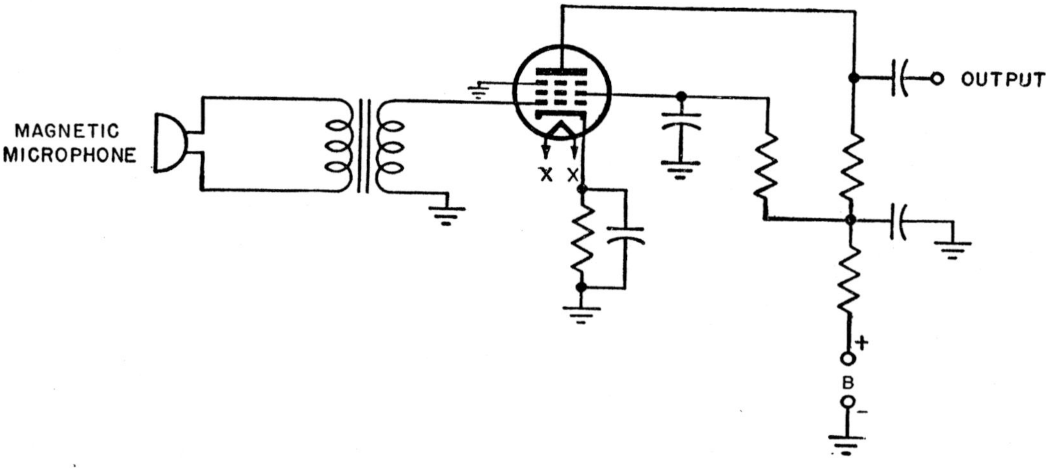

Figure 150. Typical microphone amplifiers.

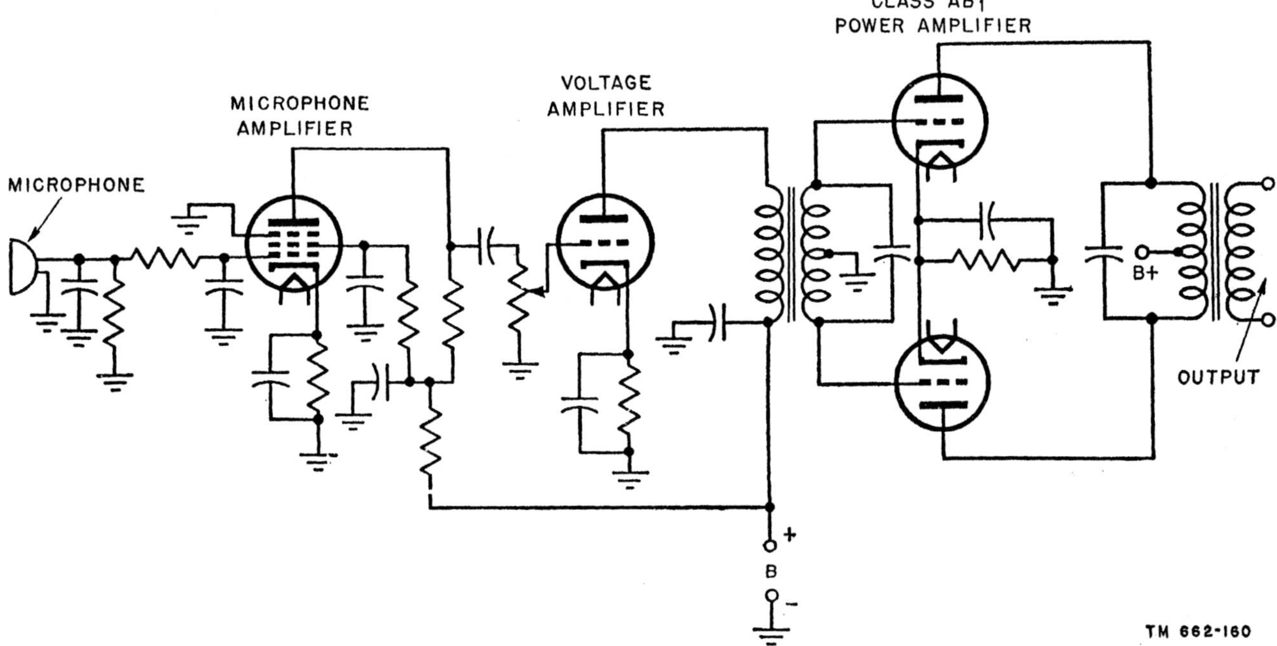

Figure 151. Speech amplifier circuit.

input and output circuits, and all high-power class B amplifiers tend to have nearly the same circuit regardless of size (fig. 152). In a class B amplifier, the grids are biased at cut-off and plate current flows only when a signal is present. Therefore, with a speech wave which is highly irregular, a wide variation of current is drawn by the stage. The power supplies for class B stages must have excellent regulation or distortion will result.

(2) In some tubes designed specifically for class B audio service, such as the 811, advantage is taken of the necessity for operating the grids at cut-off, and the tube is made so that cut-off occurs with nearly zero grid bias. Therefore, no bias supply is necessary. These are called zero-bias tubes. Triodes are widely used as class B amplifiers because of the simplicity of the circuits necessary. However, tetrodes require considerably less excitation than triodes in class B and they are used occasionally. Because of the shape of the characteristic curves of beam tetrodes, they are seldom operated completely in the class B region; instead, they are used in the region of AB_2, where the grid is not quite cut off, and some plate current flows with no signal. Except for the provision of screen voltage, such tubes are used in circuits substantially similar to that shown in figure 152.

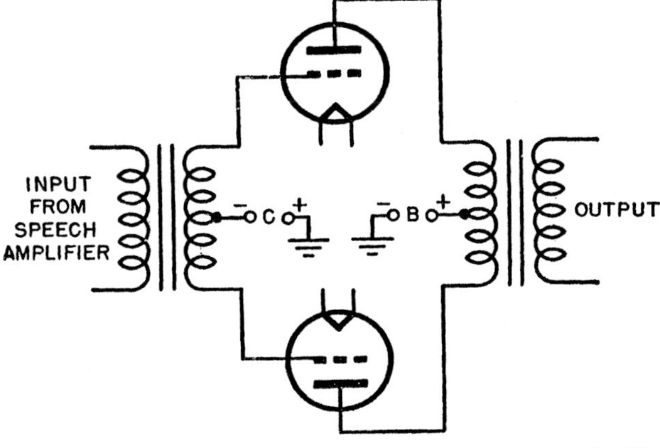

Figure 152. Class B amplifier, a high-level modulator.

(3) Radiotelephone transmitters can be modulated in a great variety of ways. When the carrier amplitude is varied in proportion to the amplitude of speech, this is termed *amplitude modulation*. When the carrier frequency is varied in proportion to the amplitude, the system is termed *frequency modulation*. When the phase of the carrier

is changed in proportion to the amplitude of the speech, the resultant modulation is termed *phase modulation*. Amplitude modulation can be introduced at either a high- or a low-power stage in a transmitter. Frequency and phase modulation always are produced in a low-level stage.

(4) When amplitude modulation is produced in a low-level stage, all subsequent stages must reproduce the amplitude of the input wave faithfully. Therefore, class B stages must be used, and no frequency multipliers can be used after the modulated stage. Modulation introduced into the final power amplifier is termed *high-level modulation*. The variation of the voltage at the grids or plate or a combination of both, proportionate to the input signal, produces amplitude modulation. If control-grid voltage is varied, the system is termed control-grid modulated. It also may be screen-modulated, **or** suppressor-, or plate-modulated.

(5) Grid modulation in its varying forms requires less power than plate modulation. However, to prevent excessive plate dissipation when the grid goes highly positive under modulation, the tube must be operated with half its normal grid-driving voltage. Since only half the plate current will be drawn and since power is proportional to the square of the current, it is clear that only one-quarter the normal power input can be used in a grid-modulated amplifier as compared to the input of a plate-modulated amplifier. However, plate modulation requires a high-power modulation transformer to couple the audio tubes to the r-f tubes. This is expensive and cumbersome, and occasionally the lower efficiency of grid modulation is tolerated because of the savings in cost and weight. Inspection of the plate-modulated amplifier in figure 153 reveals that this is a composite of the high-power triode amplifier of A of figure 148 and of the high-power class B modulator of figure 151.

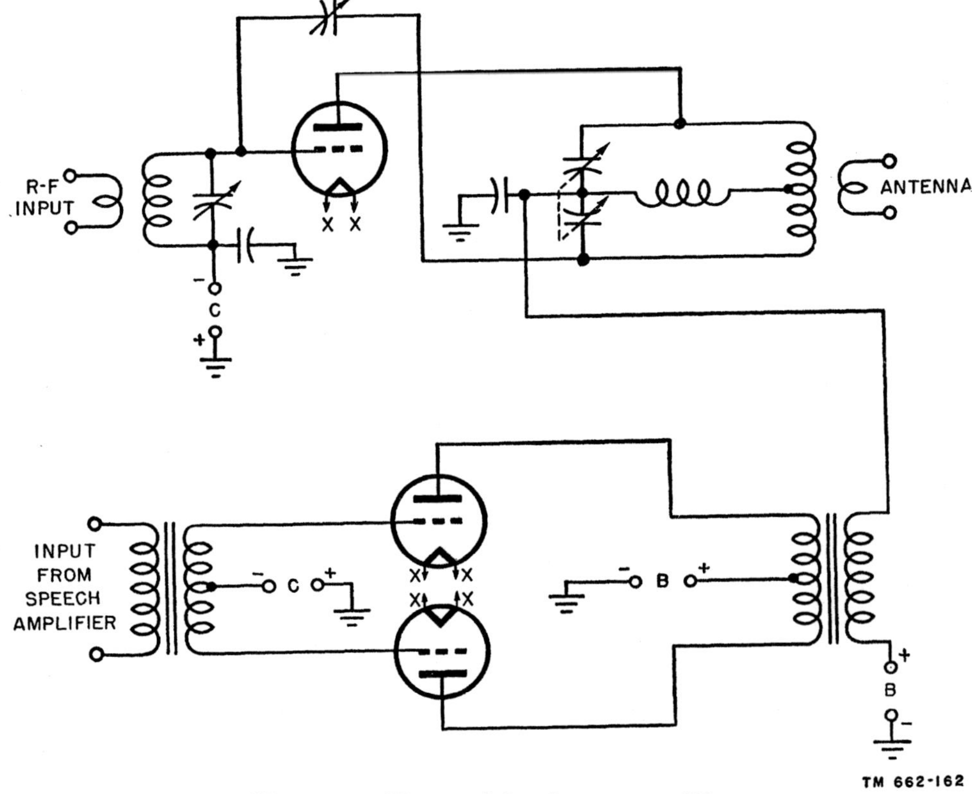

Figure 153. Plate-modulated power amplifier.

(6) In the grid-modulated amplifier in figure 154, the voltage for modulation is introduced by a small transformer which is in series with the lead to the grid-bias supply. The modulator is the push-pull AB_1 circuit of figure 151 with the triode amplifier of A of figure 148.

lation of the final amplifier is not desirable, this system is used. The major elements of high cost in a plate-modulated final amplifier are the modulation transformer and the heavy power supplies necessary for class B modulator. Engineering compromises sometimes are made to avoid these ex-

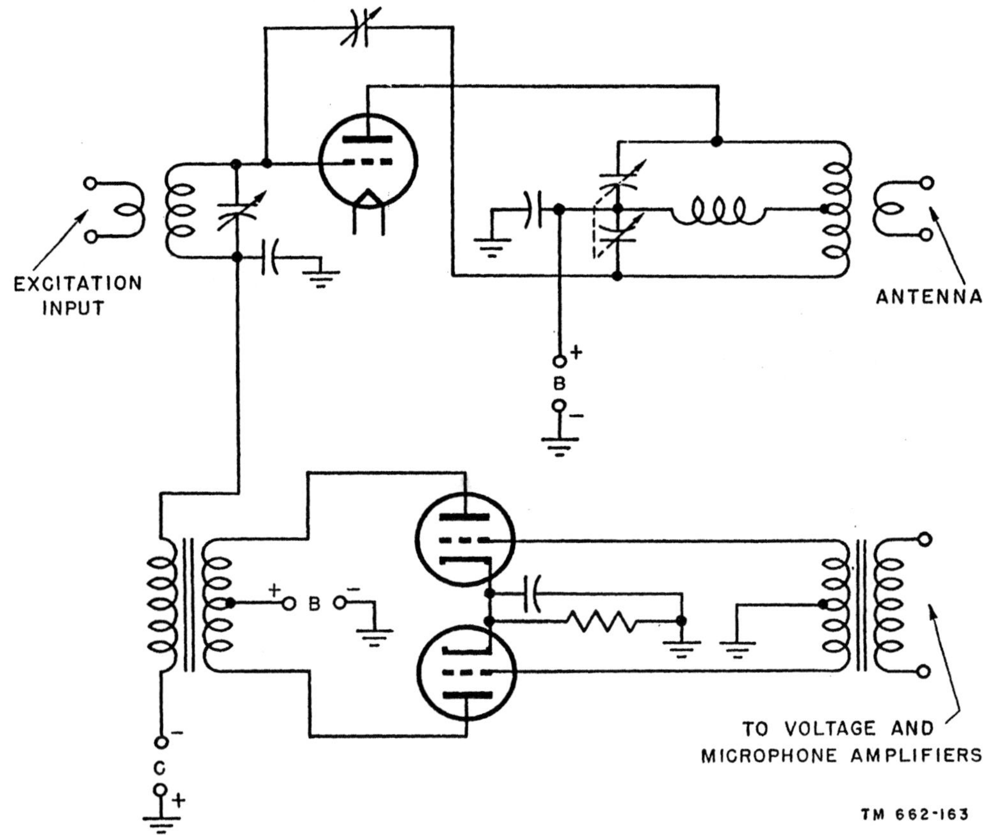

Figure 154. Grid-modulated amplifier.

(7) A low-level modulated stage is usually a plate- or grid-modulated amplifier before the final amplifier. Generally, the buffer or a stage preceding it is modulated. The circuits are not different from those used in high-level modulators, except that small tubes and components are permissible. Low-level modulation has the disadvantage of requiring class B final amplifiers to preserve the modulated waveshape, and these are not as efficient as class C tubes. However, in some instances where cost is important and grid modu-

penses. High-level plate modulation, however, is the most efficient of simple modulation systems from the standpoint of modulated-carrier output with a given final amplifier.

i. POWER SUPPLIES.

(1) Radio transmitters require the application of many different voltages and currents to the various stages. The high-power tubes used in the final amplifier and modulators operate at high voltages and currents. If an a-c line is used as the power source, the a-c

voltage is stepped up in a transformer and converted into dc by a power rectifier. A of figure 155 shows a typical high-vacuum rectifier tube, the 836, which is used to furnish high d-c voltage and current for transmitters of several hundred watts input.

(2) The important characteristics of a high-vacuum diode rectifier are maximum permissible peak current, maximum permissible average plate current, and maximum permissible inverse peak voltage. The cathode can supply only a limited amount of elec-

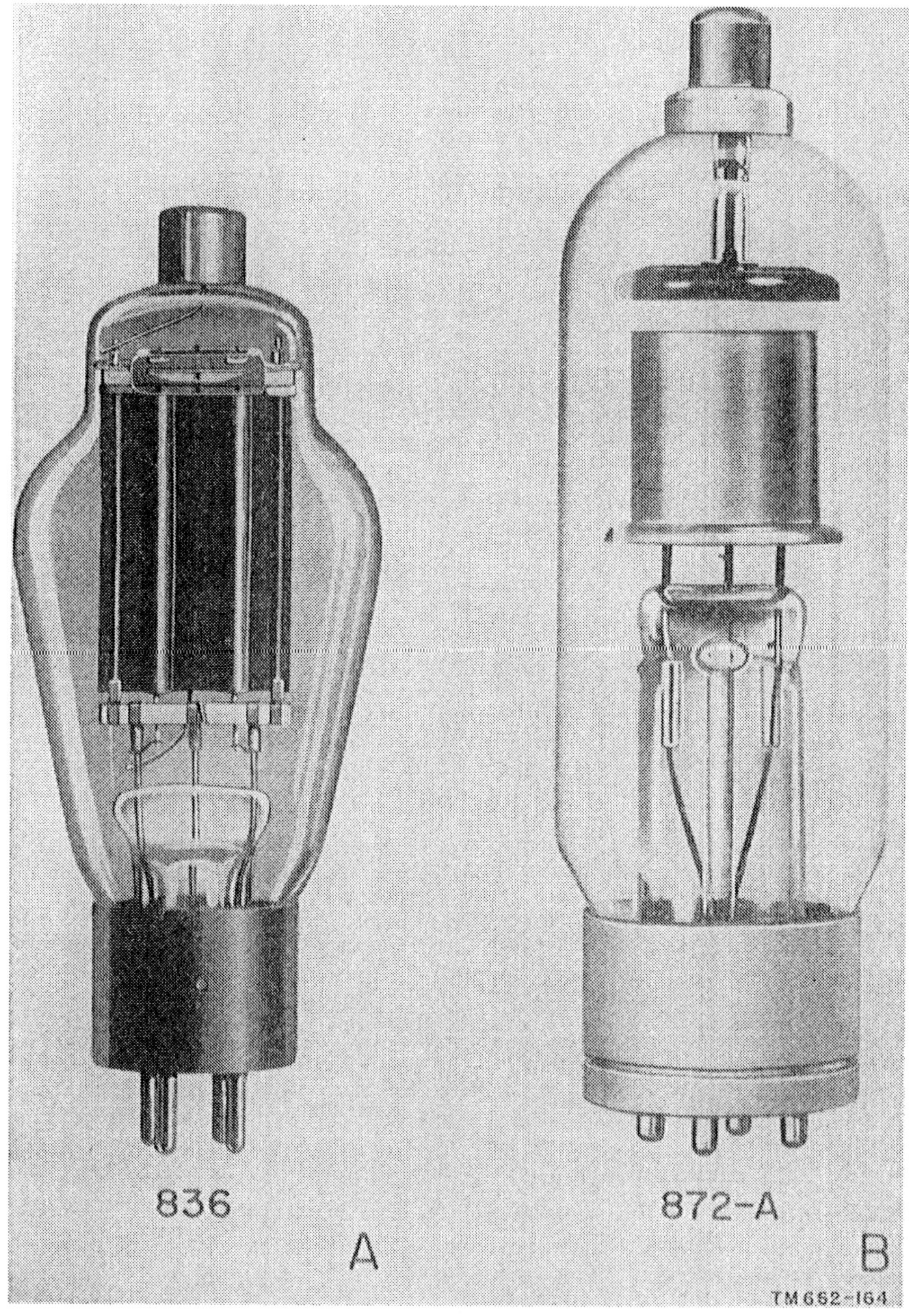

Figure 155. Typical transmitter half-wave rectifiers.

trons, and this limits the peak plate current. The maximum d-c output current that can be carried continuously without overheating the tube is reflected in the maximum permissible average plate current. The largest voltage that can be applied to the plate on the negative half of the a-c cycle without causing arcing or internal breakdown is called the maximum inverse peak voltage.

(3) Materials and construction in high-vacuum rectifiers are similar to those used in other transmitting tubes. High-vacuum rectifiers are used for developing d-c output voltages up to several thousand volts, but, in general, their use is restricted to current ranges below 400 ma. The hot-cathode mercury-vapor rectifier is a diode, such as the 872A in B of figure 155, capable of passing currents up to several amperes when the mercury vapor is ionized by collision with the electrons produced by the cathode. The result is a low potential drop across the tube of approximately 15 volts, which permits good voltage regulation even when the spacing between the plate and cathode is large. The low voltage drop, existing in the tube, permits the use of oxide-coated cathodes with correspondingly high emissivity. The same rating system applies to mercury rectifiers as applies to high-vacuum rectifiers. They require more care in operation, however. The cathode must be brought to normal operating temperature before plate voltage is applied. The tube must operate in a certain temperature range; otherwise the mercury does not vaporize properly. It must be protected from momentary overloads, because, if the peak ratings are exceeded, severe damage to the tube will result. Mercury-vapor rectifiers are used almost universally where high voltages and high currents are required. They have the advantage over high-vacuum rectifiers of much greater efficiency, better voltage regulation, and higher permissible current.

121. Tuning Procedure

a. The typical transmitter shown in the block diagram of figure 141 must be adjusted in the field so that it will perform as intended. The process of adjustment is called the tuning procedure. Usually it is needed with all new equipment and is used periodically on equipment in service. Exact tuning instructions vary from one piece of equipment to another, of course, but the same general principles are involved.

b. Basically, almost all tuning procedures consist of bringing the resonant circuits associated with the transmitting tubes to their proper frequencies. This is done by observing the variation in a current or a voltage measured at the grid, plate, cathode, or other element in the stage being adjusted, or in a subsequent stage. This means that either the inductive or the capacitive elements of the resonant circuits, or both, must be adjustable over a range sufficient to allow adjustment with different operational conditions and small variations in individual tubes.

c. If the output circuit of a class C amplifier is tuned through resonance, there is a change in the current through the tube because the plate-load impedance is changing. Because the plate current is in the form of positive pulses, an average positive direct current is associated with these pulses of plate current that also will vary with tuning. Measuring instruments for dc are much easier to use and construct than those for rf and ac, and therefore, this change in d-c plate current in the class C amplifier is a useful indication of resonance in the plate circuit. The change is in the form of a dip in current at resonance. In a stable triode amplifier, this also corresponds to maximum power output. In tetrodes and pentodes, maximum power output may not necessarily occur at this dip. In this case, the screen current may be a more reliable indication.

d. In a class C amplifier, the grid current can be used as a means of determining the resonance of the input circuit. Maximum grid current as measured on a d-c meter is a reliable indication of resonance in the input tank circuit.

e. It sometimes is inconvenient to measure d-c plate current because of the high voltage to which the meter may be exposed. In this situation, cathode current, as measured at the center tap of the filament transformer, also serves as an indicator of plate resonance, although some additional current change is caused by the grid current. In general, however, the grid current is but a small fraction of the plate current and does not materially affect the measurement.

f. The steps below give a general tuning procedure for multistage transmitters like those in the block diagram of figure 141, based on these principles. Where the stage to be tuned is not a class C stage, the maximum power output can be determined by the effect on the grid current of the class C stage which follows it. Maximum power output and optimum coupling will be indicated by maximum grid current.

(1) Connect the dummy load in place of the antenna.

(2) Turn all filaments on.

(3) Apply power to the oscillator and first buffer stage.

(4) Tune the buffer stage for maximum grid current in the following tube.

(5) Apply plate voltage to the next stage. If the next section of the transmitter is a frequency multiplier, adjust the plate-tank capacitor for minimum plate current. Do this in each multiplier stage, starting with the lowest frequency.

(6) Check with an absorption wavemeter or other r-f indicator to make sure that the output of each stage is at the desired harmonic of the input signal.

(7) Apply power to the driver or the buffer before the final amplifier. Tune its output circuit for minimum plate current. Tune the grid-tank circuit of the final amplifier for maximum grid current. Check to see that this corresponds to the rated grid excitation. Check the fixed bias supply of the final amplifier for rated voltage. Retune the buffer output for minimum plate current. Adjust the coupling between the buffer and the final amplifier if the grid current of the final stage is not sufficient or is above the recommended value. Each time an adjustment in coupling is made, the tuning process is repeated.

(8) The next step is neutralizing the final amplifier as described in the beginning of the chapter. After the stage has been stabilized, apply final plate voltage.

(9) Adjust the final-amplifier tank capacitor for a drop in plate current, which indicates resonance. The transmitter is now ready to go on the air.

(10) Remove the final plate voltage. Remove the dummy load. Connect the antenna. Reapply final plate voltage and proceed with communication.

122. Summary

a. Transmitter tubes are classified as diode, triode, tetrode, pentode, and beam power types.

b. Large transmitter tubes produce large amounts of heat which must be removed in order to prevent damage to the tubes and associated circuits. This heat is produced mainly at the plate, but also at any grid drawing current.

c. Heat removal can be accomplished by natural air currents, forced-air cooling, or water cooling. Blackening the plate also assists in heat removal. Grids drawing large currents are constructed with internal tubing through which water is circulated.

d. Oxide-coated cathodes can be used in transmitter tubes dissipating up to about 50 watts. For high emission, the thoriated-tungsten cathode is used. A thin layer of thorium is deposited on the surface of the tungsten during the manufacturing process. For very high emission, the thorium is deposited on a layer of tungsten carbide to form a carburized filament. Multistrand carburized filaments are used in the highest power transmitter tubes.

e. The ability of a metal to emit heat as radiation is called its thermal emissivity. Among the metals used for plates, graphite has the highest thermal emissivity.

f. Typical plate materials are graphite, copper, molybdenum, nickel, tantalum, and tungsten. Nickel is used for low-power transmitter-tube plates, graphite and molybdenum for medium power, tantalum for high power, and tungsten for extremely high power.

g. Materials with a high melting point, such as molybdenum, tantalum, and tungsten, are used in the construction of grids.

h. Transmitter-tube bases usually are constructed of a plastic such as bakelite. For high power a metal-clad ceramic can be used. At very high frequencies, no base is used in order to shorten leads to the tube electrodes.

i. Hard glass with a high softening point and good electrical insulating properties is used to form the envelopes of transmitter tubes.

j. An oscillator is used in transmitters to generate a radio-frequency signal. Transmitter oscillators are basically the same as those used in receivers. The most common transmitter oscillator is the crystal-controlled oscillator with its high degree of frequency stabilization.

k. Class C amplifiers are used to amplify the output of the oscillator because of their efficiency. If triode tubes are used, they must be neutralized to prevent undesirable self-oscillation.

l. Buffers are used to prevent a power stage from loading down the stages preceding it. For example, a buffer is inserted between an oscillator and a power amplifier to prevent the power stage from changing the operating frequency of the oscillator.

m. A frequency multiplier is used to raise the output frequency of a crystal oscillator. It operates at some multiple of the oscillator frequency. It can be a frequency doubler, tripler, or quadrupler.

n. Power amplifiers are used in the output stages of transmitters to strengthen the modulated carrier for transmission. Usually, they are operated class C. Class B can be used also in single-ended operation.

o. Speech amplifiers usually are operated class A in order to keep distortion at a minimum. High voltage gain, low circuit noise, and low hum pick-up are desirable.

p. Three principle types of microphones are used with speech amplifiers. They are the carbon, the crystal, and the magnetic microphones. The carbon has the highest output and the lowest fidelity; the magnetic has the lowest output and the highest fidelity.

q. The modulator builds up the audio signal and modulates the r-f carrier with its output. It usually is operated class B for relatively high power and fidelity.

r. If the audio-signal output of the modulator is injected into the final power amplifier, the process is called high-level modulation. If it is injected into some preceding stage, the process is called low-level modulation. The audio signal can be injected at any grid or at the plate of the modulation stage.

s. High-vacuum diode rectifiers and hot-cathode mercury-vapor rectifiers are used to convert a-c power to the d-c power required by the transmitter.

t. Tuning a transmitter usually consists of adjusting resonant circuits to their proper operating frequencies. Meters usually are incorporated in various transmitter circuits to give resonant indications.

123. Review Questions

a. Why do transmitting tubes differ from receiving tubes, though of the same general type?

b. State several methods for removing heat losses.

c. What are the advantages and disadvantages of each?

d. Why must good insulation be used between the various parts of the tube?

e. What types of cathode are used in small transmitting tubes?

f. Why can they not be used in higher-power tubes?

g. What is a thoriated-tungsten cathode?

h. What are its advantages and disadvantages?

i. How is it reactivated?

j. How can the life of tungsten filaments be prolonged, and why should the initial surge of filament current be limited?

k. What are the advantages of water cooling, and what precautions must be observed in the operation of water-cooled tubes?

l. What is meant by thermal emissivity?

m. What determines the choice of plate material?

n. What are the advantages and disadvantages of the following as plate materials: nickel, graphite, copper, tantalum, molybdenum?

o. Why must materials that can withstand high temperatures be used for the grid?

p. What danger is inherent in high secondary emission at the grid?

q. What types of glass and metal construction are used in transmitting tubes? Why?

r. Describe briefly, with reference to the block diagram of figure 141, the function of each part of the transmitter.

s. What are the radio frequencies used for communications?

t. What are the requirements for tubes used as oscillators in transmitters?

u. What are the advantages of crystal-controlled oscillators?

v. What type of amplifier generally is used for power amplifiers?

w. Why must triodes be neutralized when used as r-f amplifiers?

x. How is neutralization accomplished?

y. What are the requirements placed upon a stage which provides excitation to a class C amplifier?

z. Why are frequency-multiplying circuits used?

aa. What is the difference in the waveshape of the plate current of a frequency multiplier and a power amplifier using the same tube and circuit?

ab. Why does the output circuit of a frequency multiplier contain rf of a frequency higher than that of the input circuit?

ac. Is a frequency multiplier less efficient than a power amplifier? Why?

ad. When are class B power amplifiers used?

ae. What is the difference between a single-ended circuit and a push-pull circuit with regard to harmonic output?

af. Describe some methods of providing grid bias to tubes in push-pull and single-ended amplifiers operating class C.

ag. Why is a split-stator capacitor or a grounded center-tapped coil used in a push-pull circuit?

ah. What are the requirements for the tubes used as microphone amplifiers?

ai. What types of tubes are used for the generation of large amounts of audio power? In what class do they run?

aj. What are the relative advantages of grid and low-level modulation?

ak. What features are important in a high-vacuum rectifier tube?

al. Why are mercury-vapor rectifiers used in preference to high-vacuum rectifiers?

am. What operating precautions must be observed with mercury-vapor rectifiers?

an. What are the indications that the plate circuits of an amplifier are tuned to resonance?

ao. What are the indications that the grid circuit is tuned to resonance?

ap. When tuning a transmitter, when should the final amplifier be neutralized?

CHAPTER 11

OTHER TUBES

124. Gas-filled Tubes

a. GENERAL. In high-vacuum tubes, discussed previously, the possibility of an electron colliding with a gas molecule is relatively small. The following paragraphs discuss tubes in which the flow of current takes place through relatively dense gas. A relatively dense gas is meant to be a gas of approximately one ten-thousandths of normal atmospheric pressure.

b. CONDUCTION.

(1) When an electron collides with a gas molecule, the energy imparted by the impact can cause the molecule to release an electron. This molecule is known as an ion. A gas or vapor containing no ions is an almost perfect insulator. If two electrodes are placed in such a gas, no current flows between them. However, gases always have some residual ionization due to cosmic rays, radioactive materials in the walls of the container, and the action of light. If a potential is applied between two electrodes in such a gas, the ions migrate between them, giving the effect of a current flow. This current is called the *dark current* because no visible light is associated with it. It is usually about 1 μa.

(2) If the voltage on the electrodes is increased, the current begins to rise. At a particular value, called the *threshold current,* usually about 2μa, the current suddenly begins to rise without any increase in applied voltage. If there is sufficient resistance in the external circuit to prevent the current from rising quickly, the voltage drops suddenly to a lower value, and *breakdown* occurs. This abrupt change takes place as the result of the ionization of the gas by electron collision. The electrons released by the ionized gas join the stream of electrons and can liberate other electrons. The process then is cumulative. The *breakdown voltage* at which this change takes place is determined by the type of gas, the materials used for the electrodes, their size and spacing, and other factors.

(3) Once ionization takes place, the current can rise to 50 ma or more with little change in the applied voltage. If the voltage is raised still farther, the current becomes higher and the electrode acting as the cathode becomes heated by the bombardment of the ions which strike it. If it gets hot enough, it emits electrons by thermionic emission. This emission reduces the voltage drop in the tube, causing further increase in current and greater emission and ionization. The cumulative action results in a sudden decrease in voltage drop across the tube, and the current rises to the extremely high value of several amperes. Unless the tube is designed specifically to operate under this condition, it can be destroyed by the heavy current. The mechanism just described is the basic process for the formation of an arc, and tubes which operate at these high currents are called arc tubes. In the region up to 50 ma, the tube usually is small and is called a *glow tube,* from the colored light it produces. The familiar neon sign is such a glow tube with neon as the gas.

(4) In figure 156 the dark-current region described above is shown from A to B. The breakdown-voltage point is at B. The drop in voltage and the sudden rise in current (threshold current) with little change in voltage are shown between C and D. The current continues to rise until an arc takes place at E, with a sudden drop in voltage and a great increase in current in the arc from F to G. The many varieties of arc, glow, and other gas tubes, all operate on some portion of the complete curve.

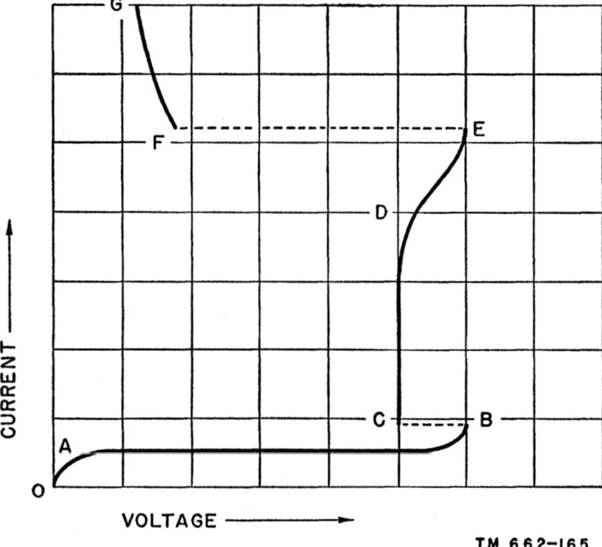

Figure 156. Graphical relations in gaseous conduction.

c. DIODES.

(1) A gas-filled tube in which two electrodes are inserted is called a gas diode. The electrode to which the positive potential is applied is called the plate and the other is the cathode, as in vacuum tubes. The cathode in a gas diode can be an electrode like the plate, or it can be a thermionic emitter. The former is known as a *cold* cathode, and the latter as a *hot* cathode.

(2) Cold-cathode tubes are used for many purposes. Among these are voltage regulation, rectification, oscillation, circuit protection, and light production, as for neon signs.

(3) Referring to figure 156, the small change of voltage from D to E is utilized in a voltage-regulator circuit. In the simple circuit shown in figure 157, the black dot within the envelope signifies that the tube is filled with gas. The resistance, R, is high enough to limit the tube current to the constant-voltage range when the load current is low. When the load current increases, the voltage drop across R increases and reduces the tube voltage. A small reduction of tube voltage in this range, called the *normal* region of the glow discharge, results in a large decrease in tube current, which decreases the voltage drop across R. Therefore, small variations of load current cause compensating variation in tube current, and the voltage across the tube remains essentially constant. A typical tube used in this circuit is the OD3/VR150.

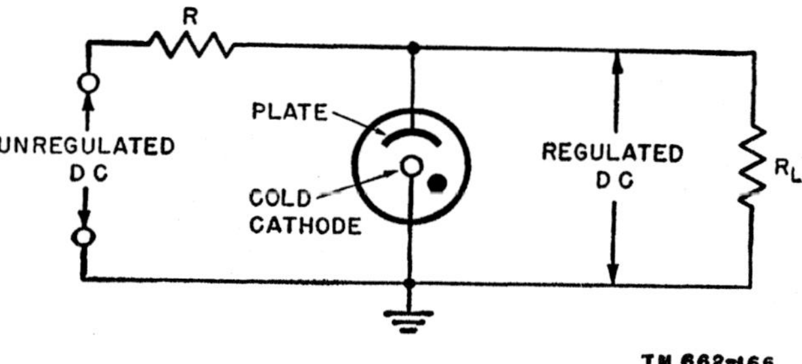

Figure 157. Simple voltage regulator circuit.

(4) Cold-cathode tubes rarely are used as rectifiers because of their high voltage drop, although a variation of the cold-cathode tube is used where filament power for a heater-cathode rectifier is difficult to obtain. An example of a cold-cathode rectifier is used in vehicular equipment. The principle of operation is based on the heating of the cathode under ionic bombardment in the region of the curve between F and G (fig. 156). The ion-heated cathode makes this tube a hot-cathode type without the application of heater voltage.

(5) If a gas diode with a cold cathode is connected as shown in A, figure 158, the resulting circuit is known as a *relaxation oscillator*. With a d-c input, capacitor C charges through the high resistance of R. The voltage across the capacitor rises as the capacitor charges. When the capacitor voltage becomes equal to the breakdown voltage, the glow tube fires, and the capacitor discharges through the tube. The capacitor discharges very quickly to a voltage equal to the extinction voltage of the tube. This is the voltage at which the tube ceases to conduct. Consequently, the capacitor begins to charge again. The output voltage is a sawtooth. Several cycles of capacitor charge and discharge generate the waveform shown in part B.

(6) Glow tubes are used widely for the protection of circuits. If a circuit may be damaged by the application of high voltages for short periods of time, a tube such as the NE–2 can be connected across it. The tube conducts during the high-voltage period, acting as a low resistance which prevents the high voltage from damaging the circuit. When the high voltage is removed, the NE–2 stops conducting and the circuit performs normally.

(7) Hot-cathode gas-filled tubes are important in power rectification. The gas most frequently used is mercury vapor, and the tubes are called mercury-vapor rectifiers, of which there are two types. One has an oxide-coated filament heated by an a-c current; the other uses a pool of mercury for a cathode.

(8) Mercury-vapor rectifiers with filaments, such as the 866A, are able to pass much higher currents than high-vacuum rectifiers because of the ionization of the mercury vapor. The mercury ions make it unnecessary to rely on the electrons produced by the filament alone. The filament is used merely to start the ionization. The heat from the filament vaporizes the small amount of liquid mercury incorporated in the tube envelope. This tube has a very high efficiency in power rectification because of its low voltage drop. The plate efficiency of a large mercury-vapor rectifier easily can approach 99 percent. Moreover, the tube drop does not vary with a varying load, and therefore the voltage regulation is better than that obtained with a high-vacuum rectifier tube.

(9) Arc tubes which use a pool of mercury as the cathode are known as mercury-pool rectifiers, an example being the 5554. This tube is capable of supplying 75-ampere dc continuously. The initial ionization is provided by an arc started between a subsidiary electrode, called the ignitor, and the cathode. The ionization results in greatly increased current. Since the ignitor must be used to start the arc, the tube can be controlled by small pulses of current to the ignitor, and thus hundreds of amperes can be controlled by a small amount of current.

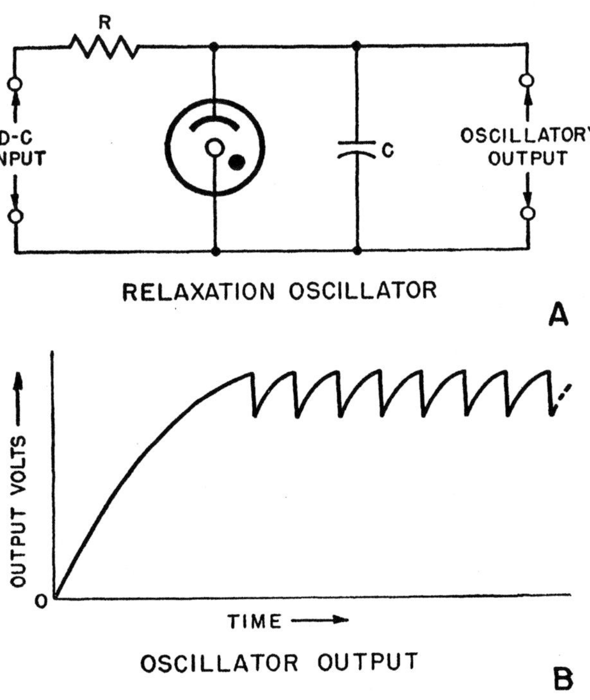

Figure 158. Relaxation oscillator and waveform produced.

d. THYRATRONS.

(1) If a grid is placed between the cathode and the plate of a gas tube, the voltage at which breakdown occurs can be controlled by the voltage on the grid. The entire plate surface in this tube usually is shielded by the grid before breakdown. The grid is placed close to the plate to prevent discharge between the two. If such a discharge does take place, it is only in the unimportant dark-current range. In a grid-controlled gas discharge tube, the plate-supply voltage exceeds the plate-cathode breakdown voltage and the grid is held either 0 or negative with relation to the cathode. Under these conditions, breakdown does not take place.

(2) If the grid voltage is raised, breakdown occurs between the grid and the cathode. This ionizes all the gas in the tube, and the discharge continues with plate-cathode current flow. Resistance in series with the grid limits its current on breakdown to a safe value. After breakdown, the grid no longer can control the discharge. If it is made negative with relation to the cathode, positive ions surround the grid wires, and electrons are repelled from them. The discharge then is shielded completely from the grid. To re-establish grid control, the plate potential must be reduced to the extinction potential of the cathode-plate discharge.

(3) This principle of grid control can be applied to almost any gas-discharge tube. It is used with cold-cathode, hot-cathode, and arc tubes. All of these types are given the generic name *thyratron.* Where a hot cathode is used, as in the 2D21 or 884, the grid acts primarily as a shield between the plate and the cathode, preventing electrons emitted by the cathode from ionizing the gas between the electrodes. By proper electrode arrangement a positive voltage on the grid is needed to start the discharge.

(4) The effectiveness of grid control is indicated by the grid-control characteristic curve for the 844 in figure 159. The curve shows the relationship between the plate voltage which must be applied with a given grid voltage to cause the tube to conduct. If the grid is below —30 volts, no amount of plate voltage can fire the tube. If the grid is above —30 volts, conduction begins within a few microseconds after the proper plate voltage is applied, depending on the gas used. Mercury vapor is common for most devices where the thyratron is used to control power, motors, etc. Hydrogen, however, is used where extremely rapid firing time is needed, as in radar applications. This is because of the lightness of its ions and the speed with which they move.

(5) The discharge of a gas tube can be controlled also by a grid external to the tube in contact with the glass. This principle is used in stroboscopic lighting for photographic purposes. A magnetic field can be used to control

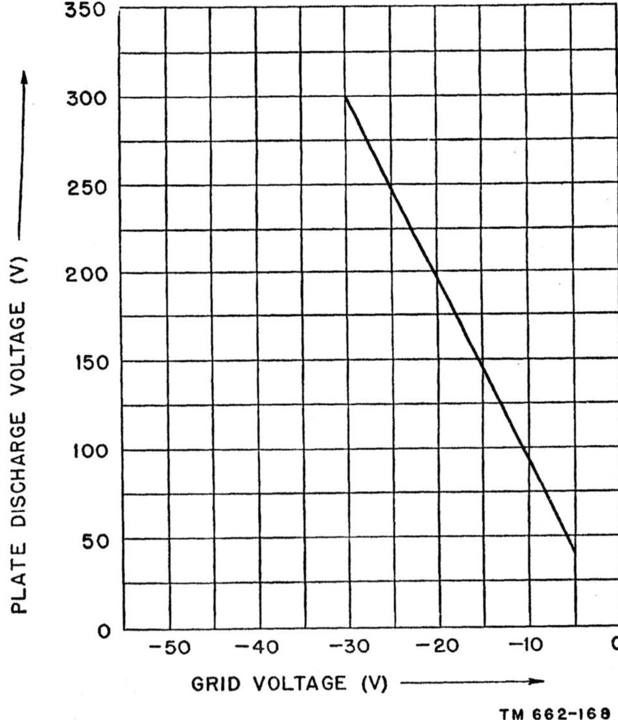

Figure 159. Grid-control characteristic of the 884.

the discharge. The field raises or lowers the firing voltage as it deflects the electrons leaving the cathode. This principle is used in the control of high-power rectifiers. The discharge between the cathode and the plate also can be initiated by means of two auxiliary electrodes that are made to start a glow discharge between them.

125. Phototubes

a. GENERAL. An important class of tubes is that in which the output current is controlled by light falling on the tube. This *photoelectric* effect is the emission of electrons from metallic surfaces under the action of light. The number of electrons released in a unit time by light of a definite wavelength is directly proportional to the intensity of this light. The energy that these electrons have is directly proportional to the frequency of the light. There is a lower limit to the frequency below which insufficient energy is imparted to the surface to cause emission. Photoelectric tubes, like the human eye, are not equally responsive to all wave-lengths, or frequencies. For this reason, the response of a phototube to any given amount of light depends on the distribution of light frequencies present.

b. CONSTRUCTION.

(1) A phototube consists basically of two electrodes in an evacuated glass bulb. One of these is the cathode, which emits electrons when light is allowed to fall on it. These electrons are drawn to the plate by application of a positive voltage.

(2) The sensitivity of a phototube depends on the frequency, or color, of the light used to excite the tube. Different phototubes are manufactured to provide different sensitivity characteristics for various applications. Some are particularly sensitive to red light, some to blue light, and some have response characteristics similiar to those of the human eye. The sensitivity of a given tube always is specified in terms of the light frequencies used to excite it.

(3) The internal construction of the typical 929 phototube (fig. 160) is almost self-evident. The half-cylinder visible inside the tube is the photosensitive cathode. It is covered with a multiple layer of the rare metal cesium overlaid on cesium oxide, which in turn lies on a layer of silver. The plate is the small rod in the center.

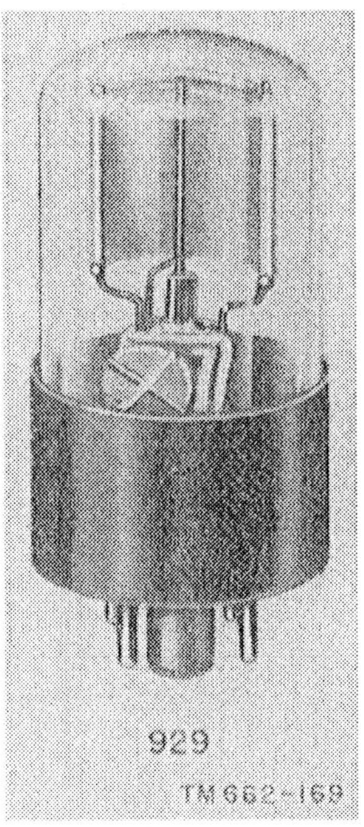

Figure 160. Typical phototube.

c. VACUUM, GAS, AND MULTIPLIER TYPES.

(1) If a small amount of gas is introduced in the evacuated envelope, the gas, by ionization, increases the amount of current that the tube is able to pass for a given amount of cathode illumination. Gas phototubes, such as the 930, have a higher sensitivity than corresponding vacuum types, such as the 929, and are used principally for reproduction of sound from sound motion pictures. However, the high-vacuum type is more stable, less easily damaged by higher than rated voltage or current, and has a higher internal resistance. The gas ions strike the cathode and produce appreciable secondary emission that also increases

the sensitivity. Consequently, the characteristic curves of vacuum and gas phototubes are noticeably different.

(2) Figure 161 shows the relationship, in the 929 and the 930, between the plate current and the plate voltage for various amounts of light. The curvature in the characteristic of the gas phototube caused by gas ionization is noticeable, and the increased current change for a given light change also is apparent. Gas phototubes must not be operated above their rated voltage or a glow discharge takes place between the plate and the cathode which can injure or destroy the delicate photosensitive surface.

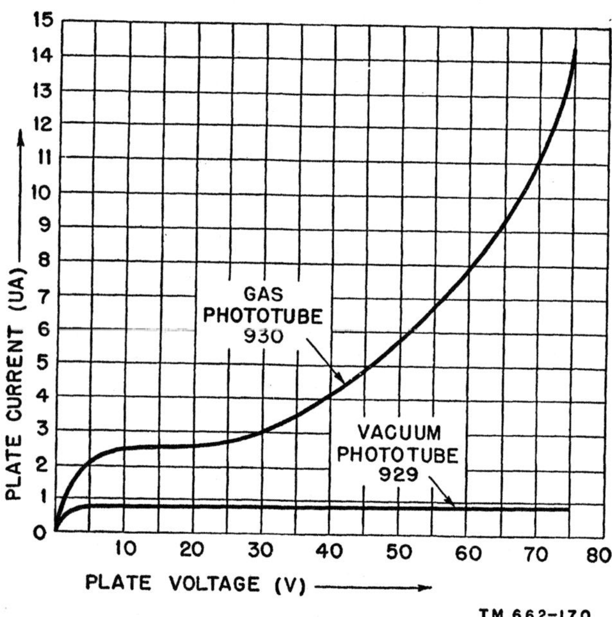

Figure 161. Comparison between control characteristics of the 929 and the 930.

(3) To increase the sensitivity of phototubes still farther, the mechanism of secondary emission is utilized in tubes like the 931–A. The initial electron emitted from the cathode is directed to strike against a series of subsidiary plates called dynodes (fig. 162). Each dynode has a surface treated for high secondary emission. As the initial electron strikes the first dynode, several

secondary electrons are released. These, in turn, strike the second dynode, each one emitting several secondary electrons. With the nine dynodes used in this tube, the amplification that can be achieved is enormous. These tubes are used wherever extremely high sensitivity, exceeding that of the human eye, is required.

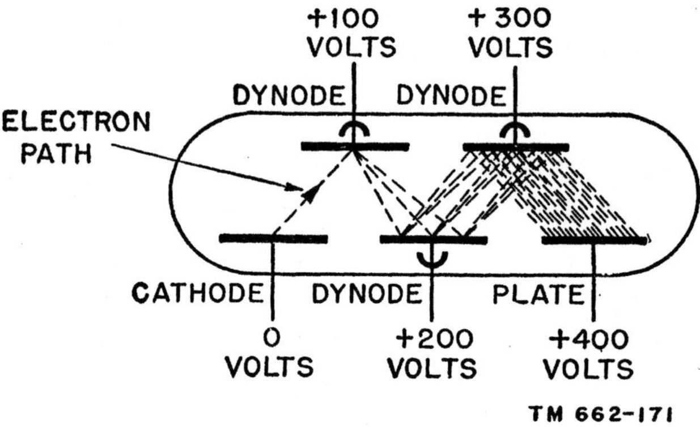

Figure 162. Operation of secondary-emission multiplier.

d. CIRCUITS AND APPLICATIONS.

(1) The current passed by a phototube is very small. Therefore, vacuum-tube amplifiers having one or more stages of amplification generally are used to build up these currents. The current change in a phototube is converted into a voltage change by passing it through a high resistance. Then this voltage is applied to the grid of a vacuum tube (fig. 163). A battery supplies the necessary plate voltage to both the amplifier and the phototube. Another voltage is necessary to provide negative bias for the amplifier tube and a return for the current from the phototube cathode. When the current in the phototube increases under the action of light, its cathode emits electrons, making the grid of the tube more positive. This increases the plate current which, in turn, operates the relay in series with the plate. The relay is provided with auxiliary contacts that actuate whatever circuit is being controlled.

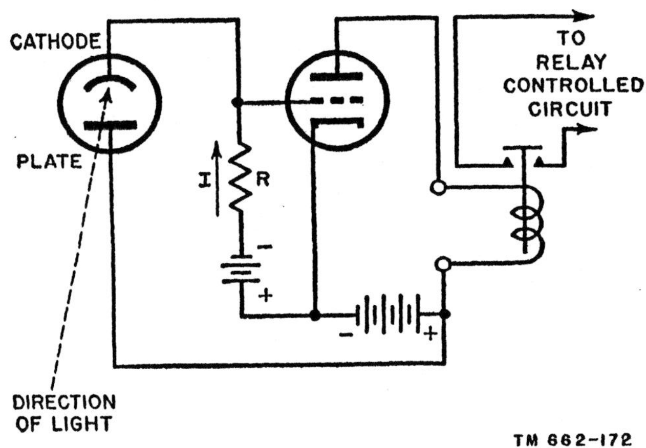

Figure 163. Basic phototube control circuit.

(2) For the production of sound from a source of light which has been modulated in intensity in proportion to the strength of the sound, the circuit of figure 164 is used. This is a conventional resistance-coupled amplifier which recovers the a-c component of the phototube output caused by the variation of the light from the cathode end of the phototube load resistor.

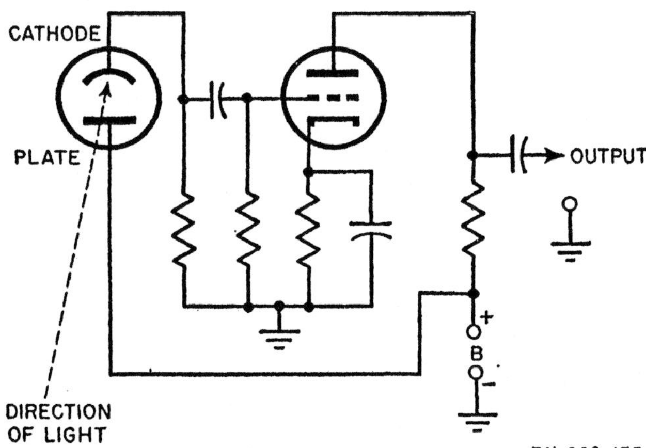

Figure 164. Phototube circuit for reproduction of sound.

126. Electron-ray Indicators

a. Electron-ray indicators (fig. 165) are used widely in radio receivers to indicate proper tuning. In addition, they find application as null indicators in bridge circuits and other apparatus in which a visual indication of small voltage changes is necessary. Most indicator, or magic-eye, tubes contain two sets of elements, one of

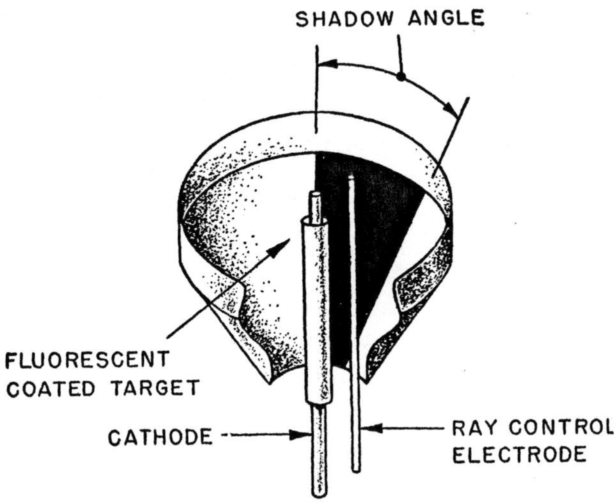

Figure 165. Cutaway view of 6E5 electron-ray indicator.

which is a triode amplifier. The other section is a cathode-ray indicator.

b. The electrons emitted by the cathode strike the conical plate, or target. This target is coated with a fluorescent paint that glows under the impact of the electrons. A small wire electrode called the ray-control electrode is parallel and close to the cathode. It deflects some of the electrons emitted from the cathode, producing a shadow on the target. This shadow is wedge-shaped, and the angle of the wedge varies with the voltage on the ray-control electrode. The plate of the internal triode amplifier is connected to the ray-control electrode, and therefore the shadow angle varies with the negative voltage applied to the grid of this triode. When the ray-control electrode is at the same potential as the target, the shadow closes completely. If the ray-control electrode is less positive than the plate, a shadow appears which is proportional in size to the difference in voltage. Since the voltage on the electrode is the same as that of the internal triode plate, the shadow angle increases with a more positive grid voltage.

127. Tubes for High Frequencies

a. As the operating frequency of an electron-tube circuit is increased, certain properties of the tube which heretofore have been negligible become important. One of the principal properties, usually neglected at low frequencies, is the capacitance between the various tube electrodes. In addition, the wires which connect the elements of the tube to the base and the tube elements themselves have small but definite amounts of inductance with appreciable reactance at higher frequencies. Also, the tube is affected by the finite time required for an electron to travel from the cathode to the plate. If this time interval, called the transit time, becomes appreciable in respect to the applied frequency, the normal phase relationships of plate and grid voltages cease to hold.

b. Special tubes at high frequencies are needed to overcome these various effects. The effect of the lead inductance can be decreased by using short leads, omitting the usual base and, in some cases, making the tube an integral part of the circuit for which it is designed. The tube shown in figure 166 with external ring electrodes illustrates the methods used for overcoming the various problems. This is known from its physical appearance as a lighthouse tube. The elements are as small as possible. They are arranged in a set of parallel planes to expose a minimal amount of surface area. Short, heavy metal rings are used instead of leads, and the elements are spaced close together so that it takes little time for an electron to travel from the cathode to the plate. The transit time, which can cause undesirable shifts in the phase of plate and grid voltages, is reduced.

c. Other high-frequency tubes incorporate similar structural refinements to achieve amplification at high frequencies. *Miniature* construction is customary for small receiving tubes in the frequency range below 200 mc. These tubes are all glass with short stem leads brought out at the base. The electrodes are small, and close spacing is used. The space-saving features of miniature construction make it desirable for many receiving applications, and the high performance of these tubes even at medium frequencies causes their use to be widespread.

128. Microwave Tubes

a. Above 3,000 mc, the limitations brought about by transit time, stray capacitance, and lead inductance become so severe that designs based upon tubes with control grids are not

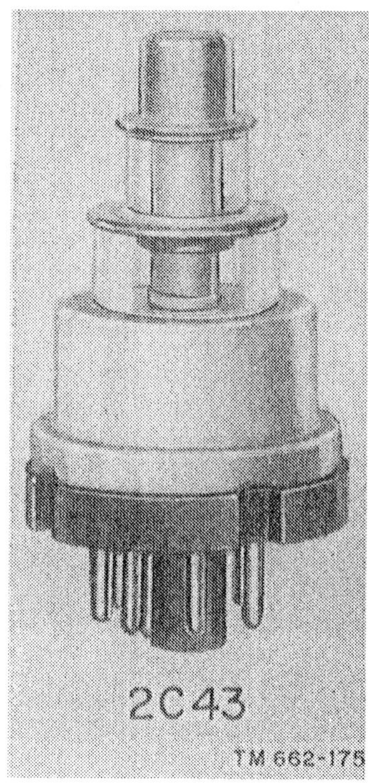

Figure 166. Lighthouse tube 2C43.

practical. Because the length of radio waves at these frequencies is less than 3 inches and as small as $\frac{1}{3}$-inch, the tubes are necessarily of a size equal to or greater than a wavelength. When these tubes are used as oscillators or amplifiers, it is necessary to incorporate tuned circuits in the tube itself. These resonant circuits usually are in the form of cylindrical cavities hollowed out of the metal structure of the tube itself.

b. Figure 167 is a cutaway view of a tube that is used as an oscillator in this range. It is called a magnetron, because it consists essentially of a diode with a strong magnetic field parallel to the axis of the cathode. This magnetic field causes the electrons which leave the cathode to move in a spiral path before striking the plate. As the clouds of electrons spiral past the resonant cavities in the plate, they cause the latter to oscillate in much the same manner as a bottle can be made to produce a musical note by blowing a stream of air across its mouth. If the dimensions of the cavities are sufficiently small, the frequency can be raised to as high as 30,000 mc. Large amounts of power can be produced by this tube at high efficiency, although it can operate only as an oscillator. Its frequency is determined mainly by the mechanical dimensions of the tube and cavities.

pulsation. Therefore, amplification takes place between the input and output resonators. The tube can be used as an oscillator if some power from the output is coupled back to the input in the proper phase.

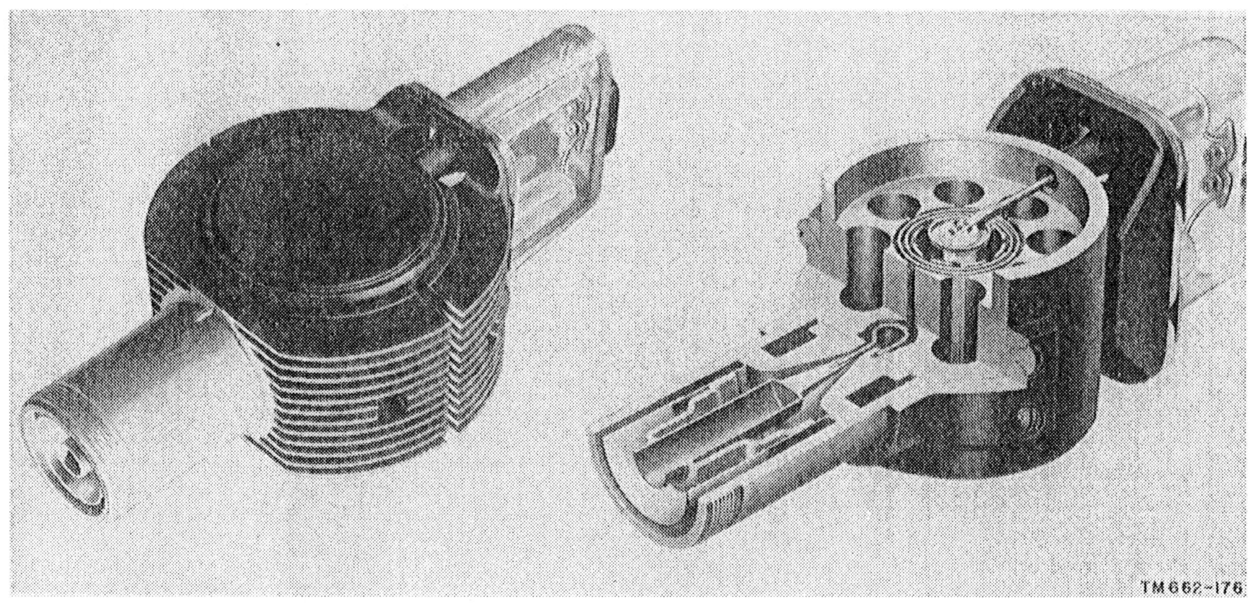

Figure 167. Operational and cutaway view of typical magnetron.

c. The magnetron is not a very flexible oscillator tube. The klystron tube in figure 168 can operate as an amplifier or oscillator over a small tunable range. Its operation depends on the changes introduced in the velocity of a stream of electrons by alternately slowing it down and speeding it up, using the transit time between two points to produce an alternating current. This current delivers power to a resonant circuit in the form of a cavity.

d. The cathode emits a stream of electrons which is smoothed out to uniform velocity by the smoother grid. A radio-frequency field is applied to the grids of the input cavity resonator. This imposes a varying velocity on the stream, retarding or speeding it up. In the drift space, the electrons that have been speeded up will overtake those that have been slowed down on an earlier cycle. This produces a still stronger pulsation in the electron stream as it passes through the grids of the output cavity resonator. The latter takes power from the stream if it is tuned to the frequency of the

e. Small tubes, called reflex klystrons, have been designed solely for use as local oscillators in microwave superheterodyne receivers. They use a single-cavity resonator, the electron stream passing through it twice—once in a

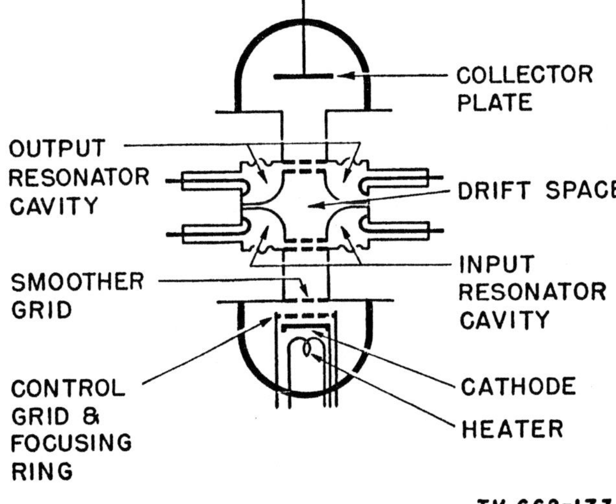

Figure 168. Operation of typical klystron.

forward direction and then reflected back through the cavity again. They can be tuned over a wide range of frequencies, but have limited power output. This is the klystron most frequently used in microwave equipment.

f. Both the klystron and the magnetron make use of electron streams or beams to operate at microwave frequencies. They transfer energy to cavity resonators through these electron streams. The cavity resonator is a device with an inherently high ratio of reactance to resistance. In other words, it is equivalent to a tuned circuit with a very high Q. At a resonant frequency of 10,000 mc a Q of 5,000 is frequent.

129. Cathode-ray Tubes

a. When electrons strike certain substances with sufficiently high velocities, the substances give off visible light. This principle is utilized in the cathode-ray tube, in which electrons strike a coated screen, causing it to fluoresce. The cathode emits electrons which are accelerated forward, formed into a thin pencil or beam, and allowed to strike a fluorescent screen.

b. Figure 169 shows two types of cathode-ray tube. In A, electrons leaving the cathode pass through a control grid. The control grid determines the number of electrons that can pass, and consequently, the intensity of light emitted from the screen. After leaving the control grid, the electrons are focused in a thin beam by the focusing anode, and the velocity of the electrons is increased by the accelerating anode. The electron stream is then directed as desired at any portion of the screen by two pairs of deflection plates, which can be compared with the plates of a capacitor. A plate with a negative charge repels the electrons, and a plate with a positive charge attracts the beam. Because this effect is caused by an electrostatic field between the plates, this cathode-ray tube is known as an electrostatic-deflection type.

c. In B, a magnetic-deflection type of cathode-ray tube is shown. The focusing anode is replaced by a focusing coil, and the deflection plates by a deflection coil. These coils are wound externally about the neck of the tube. The coils serve the same purposes as the anode and plates of the electrostatic type.

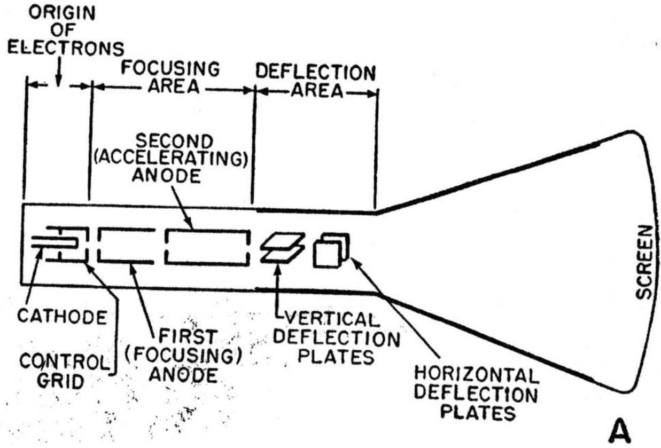

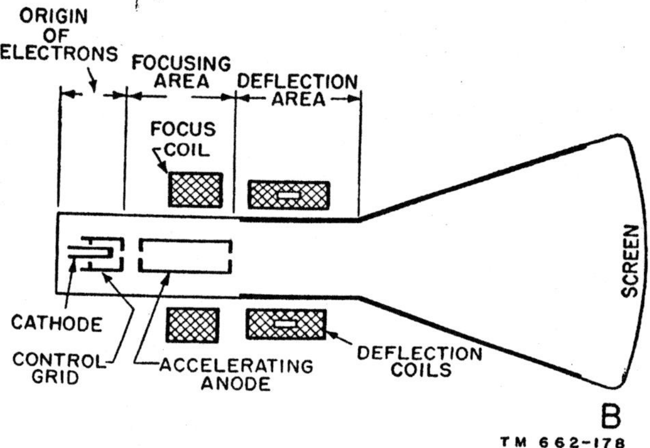

Figure 169. Basic cathode-ray tubes using electrostatic and electromagnetic deflection.

d. The usefulness of the cathode-ray tube is due to the ease with which the electron beam can be diverted from its straight path by an electrostatic or electromagnetic field. In the electrostatic tube, two small pairs of plates are located at right angles to each other. A voltage applied to one pair deflects the beam up or down; applied to the other pair, it deflects the beam right or left. This control can be accomplished magnetically by coils external to the tube whose magnetic fields deflect the beam. Electrons have little inertia and can be deflected almost instantaneously. Therefore, the beam can move back and forth or up and down at rapid rates, even up to the very high radio frequencies. When the beam moves at rapid rates, it seems to trace a line on the screen.

e. The cathode, grids, and anodes controlling the intensity and focus of the electron beam usually are assembled in one unit, called the electron gun. It is possible to put more than one electron gun in a single tube envelope. The tube then can be used for the comparison of several independent voltages at the same time.

130. Summary

a. Gases conduct electric currents by ionization of the gas molecules.

b. Cold-cathode tubes are used for the production of light, voltage regulation, rectification, protection of circuits, and as relaxation oscillators.

c. Hot-cathode and ionic-heated cathode tubes are used principally as rectifiers if they are diodes and as control tubes for large amounts of current if they are triodes of thyratrons.

d. Phototubes convert the light falling on them into a variation of current flow. The current is passed through a resistor to provide a control voltage for the operation of an amplifier.

e. The multiplier tube uses dynodes which increase the emission by releasing secondary electrons when they are struck by an electron from the light-sensitive surface.

f. Electron-ray indicators portray visually the value of a d-c voltage, by causing a shadow to appear on the surface of a fluorescent screen.

g. At very high frequencies the interelectrode capacitances, lead inductance, and transit time of standard vacuum tubes become so great that compact designs are necessary with low inductance and capacitance and with short transit time.

h. The magnetron is a diode in a strong magnetic field which causes the electron stream to move in a spiral path before it reaches the plate.

i. The klystron uses the variation in velocity produced in an electron stream passing through the field of a cavity resonator

j. Cathode-ray tubes are used in oscilloscopes for the visual display of waveforms and radar information, and in television receivers.

131. Review Questions

a. Describe the causes of current flow in a gas.

b. What happens at each stage in the current flow?

c. What are the principal uses for cold-cathode tubes?

d. Describe the function of the cold-cathode tube in each of these applications.

e. What is the most important application of hot-cathode gas-filled tubes?

f. Describe the operation of a thyratron.

g. What is the purpose of a mercury-pool rectifier, and how does it work?

h. Describe the construction and operation of a vacuum phototube.

i. How is the sensitivity of a phototube increased?

j. Describe the operation of a multiplier type phototube.

k. Describe the function of the ray-control electrode in the electron-ray indicator.

l. What factors influence the operation of tubes at high frequencies?

m. How do tubes used at high frequencies differ from those used at low frequencies?

n. What are the principal types of microwave tubes?

o. Describe the operation of a magnetron.

p. Describe the operation of a klystron.

q. Describe the principal parts and operation of a cathode-ray tube.

r. What two types of deflection are used in cathode-ray tubes?

132. Electron-tube Electrode Connections

a. The various types of electron tubes discussed in this manual all must have some means of applying potentials and making connections to the various electrodes within the envelope. The external leads take the form of tube prongs, pins, or caps. Usually, a group of prongs or pins is built into a tube base. The base material often is bakelite, although other insulating materials are used. Sometimes the connecting leads take the form of pins which are built into the tube envelope itself. Occasionally, metal caps are bonded to the tube envelope and the tube electrodes are connected to these caps through the envelope. The specific method used depends on the particular tube involved.

b. The early triode receiving tubes used a four-prong base. Two of these prongs were connected internally to the filament, and the other two were connected to the grid and the plate. The two filament prongs were slightly larger in diameter than the other prongs, so that the tube could be inserted properly in a corresponding four-hole socket. This arrangement still is used for some modern tubes. It is necessary to insert the tube properly so that the proper operating voltages can be applied to the correct electrodes. Unless some such method is used to *key* the tube, it can be damaged by improper placement in its socket. Other methods have been used to key the four-pin base. One method involves the use of a small metal projection on the tube base which permits the tube to be inserted in its socket in only one manner. Another method is to arrange the base pins in such a pattern that the tube can be inserted into its socket in only one way.

c. With the advent of more complex tubes, it became necessary to add more connecting pins to the tube base. The five-, six-, seven-, eight-, and nine-pin bases were introduced. All of these bases are keyed by means of large diameter pins, placement of the pins, or the addition of special fittings on the tube base which permit the tube to be inserted in the socket only one way (fig. 170).

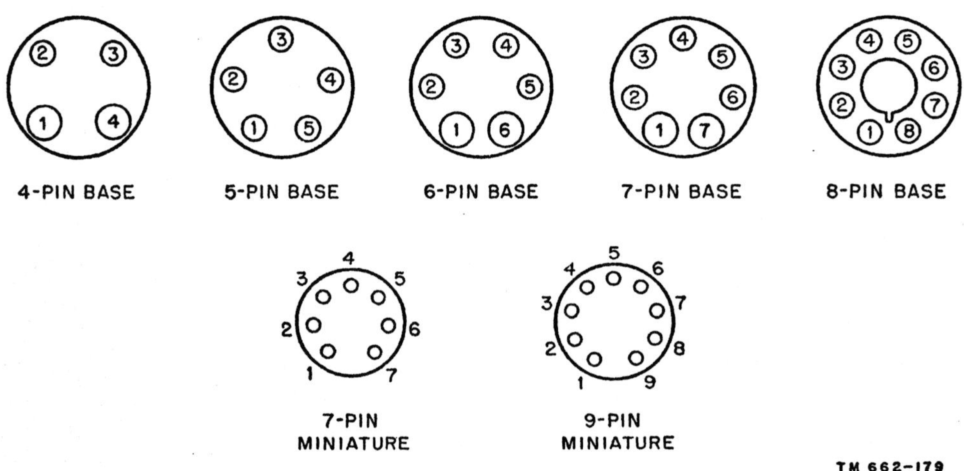

Figure 170. Common electron-tube bases showing arrangement of pins.

d. One of the most widely used of these multipin bases for receiving tubes is the eight-pin or *octal* base. In this base all the pins have the same diameter and they are uniformly spaced. However, at the center of the base is an insulating post which has a vertical ridge. This acts as a key or guide pin which fits in a keyway in the tube socket. Thus, the tube can be inserted in only one way. The original idea of the octal socket was to have similar electrodes of any type of tube connected to the same pins, so that some degree of standardization would result. If any of the pins are not used they are left off the base, or no connection is made to them.

e. A variation of the octal base is used with *lock-in* receiving tubes. The base of such tubes also has eight pins. However, the contact pins are sealed directly into the glass envelope and no insulating base is used. The bottom portion of the envelope is fitted with a metal shell and a metal key or guide pin. This guide pin has a

vertical ridge like the one used in the octal base. A groove around the bottom of the locating pin fits into a spring catch in the socket. This holds the tube firmly in the latter.

f. Another variation in the base of receiving tubes is used with miniature glass tubes. These tubes are becoming popular in modern electronic equipment because of their small size and many other desirable characteristics. Contact pins of these tubes are sealed directly into the glass envelope. Either seven or nine pins generally are used. Because of the additional spacing between two of these pins (fig. 170), the tubes cannot be inserted improperly in their sockets. Figure 171 illustrates six different types of electron-tube bases and their corresponding sockets.

g. Transmitting and special purpose tubes use sockets and methods of connection which are subject to considerable variation. Some small transmitting tubes use a base structure

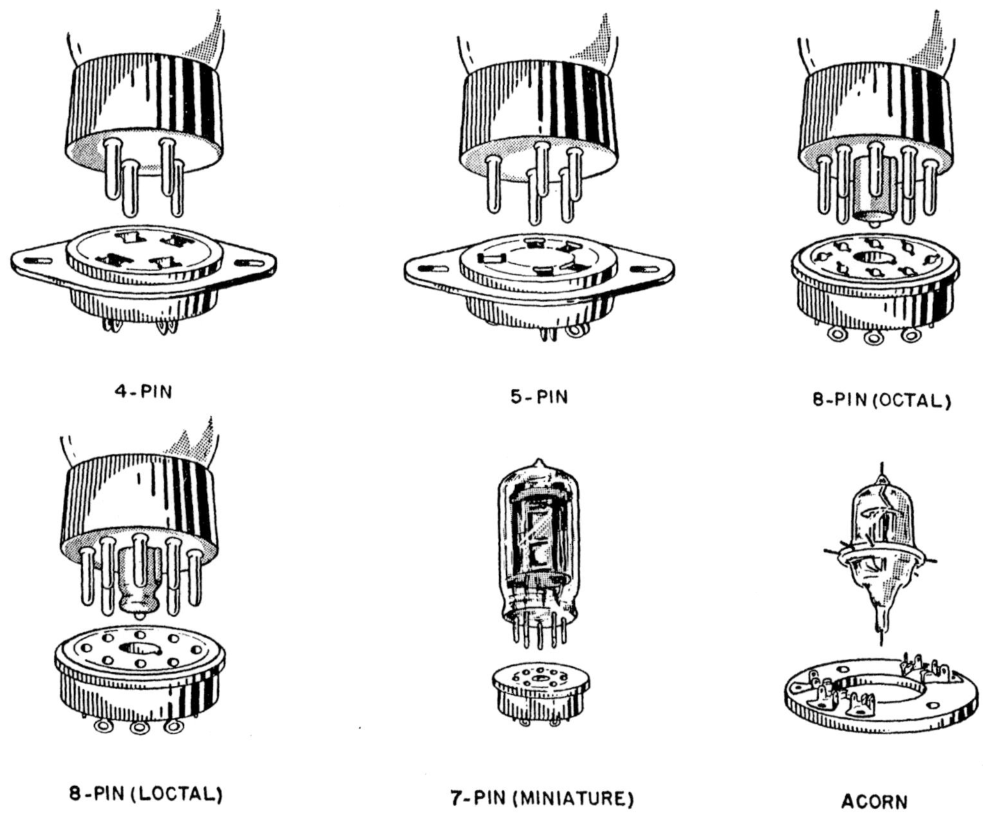

Figure 171. Receiving-type electron-tube bases and corresponding sockets.

similar to that used for receiving tubes. However, the larger types use special connections and terminals which are not at all standardized. Special high-frequency tubes use connection methods which conform with their special requirements. Cathode-ray tubes may use conventional octal sockets or sockets which have more than eight pins. A few commonly used bases for these tubes are the magnal (eleven-pin), duodecal (twelve-pin), and the diheptal (fourteen-pin).

h. A standard system has been set up for numbering the base pins of the common tube bases. The pins are numbered consecutively in a *clockwise* direction looking up at the bottom of the tube base. When fewer than eight pins are required, the unnecessary ones are omitted and the spacing and the numbering of the remaining pins are unchanged. In the octal and lock-in types, pin 1 is the pin directly clockwise of the ridge on the guide pin, as in figure 170. In the miniature types, pin 1 is the clockwise pin of the two widely spaced pins. Other designations are as shown in the figure.

i. Some attempts at electrode connection standardization have been attempted by electron-tube manufacturers. These were only partly successful, because of the tremendous variety of tubes which have been manufactured. A good many receiving tubes, however, do show a degree of uniformity in pin connections worth noting. For example, in the four-pin base, pins 1 and 4 usually are connected to the filament, pin 2 is connected to the plate, and pin 3 is connected to the control grid. In the five-pin base, pins 1 and 5 frequently are connected to the heater, pin 2 is connected to the plate, pin 3 is connected to the control grid, and pin 4 is connected to the cathode. When a five-pin base is used for a pentode tube, it is common practice to make the same connection as above except that the screen grid is connected to pin 3, and the control grid is connected to a grid cap at the top of the tube. The suppressor grid is connected internally to the cathode. In the six- and seven-pin bases, pins 1 and 6 and pins 1 and 7 frequently are used as the heater connections. In the octal base, pin 1 usually is connected to the metal envelope or internal shield, pins 2 and 7 are connected to the heater, pin 3 is connected to the plate, pin 4 is connected to the screen grid, pin 5 is connected to the control grid, and pin 8 is connected to the cathode and the suppressor grid. In the lock-in tube base, pins 1 and 8 usually are the heater connections. It must be emphasized that the wide variety of tube types makes it impossible to adhere rigidly to these pin connections.

133. Electron-tube Type Designation

a. Every electron tube is identified by a number or a combination of numbers and letters. In 1933 a systematic method of designation was developed. So many different types of tubes have been introduced since that time that it has become impossible to adhere rigidly to the system that was set up. However, some of the original ideas contained in this system are still being followed.

b. The type number of a tube is divided into four parts. First, a number consisting of one or more digits designates the filament or heater voltage. Second, one or more letters designate the type or function of the tube. Third, a number designates the number of useful elements in the tube. Fourth, one or more letters designate the size or construction. For example, the type 2A3 is a power triode which requires a filament voltage of about 2 volts (actually 2.5 volts), it is an amplifier tube, and it has three useful elements. The fourth part of the designation is omitted. The type 5Y3–G is a duodiode which requires a filament voltage of 5 volts, it is a rectifier (letters from U to Z are used for rectifiers), and it has three useful elements. The letter G indicates that the tube has a glass envelope. The type 50L6–GT requires a heater voltage of 50 volts, it is a beam power amplifier (the letter L is used for such tubes), and it has six useful elements if the heater and cathode are considered separately. The letters GT indicate the use of a glass envelope somewhat smaller than the conventional size.

c. Because of the thousands of different types of receiving tubes that are manufactured, there are probably more exceptions to this system of designation than there are tubes which follow it completely. The situation is even more confused for transmitting tubes because each manufacturer has his own system of designation.

It is possible to find tubes of practically identical characteristics made by different manufacturers with completely different type designations.

d. The situation so far as cathode-ray tubes are concerned is somewhat better. A logical system of designation is being used by nearly every manufacturer. In this system, there is first a number consisting of one or two digits which gives the screen diameter in inches. Next, there is a letter which designates the manufacturer's order of registration of the particular tube. Finally, there are a letter and a number which indicate the specific type of screen material used. These screen materials or phosphors have been fairly well standardized as far as their operating characteristics are concerned. Sometimes, a letter is added to the designation to indicate a modification in the basic cathode-ray tube. The heater voltage of most cathode-ray tubes is standardized at 6.3 volts. As an example, consider the 5AP4. This tube has a 5-inch screen diameter and it is the first tube of its type registered by a specific manufacturer. The tube uses a type P4 phosphor, whose color of fluorescence is white and whose persistence is medium.

e. Because of the lack of standardized tube designations for both receiving and transmitting tubes, and because of the impossibility of disclosing the many operating characteristics by means of a simple designation, it often is necessary to refer to a tube manual.

134. Information in Tube Manual

a. Most manufacturers of electron tubes have available listings of their particular tubes with the characteristics and technical descriptions. In some cases, this listing is small as it includes only a few special-purpose tubes or the tubes manufactured by a small manufacturer. On the other hand, if a manufacturer who produces many different tube types prepares such a listing, a useful publication results. Several such publications do exist and they are known as tube manuals. The largest tube manuals, which include several hundred pages or more, list only receiving-type electron tubes. When *tube manual* is referred to in this chapter, the receiving tube manual is meant.

b. Although no two of these tube manuals are identical, they all contain more or less the same type of information. Some manuals are so designed that pages describing new tube types can be inserted to keep the books up-to-date. Other manuals are revised and reprinted from time to time when a sufficient number of new tubes have been produced.

c. The tubes are listed according to the numerical-alphabetical sequence of their type designations. The schematic symbol of each tube, showing the base-pin connections to the various electrodes, is given. A brief description of the tube is included in some manuals as an introduction to the tube characteristics. Next, the physical specifications are designated. These include information concerning the dimensions of the envelope, the type of base, and sometimes the preferred mounting position.

d. Following the physical specification, the electrical ratings are given. These include information regarding the filament or heater voltage and current, as well as the maximum electrical ratings of the tube. Maximum plate and screen voltages, maximum plate and screen dissipations, and peak heater-to-cathode voltage are included. In addition, the interelectrode capacitances of some types are listed. If the tube has other modes of operation (for example, a pentode operated as a triode or a pentode operated in push-pull), additional ratings and electrical specifications frequently are given.

e. Next, typical operation of the tube is shown. Figures for the following often are included: typical electrode voltages, required value of cathode bias resistor, peak signal voltage, typical electrode currents under conditions of zero and maximum signal, required value of load resistance, power output, and total harmonic distortion. In addition, values of amplification factor, transconductance, and plate resistance are supplied. If the tube commonly is operated under different conditions, a complete set of typical operating values frequently is included. For example, in one tube manual, maximum ratings and typical operating values are given for the 6L6 (beam power amplifier) under the following operating conditions: single-tube class A amplifier, single-tube class A amplifier (triode connected), push-pull class A amplifier, push-pull class AB_1 amplifier, and push-pull class AB_2 amplifier. Ratings also are given for most of the foregoing, using fixed bias or cathode bias.

f. Following the typical operating values is a section dealing with specific applications. Special installation notes having to do with the particular tube type also may be supplied, and unusual features of the tube are discussed.

g. Finally, one or more families of curves depicting the operation of the tube are shown. Usually these curves are the average plate characteristics for various values of grid voltage. Sometimes one or more load lines are drawn on the characteristic curves. In some tube manuals, average transfer characteristic curves are shown, along with curves which illustrate the variation in plate resistance, transconductance, and amplification factor at various electrode voltages.

h. If the particular tube listed happens to be a rectifier, ratings or curves are given which apply to use of the tube. The maximum peak inverse plate voltage and the maximum peak plate current are given, as well as the voltage drop across the tube at certain values of plate current. The output current for various a-c input voltages and types of filter circuits is designated. Curves frequently are shown which give the d-c output voltage and load current for various input voltages and filter circuits.

i. Many tube manuals supply additional information. A section dealing with general tube and circuit theory may be included. Sometimes the common tubes are classified as to their use and characteristics. Frequently, a section of the tube manual is devoted to the design of resistance-coupled voltage amplifiers. This section usually consists of tables for the commonly used amplifier tubes. The tables include information concerning the proper combination of plate load resistor, grid resistor, screen-grid resistor, cathode-bias resistor, and coupling and bypass capacitors for various values of plate-supply voltage. The output voltage, voltage gain, and sometimes the percentage of distortion are included also under the various conditions outlined in the table.

j. Finally, the tube manual may contain circuit diagrams which illustrate some of the more important applications of the tubes listed in the manual. Some manuals provide information on obsolete or seldom-encountered types, as well as on panel and ballast lamp specifications.

135. Uses of Tube Manual

a. The tube manual provides a listing of the characteristics and socket connections of the electron tube. In servicing electronic equipment it frequently is necessary to trace circuits, check components connected to various electrodes of tubes, and measure tube electrode voltages. Because of the wide variety of electron tubes used in modern equipment and because of the lack, on the whole, of standard base connections, it is necessary to refer to a tube manual for socket connections. It must be remembered that all views of tube bases or sockets, unless otherwise indicated, are *bottom* views.

b. The normal operating voltages shown in the manual serve as a guide to servicing personnel. The technician can compare the operating voltages given for a particular tube in the manual with the voltages measured in the equipment. If an electron tube is used for a special application, the operating voltages may not be similar to those shown in the manual. However, the measured voltages should not exceed the maximum ratings given and the filament or heater voltage should certainly correspond to the value designated in the manual.

c. The average plate characteristic curves have several uses. They show the operating conditions of the tube with various electrode potentials. These can be used to compare the actual operation of a tube in a circuit with the proper average operation. In addition, the curves serve as a basis for many useful calculations. A load line is constructed on the family of curves for the particular value of plate load used. By means of this load line and the curves shown in the manual, the power output and the percentage of distortion can be determined. These are determined by direct graphical methods; that is, actual values are read from the curves and these values are substituted in simple equations which show the power output and distortion. The curves also are used for design purposes. A given set of electrode potentials is assumed, a load line is drawn, and calculations are based on the curves. These calculations show whether the plate dissipation of the tube is exceeded, and whether the power output and the fidelity are adequate. The transfer characteristic curves can be used to determine the operating range for tubes used for detection or for avc action. Conversion characteristic curves are used in the design of converter stages, and diode load curves are useful in designing electron-tube voltmeters or avc systems.

d. The tube manual permits a comparison between tubes. Comparative characteristics of several beam-power tubes, for example, can be examined to determine which tube fits the specific application required. In addition, physical dimensions of tubes can be found. This information is of importance in the mechanical design and construction of a piece of electronic equipment.

e. The section of the manual dealing with the design of resistance-coupled amplifiers gives specific component values that can be used to achieve certain results.

f. In addition to the preceding, some tube manuals contain an excellent section on theory and application of electron tubes. Illustrations frequently are included to show the internal construction of various types of receiving tubes.

136. Summary

a. Use of a base having a number of contact pins is a common method of making connections to the various electrodes in an electron tube.

b. Methods of keying tube bases include the use of a special center post in the base, special arrangement of pins, and various pin diameters.

c. All electron tubes are identified by a number or a combination of numbers and letters. Some attempts at standardization of tube designations have been attempted, but, because of the great number of different tube types, these have not been adhered to rigidly.

d. A tube manual lists a great number of electron tubes, with their electrode connections and major characteristics.

e. Physical specifications as well as maximum and typical electrical ratings are given in the tube manual.

f. The tube manual is useful in determining tube socket connections and measuring tube electrode voltages during the servicing of electronic equipment.

137. Review Questions

a. What are the common methods used to make connections and apply voltages to various tube electrodes?

b. Why must tube bases be keyed?

c. How are the pins in a tube base numbered?

d. To which pins is the filament or heater usually connected in a four-pin base? In a six-pin base? In an octal base?

e. Describe briefly the standard system of tube type designation.

f. What is a tube manual?

g. What important information can be found in a tube manual?

h. How can the curves given in the tube manual be used?

i. Of what use are the schematic diagrams showing the tube base pin connections?

j. Give some specific uses for the tube manual.

APPENDIX

LETTER SYMBOLS

1. Plate Circuit

e_b — instantaneous total plate voltage.

e_{b1} — instantaneous total plate voltage of tube V1 (e_{b2}—of tube V2. . .etc.).

E_b — total plate voltage, average.

E_{bb} — d-c plate-supply voltage.

E_{bo} — quiescent or zero signal, average value of plate voltage.

e_p — instantaneous value of varying component of plate voltage.

e_L — instantaneous total voltage across the load resistor.

e_z — instantaneous varying component of voltage across the load impedance.

E_{Lo} — quiescent or zero signal, average value of d-c voltage across the load resistor.

E_{zo} — varying component of average voltage across the load impedance.

i_b — instantaneous total plate current.

I_b — average total plate current.

I_{bo} — quiescent or zero signal, average value of plate current.

i_p — instantaneous value of varying component of plate current.

2. Grid Circuit

$e_c\,(e_{c1})$ — instantaneous total control-grid voltage.

e_{c2} — instantaneous total screen-grid voltage.

e_{c3} — instantaneous total suppressor-grid voltage.

E_{c1} — average total control-grid voltage.

E_{cc} — control-grid d-c supply voltage.

E_{c2} — screen-grid d-c supply voltage.

e_g — instantaneous value of varying component of control-grid voltage.

$i_c\,(i_{c1})$ — instantaneous total control-grid current.

i_{c2} — instantaneous total screen-grid current.

i_{c3} — instantaneous total suppressor-grid current.

3. Cathode Circuit

e_k — instantaneous total cathode voltage.

E_k — average total cathode voltage.

i_k — instantaneous cathode current.

I_k — average d-c cathode current.

4. Miscellaneous

e_{in} — instantaneous input signal voltage.

e_{out} — instantaneous output signal voltage.

E_{max} — maximum value of d-c voltage.

I_{max} — maximum value of d-c plate current.

P_1 — input power to a system.

P_2 — output power from a system.

P_o — d-c output power.

P_i — d-c input power.

P_p — plate dissipation.

For sale by the Superintendent of Documents, U. S. Government Printing Office
Washington 25, D. C. - Price $1.25

Printed in the United States
59937LVS00005B/55-74